Mathematics in Image Processing

IAS/PARK CITY
MATHEMATICS SERIES

Volume 19

Mathematics in Image Processing

Hongkai Zhao
Editor

American Mathematical Society
Institute for Advanced Study

John C. Polking, Series Editor
Hongkai Zhao, Volume Editor

IAS/Park City Mathematics Institute runs mathematics education programs that bring together high school mathematics teachers, researchers in mathematics and mathematics education, undergraduate mathematics faculty, graduate students, and undergraduates to participate in distinct but overlapping programs of research and education. This volume contains the lecture notes from the Graduate Summer School program

2010 *Mathematics Subject Classification.* Primary 35-XX, 42-XX, 49-XX, 65-XX, 94-XX.

Library of Congress Cataloging-in-Publication Data

Mathematics in image processing / Hongkai Zhao, editor.
pages cm. — (IAS/Park City mathematics series, volume 19)
Includes bibliographical references.
ISBN 978-0-8218-9841-3 (alk. paper)
1. Image processing–Mathematics. I. Zhao, Hong-Kai, 1968–

TA1637.5.M37 2013
006.601′51–dc23 2012048985

∞ The paper used in this book is acid-free and falls within the guidelines
established to ensure permanence and durability.
Visit the AMS home page at http://www.ams.org/

10 9 8 7 6 5 4 3 2 1 18 17 16 15 14 13

Contents

Preface

The IAS/Park City Mathematics Institute (PCMI) was founded in 1991 as part of the "Regional Geometry Institute" initiative of the National Science Foundation. In mid 1993 the program found an institutional home at the Institute for Advanced Study (IAS) in Princeton, New Jersey.

The IAS/Park City Mathematics Institute encourages both research and education in mathematics and fosters interaction between the two. The three-week summer institute offers programs for researchers and postdoctoral scholars, graduate students, undergraduate students, high school teachers, undergraduate faculty, and researchers in mathematics education. One of PCMI's main goals is to make all of the participants aware of the total spectrum of activities that occur in mathematics education and research. We wish to involve professional mathematicians in education and to bring modern concepts in mathematics to the attention of educators. To that end the summer institute features general sessions designed to encourage interaction among the various groups. In-year activities at the sites around the country form an integral part of the High School Teachers Program.

Each summer a different topic is chosen as the focus of the Research Program and Graduate Summer School. Activities in the Undergraduate Summer School deal with this topic as well. Lecture notes from the Graduate Summer School are being published each year in this series. The first nineteen volumes are:

- Volume 1: *Geometry and Quantum Field Theory* (1991)
- Volume 2: *Nonlinear Partial Differential Equations in Differential Geometry* (1992)
- Volume 3: *Complex Algebraic Geometry* (1993)
- Volume 4: *Gauge Theory and the Topology of Four-Manifolds* (1994)
- Volume 5: *Hyperbolic Equations and Frequency Interactions* (1995)
- Volume 6: *Probability Theory and Applications* (1996)
- Volume 7: *Symplectic Geometry and Topology* (1997)
- Volume 8: *Representation Theory of Lie Groups* (1998)
- Volume 9: *Arithmetic Algebraic Geometry* (1999)
- Volume 10: *Computational Complexity Theory* (2000)
- Volume 11: *Quantum Field Theory, Supersymmetry, and Enumerative Geometry* (2001)
- Volume 12: *Automorphic Forms and their Applications* (2002)
- Volume 13: *Geometric Combinatorics* (2004)
- Volume 14: *Mathematical Biology* (2005)
- Volume 15: *Low Dimensional Topology* (2006)
- Volume 16: *Statistical Mechanics* (2007)
- Volume 17: *Analytic and Algebraic Geometry: Common Problems, Different Methods* (2008)
- Volume 18: *Arithmetic of L-functions* (2009)
- Volume 19: *Mathematics in Image Processing* (2010)

Volumes are in preparation for subsequent years.

Some material from the Undergraduate Summer School is published as part of the Student Mathematical Library series of the American Mathematical Society. We hope to publish material from other parts of the IAS/PCMI in the future. This will include material from the High School Teachers Program and publications documenting the interactive activities which are a primary focus of the PCMI. At the summer institute late afternoons are devoted to seminars of common interest to all participants. Many deal with current issues in education: others treat mathematical topics at a level which encourages broad participation. The PCMI has also spawned interactions between universities and high schools at a local level. We hope to share these activities with a wider audience in future volumes.

John C. Polking
Series Editor
December 2012

Introduction

Hongkai Zhao

IAS/Park City Mathematics Series
Volume 19, 2010

Introduction

Hongkai Zhao

1. The Summer School and the Theme

The theme of 2010 PCMI Summer School was Mathematics in Image Processing in a broad sense, including mathematical theory, analysis, computation algorithms and applications. Image processing can be viewed as an area in information science where information needs to be processed, extracted and analyzed from visual contents such as imageries or videos. The demands for efficient processing of overwhelming visual information from extensive imaging technologies and the internet are ever growing and challenging. These demands include standard tasks such as compression and denoising as well as high level understanding and analysis, such as recognition and classification. Mathematics has been one of the main driving forces in this field and has provided many powerful tools, such as linear algebra, probability and statistics, harmonic analysis, partial differential equations, optimization, etc. Centered around the theme of mathematics in image processing, the summer school covered a quite wide spectrum of topics in this field through lecture series with associated discussion sessions and computer lab sessions as well as a pretty long list of research seminars.

What makes the summer school more timely and exciting is the very recent advances and developments in mathematical theory and computational methods for sparse representation. Sparsity is an important property that should be investigated and explored for any signal or information. Two key mathematical questions associated with sparse representation are: (1) In what representation are a certain class of signals sparse? For examples, the sparse representation could be in a certain transform domain with a pre-determined basis, such as Fourier or wavelets, or it could be some basis or dictionary computed or learned from a training set; (2) How can one recover the sparse representation effectively from the given measurements? For example, what kind of measurements are desirable and how does one design efficient algorithms to recover the sparse signal? Quite a portion of the lectures and research seminars during the summer school presented state of the art work on sparse representation and compressive sensing from theory to algorithms and applications.

Altogether there were nine lectures series:

- Richard Baraniuk, Rice University: Compressive Sensing: Sparsity-Based Signal Acquisition and Processing
- Antonin Chambolle, École Polytechnique: Total-Variation Based Image Reconstruction

---

Department of Mathematics, University of California at Irvine
E-mail address: zhao@math.uci.edu

- Michael Elad, Israel Institute of Technology: Sparse & Redundant Representations: From Theory to Applications in Image Processing
- Anna Gilbert, University of Michigan: A Survey of Sparse Approximation
- Yann LeCun, New York University: Learning Image Feature
- Guillermo Sapiro, University of Minnesota: Dictionary Learning for Efficient Signal Representation
- Zuowei Shen, National University of Singapore: Wavelet and Wavelet Frames in Imaging Science
- Joseph M. Teran, University of California, Los Angeles: Numerical Methods for Elasticity Problems in Biomechanics
- Ross Whitaker, University of Utah: Statistical Models and Methods in Image Analysis

There were also two Clay Senior Scholars in residence during the summer program, Ingrid Daubechies (Princeton University) and Jean-Michel Morel (Ècole Normale Supérieure de Cachanin). Their presence and contributions were invaluable to the whole program. In particular, they gave two very well-received public lectures open to the whole program: Fine Art Meets Mathematics (by Daubechies) and Image Editing with the Poisson Equation: How to teach the Fourier method to undergraduates (by Morel). In addition Tony Derose from Pixar gives a stimulating public lecture about mathematics in computer animation for entertainment industry.

2. Content of the Volume

The content of the current volume consists of three lecture series which give a pretty good glimpse of the topics covered during the summer school. The lecture notes are ordered in the same way as the corresponding lectures were presented during the summer school.

1. *MRA-Based Wavelet Frames and Applications by Bin Dong and Zuowei Shen.* This lecture series includes five lectures which provide a concise, systematic and self contained overview of wavelet frames derived from a multiresolution analysis (MRA-based wavelet frames) and its applications in image analysis and restoration. It starts with an introduction to multiresolution analysis, its definition and approximation theory. Then the concept and concrete construction of tight frames are explained, including the multiresolution based wavelet tight frames and pseduo-spline tight frames. In the later part, applications to image processing based on tight frames are demonstrated. The first part focuses on the mathematical and theoretical foundations of tight frames including fast algorithms for framelet transform. The second part presents some of the very recent developments and applications in image restoration, blind deconvolution and others including a few fast optimization algorithms in these applications. For these applications, sparse representation using tight frames demonstrate various advantages in flexibility and stability due to the redundancy.

2. *Sparse and Redundant Representations Modeling of Images by Michael Elad.* This lecture series consists of five lectures on algorithms, theory and various applications for sparse and redundant representation of images, called *Sparseland.* In particular the author tries to address a few basic questions in the series. For example, how does one (1) find a good dictionary from a set of images that can provide sparse representation for images in a certain class, and (2) find a sparse

representation of an image from a given dictionary. A few formulations and algorithms for learning a dictionary from a training set are presented for question (1). Sparse-coding algorithms and approximation theory are provided for question (2). Various applications and examples are shown to demonstrate the validity of the *Sparseland* assumption and the efficiency of those algorithms introduced in the lectures.

3. *Simulation of Elasticity, Biomechanics, and Virtual Surgery by J. Teran, J. Hellrung, Jr. and J. Hegemann* This lecture series includes three lectures for the simulation of elastic materials characteristic of biological soft tissues. It starts with the modeling of elastic materials and derivations of the corresponding differential equations based on continuum mechanics and elasticity theory. The second lecture is about numerical discretization of those derived differential equations and design of efficient and stable numerical algorithms for computers. The third part provides some more details and further discussions of the implementation.

3. Acknowledgement

We would like to thank all the people who have contributed to the summer program. We are grateful to the lecturers for their effort in providing well prepared and accessible lecture series for this dynamic, broad and exciting field and to the TAs for holding discussion sessions and computer labs. Special thanks goes to those who contributed to this volume. We are sure the PCMI program and this volume will benefit the participants and the community in the years to come.

We would also like to thank Catherine Giesbrecht and Dena Vigil for their invaluable help for organizing the program. The support of the whole PCMI steering committee chaired by Richard Hain is appreciated. Our final thanks goes to John Polking for his help in preparing this volume.

MRA-Based Wavelet Frames and Applications

Bin Dong and Zuowei Shen

IAS/Park City Mathematics Series
Volume 19, 2010

MRA-Based Wavelet Frames and Applications

Bin Dong and Zuowei Shen

Introduction

This collection of lecture notes is based on a series of lectures presented by the authors at the Institute for Advanced Studies/Park City Mathematics 2010 Summer Program on "The Mathematics of Image Processing". The aim is to provide a concise, systematic and self contained overview on wavelet frames derived from a multiresolution analysis (MRA-based wavelet frames) and its applications in image analysis and restoration.

The publication of the unitary extension principle [**158**] in 1997 was the starting point of the MRA-based wavelet frames, especially the MRA-based tight wavelet frames. The unitary extension principle makes constructing wavelet frame systems with multiresolution structure painless. Wavelet frames derived from a multiresolution provide sparse approximations for piecewise smooth functions and have fast decomposition and reconstruction algorithms associated with them. These two important properties of MRA-based wavelet frames lead to many applications of wavelet frames in image processing. These lecture notes provide a detailed and comprehensive description of MRA-based wavelet frame theory and its applications.

Frame theory and its applications, notably the Gabor frames (see e.g. [**61, 99, 132**]) and wavelet frames (without a multiresolution structure) (see e.g. [**61, 132**]), were developed long before the discovery of the multiresolution analysis of [**131, 134**] and the systematic construction of the MRA-based compactly supported orthonormal wavelets of [**60**]. The concept of frame can be traced back to [**80**]. The rich literature of Gabor and wavelet frames provides a wide range of applications including time frequency analysis for signal processing, coherent state in quantum mechanics, filter bank design in electrical engineering, edge and singularity detection in image processing, and etc. It is not an intention of these notes to give a complete survey on frame theory and applications. Details on the earlier development and applications of Gabor and wavelet frames are provided in [**61, 99, 132, 134, 135**] and the references therein. Instead, the purpose here is to summarize some of contributions of the authors and their collaborators on the MRA-based wavelet frames and their applications. Special attention is given to applications of MRA-based tight wavelet frames in image restorations and analysis; a theme that is easily followed. We hope that these notes can be used for a course on this subject in future. A short note of such is given in [**167**].

Department of Mathematics, The University of Arizona, 617 N. Santa Rita Ave., Tucson, AZ, 85721-0089

E-mail address: dongbin@math.arizona.edu

Department of Mathematics, National University of Singapore, Singapore 119076

E-mail address: matzuows@nus.edu.sg

MRA-based wavelet frames, especially, MRA-based wavelet tight frames, can be viewed as a generalization of the MRA-based orthonormal wavelet of [**60, 131**]. Since the publication of [**60, 131**], especially after the compactly supported MRA-based orthonormal wavelets being constructed by [**60**], wavelet analysis and its applications have been one of the most active research areas in applied mathematics over the past two decades. Algorithms and methods based on wavelet analysis have become powerful tools in image/signal processing and analysis. One well known application of MRA-based wavelets is image compression (see e.g. [**132**]) using the orthonormal or bi-orthogonal MRA-based wavelet bases of [**56, 60**]. Tight wavelet frames derived from over sampled orthonormal wavelet basis are already used in noise removal by [**57, 77**]. The publication of the unitary extension principle of [**158**] initiated a new wave of theoretical development, as well as exploration of new applications of MRA-based tight wavelet frames. Further theoretical developments on MRA-based wavelet frames can be found in e.g. [**52, 65, 106**] and the references therein. In order to be useful in applications, tight wavelet frames that have a multiresolution structure are preferred because this guarantees the existence of fast decomposition and reconstruction algorithms. Recently, MRA-based tight wavelet frames have been used in image inpainting, image denoising, image deblurring, blind deblurring, and image decompositions (see e.g. [**14, 15, 16, 22, 23, 25, 32, 34, 36, 38**]). More recently in [**71**], the MRA-based tight wavelet frame is used to develop algorithms for image segmentation. In some of applications mentioned above, the unitary extension principle is used to design a tight wavelet frame system adapted to the real life problems in hand, see, e.g. [**14, 25, 32, 36, 38**]. In other applications, simple tight wavelet frames derived from the unitary extension principle are used. Frame based algorithms for $3D$ surface reconstruction from scattered points and medical imaging (e.g. tomography) are currently being explored.

The community's effort to develop redundant wavelet systems that have sparse approximations for various classes of functions has led to the development of the MRA-based wavelet frames. The properties of redundancy, sparse approximation and the existence of fast decomposition and reconstruction algorithms are the key factors for the wide usage of wavelet frame systems. There are a few other redundant wavelet systems that have been developed and widely used in image/signal processing and analysis. Such redundant systems include, for example, bi-frames of [**52, 65, 101, 157**], bandlets of [**132**], ridgelets of [**78**], curvelets of [**26, 27**], and shearlets of [**102, 123**]. We forgo discussing any other redundant wavelet systems in order to have a clear focus in these lecture notes. Interested readers should consult the relevant references for further details.

The lecture notes are organized as follows. In Lecture 1, we present the multiresolution analysis, which lays the foundation of MRA-based wavelets and tight wavelet frames and provides a platform for multi-level analysis of images/signals. In Lecture 2, based on the MRA structure, we introduce the unitary extension principle of [**158**], which provides a general and convenient framework to construct MRA-based tight frames (framelets) of $L_2(\mathbb{R}^d)$. The MRA-based tight frames derived from the unitary extension principle have fast decomposition and reconstruction algorithms with the same complexity as convolutions. Details of fast algorithms for framelet decomposition and reconstruction are also given. In Lecture 3, we introduce a general class of refinable functions, called pseudo-splines [**65, 165, 75**], which includes B-splines, Daubechies orthogonal refinable function and interpolatory refinable functions as special cases. Comprehensive regularity analysis of pseudo-splines, as well as the construction of (anti)symmetric pseudo-spline tight frames is provided. In Lecture 4, we present tight frame based models namely, balanced, synthesis based and analysis based approach, as well as the corresponding fast algorithms for image restoration problems, e.g. image deblurring, denoising and inpainting. In Lecture 5, we consider some other interesting and important applications of tight frames. We discuss the model proposed in [**18**] on blind deblurring (motion deblurring to be specific) problems. The major difficulty of blind deblurring problems is that the convolution kernel is unknown and needs to be solved simultaneously with the underlying images. Then, we present a frame based image segmentation model with a fast algorithm for the general image segmentation problems of [**71**]. At the end of this lecture, we discuss the model proposed by [**112**] on reconstruction of scenes (visible surfaces) from scattered, noisy and possibly sparse range data (point clouds). High-quality 3D scene modeling has long been an important research topic in computer vision, robotic navigation, computer graphics and animation.

LECTURE 1

Multiresolution analysis

The concept of multiresolution analysis (MRA) was first introduced by Mallat [**131**] and Meyer [**134**]. It is a general framework that makes constructing orthonormal wavelet bases for $L_2(\mathbb{R})$ very easy. The MRA-based compactly supported orthonormal wavelet systems were constructed by Daubechies [**60**]. Furthermore, the MRA structure grants fast implementation of wavelet decomposition and reconstruction which makes wavelets a very practical tool for image/signal processing and analysis.

1. Definitions and basics

Here, we adopt a more general MRA structure as proposed in [**68**], rather than the structure originally proposed by Mallat and Meyer.

For a given function $\phi \in L_2(\mathbb{R})$, we define the *shift-invariant subspace* $V(\phi) \subset L_2(\mathbb{R})$ generated by ϕ as

$$V(\phi) := \overline{\operatorname{span}\{\phi(\cdot - k),\ k \in \mathbb{Z}\}},$$

and denote V_n as the 2^n-dilate of $V(\phi)$, i.e.

$$V_n = \overline{\operatorname{span}\{\phi(2^n \cdot -k),\ k \in \mathbb{Z}\}},\ n \in \mathbb{Z}. \tag{1.1}$$

We have $V = V_0$. A subspace $S \subset L_2(\mathbb{R})$ is called *translation-invariant* if for any $t \in \mathbb{R}$ and $f \in S$, we have $f(\cdot - t) \in S$. The subspace S is called *s-shift-invariant* if for any $k \in \mathbb{Z}$ and $f \in S$, we have $f(\cdot - sk) \in S$, and in particular if $s = 1$, we call S a *shift-invariant* subspace.

Now for a given sequence of subspaces $\{V_n\}_{n\in\mathbb{Z}}$, we say that $\{V_n\}$ forms a *multiresolution analysis (MRA)* for $L_2(\mathbb{R})$, if the following conditions are satisfied:

$$V_n \subset V_{n+1},\ n \in \mathbb{Z}, \tag{1.2}$$

$$\overline{\cup_n V_n} = L_2(\mathbb{R}), \tag{1.3}$$

$$\cap_n V_n = \{0\}. \tag{1.4}$$

Here we call ϕ the *generator* of the MRA. The major tasks of Lecture 1 are to investigate which conditions ϕ must satisfy in order for the corresponding sequence of subspaces $\{V_n\}_{n\in\mathbb{Z}}$ to form an MRA, and to study the approximation properties of an MRA.

To make (1.2) hold, the generator $\phi \in L_2(\mathbb{R})$ must be in V_1. Once $\phi \in V_1$, the shifts of ϕ are in V_1 as well since V_1 is $\frac{1}{2}$-shift invariant. Therefore, to have $V_0 \subset V_1$, it is natural to require ϕ to be refinable. A function $\phi \in L_2(\mathbb{R})$ is said to be *refinable* if

$$\phi(x) = 2\sum_{k\in\mathbb{Z}} h_0[k]\phi(2x - k) \tag{1.5}$$

for some $h_0 \in \ell_2(\mathbb{Z})$. The sequence h_0 is called the *refinement mask* of ϕ. In Fourier domain, the definition of refinability of ϕ can be rewritten as

$$\widehat{\phi}(2\cdot) = \widehat{h}_0\widehat{\phi}. \tag{1.6}$$

Here $\widehat{\phi}$ denotes the Fourier transform of ϕ, and $\widehat{h}_0$ denotes the Fourier series of sequence h_0. Recall that the *Fourier transform* of a function $f \in L_1(\mathbb{R})$ is defined by

$$\widehat{f}(\xi) := \int_{\mathbb{R}} f(t)e^{-i\xi t}\mathrm{d}t, \qquad \xi \in \mathbb{R},$$

which can be extended to more general function spaces (e.g. $L_2(\mathbb{R})$ or space of tempered distributions) in a natural manner. The *Fourier series* of a sequence $h_0 \in \ell_2(\mathbb{Z})$ is defined as

$$\widehat{h}_0(\xi) := \sum_{j\in\mathbb{Z}} h_0[j]e^{-ij\xi}, \qquad \xi \in \mathbb{R}.$$

It follows from [**69**, Theorem 2.14] (also see [**68**]) that $\phi \in V_1$ whenever (1.6) holds with $\widehat{h}_0$ being a 2π periodic measurable function. Therefore, we generally do not need to require $h_0 \in \ell_2(\mathbb{Z})$. However, we still assume that $h_0 \in \ell_2(\mathbb{Z})$ for convenience. Indeed, when the unitary extension principle is applied to construct tight wavelet frames, we normally assume that ϕ and its shifts form a *Bessel sequence*, i.e. there exist $C < \infty$, such that for any sequence $b \in \ell_2(\mathbb{Z})$,

$$\|\sum_{k\in\mathbb{Z}} b[k]\phi(\cdot - k)\|_{L_2(\mathbb{R})} \leq C\|b\|_{\ell_2(\mathbb{Z})}. \tag{1.7}$$

When ϕ and its shifts form a Bessel sequence, it is easy to prove that $\phi \in V_1$ under the assumption that ϕ is refinable with its mask $h_0 \in \ell_2(\mathbb{Z})$. Finally, we remark that when $\phi \in L_2(\mathbb{R})$ is compactly supported, ϕ automatically satisfies (1.7) (see Lemma 3.1 in Lecture 3).

Examples of refinable functions include B-splines or more generally pseudo-splines. Here, we give examples of refinable B-splines, and postpone detailed discussion on pseudo-splines until Lecture 3. A (centered) B-spline with order m, denoted as B_m, is defined in the Fourier domain by

$$\widehat{B}_m(\xi) = e^{-ij\frac{\xi}{2}}\left(\frac{\sin(\xi/2)}{\xi/2}\right)^m, \tag{1.8}$$

where $j = 0$ when m is even, $j = 1$ when m is odd. The corresponding refinement mask of B_m is

$$\widehat{a}(\xi) = e^{-ij\frac{\xi}{2}}\cos^m(\xi/2).$$

The B-spline B_m is a compactly supported function in $C^{m-2}(\mathbb{R})$ with m as the length of its support. For detailed discussions about B-splines, one may refer to [**67**].

It turns out that we only need to place very mild requirements on $\phi \in L_2(\mathbb{R})$ for conditions (1.3) and (1.4) be satisfied. For example, any compactly supported refinable function $\phi \in L_2(\mathbb{R})$ with $\widehat{\phi}(0) \neq 0$ will generate an MRA. In particular, any B-spline of (1.8) generates an MRA. Details will be discussed in the next two sections.

2. Density of the union of V_n

In the following discussions we denote $\phi_n := 2^{n/2}\phi(2^n\cdot)$. Then V_n, defined in (1.1), is the 2^{-n}-shift-invariant subspace generated by ϕ_n. We know from [**69**, Theorem 2.14] that a function s belongs to V_n if and only if there exists some $2^{n+1}\pi$-period measurable function τ such that

$$\hat{s} = \tau\widehat{\phi_n}, \tag{1.9}$$

and $\tau\widehat{\phi_n} \in L_2(\mathbb{R})$.

Before we give a complete characterization of the density property (1.3) , we first recall two classical results of [**163**].

PROPOSITION 1.1 ([**163**, Theorem 9.5]). *For any function $f \in L_2(\mathbb{R})$ and every $t \in \mathbb{R}$, let f_t be the translate of f defined by*

$$f_t(x) = f(x - t) \quad x \in \mathbb{R}.$$

Then the mapping

$$t \mapsto f_t$$

is a uniformly continuous mapping of $\mathbb{R}$ into $L_2(\mathbb{R})$.

PROPOSITION 1.2 ([**163**, Theorem 9.17] or [**181**, page 100]). *Let X be a closed translation-invariant subspace of $L_2(\mathbb{R})$ with its Fourier transforms defined as*

$$\widehat{X} := \{\hat{f} : f \in X\}.$$

Then we have $\widehat{X} = L_2(\Omega)$ for some measurable set $\Omega \in \mathbb{R}$. Let X_1 and X_2 are two closed translation-invariant subspaces of $L_2(\mathbb{R})$ with $\widehat{X_1} = L_2(\Omega_1)$ and $\widehat{X_2} = L_2(\Omega_2)$. Then $X_1 = X_2$ if and only if $\Omega_1 = \Omega_2$ modulo a null-set.

REMARK 1.1. In particular, when one takes the Ω_2 in Proposition 1.2 to be $\mathbb{R}$, then we have that $\widehat{X_1} = L_2(\Omega_1)$ and $\widehat{X_2} = L_2(\mathbb{R})$. Proposition 1.2 tells us that $X_1 = X_2$ if and only if $\Omega_1 = \mathbb{R}$ modulo a null-set, where we note that $X_2 = L_2(\mathbb{R})$.

Now, we begin with the following lemma of [**68**]

LEMMA 1.1. *Let $\{V_n\}_{n\in\mathbb{Z}}$ be a nested sequence, i.e. (1.2) is satisfied. Then $\overline{\cup_n V_n}$ is a closed translation-invariant subspace of $L_2(\mathbb{R})$.*

PROOF. Let $X := \cup V_n$. Then $\overline{X}$ is certainly closed. We first show that if $f \in X$, and then for any $t_0 \in \mathbb{R}$, $f_{t_0} \in \overline{X}$. We take $f \in X$. Since V_n is a nested sequence, i.e. $V_n \subset V_{n+1}$, then for n sufficiently large we have $f \in V_n$ (i.e. $\exists K > 0$, s.t. for all $n > K$, $f \in V_n$). Since V_n is 2^{-n}-shift-invariant, $f_t = f(\cdot + t)$ is in X for any $t = 2^{-n}k$, $n, k \in \mathbb{Z}$, which means that f_t is in X for all dyadic $t = 2^{-n}k$, $k, n \in \mathbb{Z}$. Proposition 1.1 tells us that translation is a continuous operation in $L_2(\mathbb{R})$, i.e. for any $\varepsilon > 0$, $\exists\delta > 0$ and $|s - t| < \delta$, such that $\|f_s - f_t\|_{L_2(\mathbb{R})} < \varepsilon$. It is also well known that the dyadic set $A := \{2^k j\}_{k,n\in\mathbb{Z}}$ is dense in $\mathbb{R}$, which means for any $t_0 \in \mathbb{R}$ we can choose $t \in A$ s.t. $|t - t_0| < \delta$ for any given $\delta > 0$. Then by the continuity of translation of L_2 functions, for any $t_0 \in \mathbb{R}$, we can always find $t \in A$ such that for any $\varepsilon > 0$, $\|f_t - f_{t_0}\|_{L_2(\mathbb{R})} < \varepsilon$. Since $\overline{X}$ is closed, we have $f_{t_0} \in \overline{X}$.

Now, we show that for $g \in \overline{X}$ and $t_0 \in \mathbb{R}$, $g_{t_0} \in \overline{X}$. We take $g \in \overline{X}$. Then there always exists a function $f \in X$, s.t. $\|g - f\|_{L_2(\mathbb{R})} < \varepsilon$, $\forall\varepsilon > 0$. We take $t_0 \in \mathbb{R}$, and then the equality $\|g_{t_0} - f_{t_0}\|_{L_2(\mathbb{R})} = \|g - f\|_{L_2(\mathbb{R})}$ and the closeness of $\overline{X}$ show that

$g_{t_0} \in \overline{X}$, since we have already proved that $f_{t_0} \in \overline{X}$. This concludes that $\overline{X} = \overline{\cup V_n}$ is a closed translation-invariant subspace of $L_2(\mathbb{R})$. ∎

Lemma 1.1 gives us the translation-invariance of $\overline{\cup V_n}$ which allows us to use Proposition 1.2 to prove the important theorem below which is first proven in [**68**] (also see [**115**]).

THEOREM 1.1. *Let $\{V_n\}_{n\in\mathbb{Z}}$ be a nested sequence, i.e.* (1.2) *is satisfied. Then $\overline{\cup_n V_n} = L_2(\mathbb{R})$ if and only if $\Omega_0 := \cup_n \text{supp}\, \widehat{\phi_n} = \mathbb{R}$ (modulo a null-set).*

PROOF. Let $X := \overline{\cup V_n}$. From Proposition 1.2 we have $\widehat{X} = L_2(\Omega)$ for some measurable set $\Omega \subset \mathbb{R}$. Moreover, Proposition 1.2 tells us that $X = L_2(\mathbb{R})$ if and only if $\Omega = \mathbb{R}$ modulo a null-set. We verify that $\Omega = \Omega_0$ modulo a null-set which will complete the proof. Since each ϕ_n is in X, we must have supp $\widehat{\phi_n} \subset \Omega$ modulo a null-set, and so, $\Omega_0 \subset \Omega$ modulo a null-set. Now we suppose that $\Omega \setminus \Omega_0$ contains a set Ω_1 of positive measure. From (1.9), we have supp $\hat{s} \subset$ supp $\widehat{\phi_n}$ for any $s \in V_n$. Since $\widehat{\phi_n}$ vanishes on Ω_1 for all $n \in \mathbb{Z}$, so does the Fourier transform of each element in V_n, $n \in \mathbb{Z}$. Hence, each element in $\cup V_n$ has Fourier transform which vanishes on Ω_1. Lastly, if we show that by taking the closure, each element in X has Fourier transform which vanishes on Ω_1, then it contradicts with the fact that $\widehat{X} = L_2(\Omega)$ contains $L_2(\Omega_1)$. Thus we will have $\Omega \setminus \Omega_0$ is of measure zero, which then implies that $\Omega = \Omega_0$ modulo a null-set.

Now take $f \in X$, there exists $g \in \cup V_n$ with $\hat{g}$ vanishes on Ω_1, such that for any $\varepsilon > 0$,

$$\|f - g\|_{L_2(\mathbb{R})} < \varepsilon.$$

Applying Parseval's identity we have

$$\|f - g\|_{L_2(\mathbb{R})} = \frac{1}{\sqrt{2\pi}}\|\hat{f} - \hat{g}\|_{L_2(\mathbb{R})}.$$

By the fact that $\hat{g}$ vanishes on Ω_1 we have

$$\begin{aligned}\frac{1}{\sqrt{2\pi}}\|\hat{f} - \hat{g}\|_{L_2(\mathbb{R})} \geq \frac{1}{\sqrt{2\pi}}\|\hat{f} - \hat{g}\|_{L_2(\Omega_1)} &\geq \frac{1}{\sqrt{2\pi}}\|\hat{f}\|_{L_2(\Omega_1)} - \frac{1}{\sqrt{2\pi}}\|\hat{g}\|_{L_2(\Omega_1)} \\ &= \frac{1}{\sqrt{2\pi}}\|\hat{f}\|_{L_2(\Omega_1)}.\end{aligned}$$

Thus we have

$$\frac{1}{\sqrt{2\pi}}\|\hat{f}\|_{L_2(\Omega_1)} < \varepsilon.$$

Therefore $\hat{f}$ also vanishes on Ω_1 and this concludes the proof of this theorem. ∎

THEOREM 1.2 ([**115**]). *Let $\phi \in L_2(\mathbb{R})$ and V_n defined by* (1.1). *Assume that ϕ is refinable. Then,*

$$\overline{\cup_n V_n} = L_2(\mathbb{R})$$

if and only if

$$\cap_n 2^n Z(\hat{\phi})$$

is a set of measure zero, where $Z(\hat{\phi}) := \{\xi \in \mathbb{R} : \hat{\phi}(\xi) = 0\}$.

PROOF. We first note that (1.2) holds because ϕ is refinable. Indeed, from the refinement equation (1.6), we have

$$\widehat{\phi}(2^{-n+1}\xi) = \widehat{h}_0(2^{-n}\xi)\widehat{\phi}(2^{-n}\xi).$$

Since $\widehat{h}_0$ is 2π-periodic, $\widehat{h}_0(2^{-n}\cdot)$ is $2^{n+1}\pi$-periodic and hence $\phi(2^{n-1}\cdot) \in V_n$ by (1.9). Since V_n is 2^{-n}-shift-invariant which is obviously also 2^{-n+1}-shift-invariant, we have $\{\phi(2^{n-1}\cdot -k) : k \in \mathbb{Z}\} \subset V_n$ which implies $V_{n-1} \subset V_n$.

From Theorem 1.1 we know that $\overline{\cup V_n} = L_2(\mathbb{R})$ if and only if $\Omega_0 := \cup \operatorname{supp} \widehat{\phi_n} = \mathbb{R}$ modulo a null-set. We will show that $\Omega_0 = \mathbb{R}$ modulo a null-set if and only if $m(\cap 2^n Z(\hat{\phi})) = 0$ which will complete the proof. The fact that $\cup \operatorname{supp} \widehat{\phi_n} = \mathbb{R}$ modulo a null-set is equivalent to

$$m\Big(\big(\cup \operatorname{supp} \widehat{\phi_n}\big)^c\Big) = 0.$$

Moreover we have

$$\big(\cup \operatorname{supp} \widehat{\phi_n}\big)^c = \cap(\operatorname{supp} \widehat{\phi_n})^c = \cap Z(\widehat{\phi_n}),$$

modulo a null-set. Last, we show that $Z(\widehat{\phi_n}) = 2^n Z(\hat{\phi})$. We first note that $\widehat{\phi_n} = 0$ is equivalent to $\hat{\phi}(2^{-n}\cdot) = 0$. Now, we take any $\xi_0 \in Z(\widehat{\phi_n}) = \{\xi \in \mathbb{R} : \hat{\phi}(2^{-n}\xi) = 0\}$. We have $2^{-n}\xi_0 \in Z(\hat{\phi})$ and that means $\xi_0 \in 2^n Z(\hat{\phi})$. We take $\xi_0 \in 2^n Z(\hat{\phi})$ which means that $\xi_0 = 2^n \xi_1$, where $\hat{\phi}(\xi_1) = 0$. Since $\hat{\phi}(2^{-n}\xi_0) = \hat{\phi}(\xi_1) = 0$, we obtain $\xi_0 \in Z(\widehat{\phi_n})$. This shows that $Z(\widehat{\phi_n}) = 2^n Z(\hat{\phi})$. Now we conclude that

$$m\Big(\big(\cup \operatorname{supp} \widehat{\phi_n}\big)^c\Big) = m\Big(\cap 2^n Z(\widehat{\phi})\Big).$$

Therefore, $\cup \operatorname{supp} \widehat{\phi_n} = \mathbb{R}$ modulo a null-set if and only if $m\Big(\cap 2^n Z(\widehat{\phi})\Big) = 0$. This concludes the proof of the theorem. ∎

COROLLARY 1.1. *Let $\phi \in L_2(\mathbb{R})$ and V_n defined by* (1.1). *Assume that $\{V_n\}_n$ is nested. If $\hat{\phi}$ is non-zero almost everywhere (a.e.) in some neighborhood of the origin, we have $\overline{\cup_n V_n} = L_2(\mathbb{R})$. In particular, assume that $\hat{\phi}$ is continuous at the origin and $\hat{\phi}(0) \neq 0$. Then we have $\overline{\cup_n V_n} = L_2(\mathbb{R})$.*

PROOF. Since $\widehat{\phi_n} = 2^{-n/2}\hat{\phi}(2^{-n}\cdot)$, $\hat{\phi}$ being non-zero a.e. on Ω implies $\widehat{\phi_n}$ is non-zero on $2^n\Omega$. Now if Ω is some neighborhood of the origin, we obtain that $\cup \operatorname{supp} \widehat{\phi_n} = \mathbb{R}$, since $\cup 2^n \Omega = \mathbb{R}$. By Theorem 1.1 we have $\overline{\cup V_n} = L_2(\mathbb{R})$. ∎

COROLLARY 1.2. *Let $\phi \in L_2(\mathbb{R})$ with $\widehat{\phi}(0) = 1$ be a compactly supported refinable function, and V_n is defined as in* (1.1). *Then $\overline{\cup_n V_n} = L_2(\mathbb{R})$.*

PROOF. Since ϕ is compactly supported, the set of zero points of $\hat{\phi}$, i.e. $Z(\hat{\phi})$ is of measure zero. Then this corollary follows directly from Theorem 1.2. ∎

3. Triviality of the intersections of V_n

In this section, we study which requirements placed on ϕ guarantee that (1.4) is satisfied. Let us first introduce some basic concepts and notations that are used in the last section of this lecture as well.

We define the *bracket product* of two functions $f, g \in L_2(\mathbb{R})$, denoted as $[f, g]$, as follows:

$$[f, g](\xi) := \sum_{k \in \mathbb{Z}} f(\xi + 2\pi k)\overline{g(\xi + 2\pi k)}.$$

It is easy to verify that if $f, g \in L_2(\mathbb{R})$, then $[f, g] \in L_1([-\pi, \pi])$, and the Cauchy-Schwartz inequality gives that

$$\left|[f, g]\right|^2 \leq [f, f][g, g], \tag{1.10}$$

where the right hand side is finite a.e. (see e.g. [**114**]. Inequality (1.10) shows that bracket products for functions in $L_2(\mathbb{R})$ are well defined.

Using the notation of bracket products, one can easily show that (see Lemma 3.1 in Lecture 3) the shifts of ϕ forming an orthonormal set, i.e.

$$\langle \phi, \phi(\cdot - k) \rangle = \delta_{k,0},$$

with $\delta_{k,0} = 0$ if $k \neq 0$ and $\delta_{0,0} = 1$, is equivalent to

$$[\widehat{\phi}, \widehat{\phi}] = 1 \quad \text{a.e.}$$

We need the following lemma to prove the main theorem of this section.

LEMMA 1.2. *For a given $\phi \in L_2(\mathbb{R})$, one can always find $\psi \in L_2(\mathbb{R})$ such that $V(\phi) \subset V(\psi)$ and the shifts of ψ are orthonormal.*

PROOF. For given $\phi \in L_2(\mathbb{R})$, we define ψ as follows:

$$\hat{\psi}(\xi) := \begin{cases} \frac{\hat{\phi}(\xi)}{\sqrt{[\hat{\phi}, \hat{\phi}](\xi)}}, & \text{if } [\hat{\phi}, \hat{\phi}](\xi) > 0, \\ 1, & \text{if } [\hat{\phi}, \hat{\phi}](\xi) = 0 \text{ and } \xi \in [-\pi, \pi), \\ 0, & \text{elsewhere.} \end{cases}$$

Since $[\hat{\phi}, \hat{\phi}] \in L_1(\mathbb{R})$, it is finite almost everywhere, and thus

$$[\hat{\psi}, \hat{\psi}](\xi) = 1 \quad \text{for almost all } \xi \in \mathbb{R}.$$

Hence $\psi \in L_2(\mathbb{R})$ and the shifts of ψ are orthonormal. Moreover,

$$\hat{\phi} = \sqrt{[\hat{\phi}, \hat{\phi}]}\hat{\psi}.$$

Then (1.9) implies that $\phi \in V(\psi)$, and hence $V(\phi) \subset V(\psi)$. ∎

The following result was first proven in [**68**] and the proof given here is from [**115**]).

THEOREM 1.3. *Given any $\phi \in L_2(\mathbb{R})$, we have $\cap_n V_n = \{0\}$.*

PROOF. By lemma 1.2, it suffices to prove the theorem for the case when $\{\phi(\cdot - k), k \in \mathbb{Z}\}$ is an orthonormal set. We note that whenever $\{\phi(\cdot - k), k \in \mathbb{Z}\}$ forms an orthonormal set, so does $\{2^{n/2}\phi(2^n \cdot -k), k \in \mathbb{Z}\}$. Here we denote $\phi_{n,k} := 2^{n/2}\phi(2^n \cdot -k)$.

Let P_n be the orthogonal projector from $L_2(\mathbb{R})$ onto V_n. The theorem will be established if we can prove

$$P_n f \to 0 \quad \text{as} \quad n \to -\infty \tag{1.11}$$

for every $f \in L_2(\mathbb{R})$. Note that compactly supported functions are dense in $L_2(\mathbb{R})$, which means there always exists $\tilde{f} \in L_2(\mathbb{R})$ which is compactly supported and

$$\|f - \tilde{f}\|_{L_2(\mathbb{R})} < \varepsilon.$$

Hence suppose we have $P_n\tilde{f} \to 0, \quad n \to -\infty$, then

$$\begin{aligned} \|P_n f\|_{L_2(\mathbb{R})} - \|P_n\tilde{f}\|_{L_2(\mathbb{R})} &\leq \|P_n f - P_n\tilde{f}\|_{L_2(\mathbb{R})} \\ &= \|P_n(f-\tilde{f})\|_{L_2(\mathbb{R})} \\ &\leq \|f-\tilde{f}\|_{L_2(\mathbb{R})} \leq \varepsilon \end{aligned}$$

which means that $P_n\tilde{f} \to 0$ implies $P_n f \to 0$. So it is sufficient to show that (1.11) is true for any compactly supported function $f \in L_2(\mathbb{R})$. Let f be such a function. Then $P_n f$ can be expressed as

$$P_n f = \sum_{k\in\mathbb{Z}} c_n[k]\phi_{n,k},$$

where

$$c_n[k] = \int_{\mathbb{R}} f(x)\overline{\phi_{n,k}(x)}dx$$

are sequences in $\ell_2(\mathbb{Z})$. Since the shifts of ϕ are orthonormal, we deduce from the equation of $P_n f$ that

$$\begin{aligned} \|P_n f\|^2_{L_2(\mathbb{R})} &= \|c_n\|^2_{\ell_2(\mathbb{Z})} \\ &= \sum_{k\in\mathbb{Z}} \Big|\int_{\mathbb{R}} f(x)\overline{\phi_{n,k}(x)}dx\Big|^2 \\ &\leq \sum_{k\in\mathbb{Z}} \Big(\int_{\mathbb{R}} |f(x)\phi_{n,k}(x)|dx\Big)^2 \\ &= \sum_{k\in\mathbb{Z}} \Big(\int_{\mathbb{R}} |f(x)2^{n/2}\phi(2^n x - k)|dx\Big)^2. \end{aligned}$$

The first identity follows by Plancherel formula for the orthonormal basis $\{\phi_{n,k}\}_{k\in\mathbb{Z}}$. If f is supported in the interval $[-R, R]$, then by the Cauchy-Schwartz inequality we have,

$$\|P_n f\|^2_{L_2(\mathbb{R})} \leq \|f\|^2_{L_2(\mathbb{R})} \sum_{k\in\mathbb{Z}} \int_{\{2^n[-R,R]-k\}} |\phi(x)|^2 dx.$$

Now if $n < 0$ and $|n|$ is sufficiently large, we have that

$$\{2^n[-R,R]-k_1\} \cap \{2^n[-R,R]-k_2\} = \emptyset, \quad k_1 \neq k_2.$$

Therefore,

$$\|P_n f\|^2_{L_2(\mathbb{R})} \leq \|f\|^2_{L_2(\mathbb{R})} \int_{E_n} |\phi(x)|^2 dx = \|f\|^2_{L_2(\mathbb{R})} \int_{\mathbb{R}} \chi_{E_n}|\phi(x)|^2 dx, \tag{1.12}$$

where

$$E_n = \cup_{k\in\mathbb{Z}}(2^n[-R,R]-k).$$

Since when $n \to -\infty$, $\chi_{E_n}|\phi(x)|^2 \to 0$ for almost every $x \in \mathbb{R}$, and $\chi_{E_n}|\phi(x)|^2 \leq |\phi(x)|^2 \in L_1(\mathbb{R})$, then by the dominated convergence theorem we have

$$\lim_{n\to-\infty} \|f\|^2_{L_2(\mathbb{R})} \int_{\mathbb{R}} \chi_{E_n}|\phi(x)|^2 dx = \|f\|^2_{L_2(\mathbb{R})} \int_{\mathbb{R}} \lim_{n\to-\infty} \chi_{E_n}|\phi(x)|^2 dx = 0.$$

Thus $\lim_{n\to-\infty} \|P_n f\|^2_{L_2(\mathbb{R})} = 0.$ ∎

Now we summarize the results we get in this section and the previous one in the following theorem.

THEOREM 1.4. *Let $\phi \in L_2(\mathbb{R})$, and V_n defined as in* (1.1). *Then,*

(1) $\cap_n V_n = \{0\}$;
(2) *Assume, in addition, that ϕ is refinable. Then,*

$$\overline{\cup_n V_n} = L_2(\mathbb{R})$$

if and only if

$$\cap_n 2^n Z(\hat{\phi})$$

is a set of measure zero, where $Z(\hat{\phi}) := \{\xi \in \mathbb{R} : \hat{\phi}(\xi) = 0\}$.

In particular, the sequence of subspaces $\{V_n\}_{n\in\mathbb{Z}}$ generated by $\phi \in L_2(\mathbb{R})$ forms an MRA if either of the following condition is satisfied:

(1) *ϕ is a compactly supported refinable function with $\widehat{\phi}(0) \neq 0$;*
(2) *ϕ is refinable and $\widehat{\phi}$ is continuous at 0 with $\widehat{\phi}(0) \neq 0$.*

4. Approximation

In order to obtain a function in V_n that provides a good approximation to a given function f, one needs an approximation scheme. One of the commonly used approximation schemes is the quasi-interpolation scheme. There are many quasi-interpolatory schemes that give good approximations. We start by defining the quasi-interpolatory scheme that is used in MRA-based tight wavelet frame approximation.

For any function $f \in L_2(\mathbb{R})$, the dyadic dilation operator $\mathcal{D}$ is defined by $\mathcal{D}f(x) := \sqrt{2}f(2x)$ and the translation operator T is defined by $T_a f(x) := f(x-a)$ for $a \in \mathbb{R}$. Given $j \in \mathbb{Z}$, we have $T_a\mathcal{D}^j = \mathcal{D}^j T_{2^j a}$. Define $\phi_{n,k} := D^n T_k \phi$.

For a given $\phi \in L_2(\mathbb{R})$ that generates an MRA $\{V_n\}_n$, the *quasi-interpolatory operator* used here is defined as

$$\mathcal{P}_n : f \mapsto \sum_{k\in\mathbb{Z}} \langle f, \phi_{n,k}\rangle \phi_{n,k}, \tag{1.13}$$

for an arbitrary $f \in L_2(\mathbb{R})$.

We first provide the following proposition which will become handy for our analysis.

PROPOSITION 1.3. *Let $f, \phi \in L_2(\mathbb{R})$. Assume that $[\widehat{\phi}, \widehat{\phi}]$ is essentially bounded. Then*

(1) *The shifts of ϕ form a Bessel system, i.e.* (1.7) *is satisfied. Furthermore, the Bessel bound, C in* (1.7), *is equal to $\|[\widehat{\phi}, \widehat{\phi}]\|^{\frac{1}{2}}_{L_\infty([-\pi,\pi])}$.*
(2) *The 2π-periodic function $[\widehat{f}, \widehat{\phi}] \in L_1([-\pi,\pi])$ is the Fourier series of the sequence $\{\langle f, \phi(\cdot - k)\rangle : k \in \mathbb{Z}\}$. As a consequence,*

$$\sum_{k\in\mathbb{Z}} \langle f, \phi(\cdot - k)\rangle e^{-ik\xi} = [\widehat{f}, \widehat{\phi}](\xi)$$

almost everywhere.
(3) *The operator $\mathcal{P}_n$ is bounded with bound $\|[\widehat{\phi}, \widehat{\phi}]\|_{L_\infty([-\pi,\pi])}$. Furthermore, the Fourier transform of $\mathcal{P}_n f$ satisfies*

$$\widehat{\mathcal{P}_n f} = [\widehat{f}(2^n\cdot), \widehat{\phi}](2^{-n}\cdot)\widehat{\phi}(2^{-n}\cdot) \tag{1.14}$$

PROOF. For (1), we denote $\tilde{\phi} := [\widehat{\phi}, \widehat{\phi}]^{\frac{1}{2}}$. The Plancherel theorem says that for all finite sequences $b \in \ell_2(\mathbb{Z})$ we have

$$2\pi \left\| \sum_{k\in\mathbb{Z}} b[k]\phi(\cdot - k) \right\|_{L_2(\mathbb{R})}^2 = \|\widehat{b}\widehat{\phi}\|_{L_2(\mathbb{R})}^2.$$

By breaking the real line into $\mathbb{R} = \cup_{k\in\mathbb{Z}}\{[-\pi,\pi] + 2k\pi\}$ and noticing that $\widehat{b}$ is 2π-periodic, we have

$$\begin{aligned}
\|\widehat{b}\widehat{\phi}\|_{L_2(\mathbb{R})}^2 &= \int_{\mathbb{R}} |\widehat{b}(\xi)|^2 |\widehat{\phi}(\xi)|^2 \mathrm{d}\xi \\
&= \sum_{k\in\mathbb{Z}} \int_{[-\pi,\pi]+2k\pi} |\widehat{b}(\xi)|^2 |\widehat{\phi}(\xi)|^2 \mathrm{d}\xi \\
&= \sum_{k\in\mathbb{Z}} \int_{-\pi}^{\pi} |\widehat{b}(\xi)|^2 |\widehat{\phi}(\xi + 2k\pi)|^2 \mathrm{d}\xi \\
&= \|\widehat{b}\tilde{\phi}\|_{L_2([-\pi,\pi])}^2.
\end{aligned}$$

Therefore, we obtain the following identity

$$2\pi \left\| \sum_{k\in\mathbb{Z}} b[k]\phi(\cdot - k) \right\|_{L_2(\mathbb{R})}^2 = \|\widehat{b}\tilde{\phi}\|_{L_2([-\pi,\pi])}^2. \tag{1.15}$$

Suppose $\tilde{\phi}$ is essentially bounded. Then (1.15) is satisfied for all $b \in \ell_2(\mathbb{Z})$. Therefore,

$$\|\widehat{b}\tilde{\phi}\|_{L_2([-\pi,\pi])} \leq C\|\widehat{b}\|_{L_2([-\pi,\pi])},$$

with $C = \|\tilde{\phi}\|_{L_\infty([-\pi,\pi])}$ and thus (1.7) follows.

For (2), using Fubini's theorem and the fact that $e^{ik\xi}$ is 2π-periodic, we have,

$$\begin{aligned}
\int_{-\pi}^{\pi} [\widehat{f}, \widehat{\phi}](\xi) e^{ik\xi} \mathrm{d}\xi &= \int_{-\pi}^{\pi} \sum_{k\in\mathbb{Z}} \widehat{f}(\xi + 2k\pi)\overline{\widehat{\phi}(\xi + 2k\pi)} e^{ik\xi} \mathrm{d}\xi \\
&= \sum_{k\in\mathbb{Z}} \int_{-\pi}^{\pi} \widehat{f}(\xi + 2k\pi)\overline{\widehat{\phi}(\xi + 2k\pi)} e^{ik\xi} \mathrm{d}\xi \\
&= \sum_{k\in\mathbb{Z}} \int_{[-\pi,\pi]+2k\pi} \widehat{f}(\xi)\overline{\widehat{\phi}(\xi)} e^{ik\xi} \mathrm{d}\xi \\
&= \int_{\mathbb{R}} \widehat{f}(\xi)\overline{\widehat{\phi}(\xi)} e^{ik\xi} \mathrm{d}\xi,
\end{aligned}$$

for any $k \in \mathbb{Z}$. Now by the Plancherel theorem, we have

$$\int_{\mathbb{R}} \widehat{f}(\xi)\overline{\widehat{\phi}(\xi)} e^{ik\xi} \mathrm{d}\xi = \int_{\mathbb{R}} \widehat{f}(\xi)\overline{\widehat{\phi}(\xi) e^{-ik\xi}} \mathrm{d}\xi = 2\pi \int_{\mathbb{R}} f(x)\overline{\phi(x-k)} \mathrm{d}x.$$

Therefore, we have

$$\frac{1}{2\pi} \int_{-\pi}^{\pi} [\widehat{f}, \widehat{\phi}](\xi) e^{ik\xi} \mathrm{d}\xi = \langle f, \phi(\cdot - k)\rangle.$$

This means that $[\widehat{f}, \widehat{\phi}](\xi)$ is the Fourier series of $\{\langle f, \phi(\cdot - k)\rangle : k \in \mathbb{Z}\}$.

For (3), the boundedness of $\mathcal{P}_n$ follows directly from the facts that $\|\mathcal{P}_n\| = \|\mathcal{P}_0\|$ for all $n \in \mathbb{Z}$ and the shifts of ϕ form a Bessel system. Now, we show the identity

(1.14) in part (3). Notice that $\mathcal{P}_n = \mathcal{D}^n\mathcal{P}_0\mathcal{D}^{-n}$. Therefore, all we need to show is that

$$\widehat{\mathcal{P}_0 f} = [\widehat{f}, \widehat{\phi}]\widehat{\phi}. \tag{1.16}$$

Indeed, we have

$$\begin{aligned}
\widehat{\mathcal{P}_0 f}(\xi) &= \int_{\mathbb{R}} \sum_{k\in\mathbb{Z}} \langle f, \phi(\cdot - k)\rangle \phi(x-k)e^{-i\xi x}\mathrm{d}x\\
&= \sum_{k\in\mathbb{Z}} \langle f, \phi(\cdot - k)\rangle e^{-ik\xi} \int_{\mathbb{R}} \phi(x-k)e^{-i\xi(x-k)}\mathrm{d}x\\
&= \sum_{k\in\mathbb{Z}} \langle f, \phi(\cdot - k)\rangle e^{-ik\xi}\widehat{\phi}(\xi)\\
&= [\widehat{f}, \widehat{\phi}](\xi)\widehat{\phi}(\xi),
\end{aligned}$$

where the switching of orders of summation and integration is justified by the fact that the summation converges in the sense of $L_2(\mathbb{R})$ and Fourier transform is continuous on $L_2(\mathbb{R})$. ■

We say that $\mathcal{P}_n$ provides approximation order m if, for all $f \in \mathcal{W}_2^m(\mathbb{R})$,

$$\|f - \mathcal{P}_n f\|_{L_2(\mathbb{R})} = O(2^{-nm}).$$

Here $\mathcal{W}_2^m(\mathbb{R})$, $m \in \mathbb{R}_+$, are Sobolev spaces defined by

$$\mathcal{W}_2^m(\mathbb{R}) := \{f \in L_2(\mathbb{R}) : \|f\|_{\mathcal{W}_2^m(\mathbb{R})} := \sqrt{2\pi}\|(1+|\cdot|)^m \widehat{f}\|_{L_2(\mathbb{R})} < \infty\}.$$

The rest of this section is devoted to the analysis of approximation orders of $\mathcal{P}_n$. As one will see, the approximation order of $\mathcal{P}_n$ is closely related to the property of $1-|\widehat{\phi}|^2$ and $[\widehat{\phi}, \widehat{\phi}] - |\widehat{\phi}|^2$ at the origin.

In the following discussions, we denote $|f| = O(|\cdot|^m)$ whenever $\lim_{x\to 0}\frac{|f|(x)}{|x|^m} \le C$ with $C > 0$. We say that a function has zero of order m if $f^{(j)}(0) = 0$ for $j = 0, 1, \ldots, m-1$. It is clear that whenever $f \in C^m$ and it has zero of order m, then $|f| = O(|\cdot|^m)$. For convenience of arguments, whenever we say that f has zero of order m, we implicitly assume that $f \in C^m$ (although we only require the existence of $f^{(m-1)}$ in the original definition).

The following theorem, whose proof follows similar ideas of [**69**], shows the approximation order of the operator $\mathcal{P}_n$ depends on the behavior of the function $[\widehat{\phi}, \widehat{\phi}] - |\widehat{\phi}|^2$ at the origin.

THEOREM 1.5. *Let $\mathcal{P}_n$ be defined by* (1.13) *for a given function $\phi \in L_2(\mathbb{R})$ satisfying*

(1) *$[\widehat{\phi}, \widehat{\phi}]$ is essentially bounded;*
(2) *$[\widehat{\phi}, \widehat{\phi}] - |\widehat{\phi}|^2 = O(|\cdot|^{2m})$;*
(3) *$1 - |\widehat{\phi}|^2 = O(|\cdot|^{2m_0})$.*

Then, $\mathcal{P}_n$ provides approximation order $m_1 = \min\{m, 2m_0\}$.

PROOF. Part (3) of Proposition 1.3 gives us

$$\begin{aligned}
\widehat{\mathcal{P}_n f} &= [\widehat{f}(2^n\cdot), \widehat{\phi}](2^{-n}\cdot)\widehat{\phi}(2^{-n}\cdot)\\
&= \sum_{k\in\mathbb{Z}} \widehat{f}(\xi + 2^{n+1}k\pi)\overline{\widehat{\phi}(2^{-n}\xi + 2k\pi)}\widehat{\phi}(2^{-n}\xi).
\end{aligned}$$

Then we have

$$\begin{aligned}\|\widehat{\mathcal{P}_n f} - \widehat{f}\|_{L_2(\mathbb{R})}^2 &= \int_{\mathbb{R}} \left|[\widehat{f}(2^n\cdot), \widehat{\phi}](2^{-n}\xi)\widehat{\phi}(2^{-n}\xi) - \widehat{f}(\xi)\right|^2 \mathrm{d}\xi \\ &= \int_{\mathbb{R}} \left|[\widehat{f}(2^n\cdot), \widehat{\phi}](2^{-n}\xi)\right|^2 \left|\widehat{\phi}(2^{-n}\xi)\right|^2 \mathrm{d}\xi \\ &- \int_{\mathbb{R}} [\widehat{f}(2^n\cdot), \widehat{\phi}](2^{-n}\xi)\widehat{\phi}(2^{-n}\xi)\overline{\widehat{f}(\xi)}\mathrm{d}\xi \\ &- \int_{\mathbb{R}} \overline{[\widehat{f}(2^n\cdot), \widehat{\phi}](2^{-n}\xi)\widehat{\phi}(2^{-n}\xi)}\widehat{f}(\xi)\mathrm{d}\xi + \|\widehat{f}\|_{L_2(\mathbb{R})}^2.\end{aligned}$$

Since $[\widehat{f}(2^n\cdot), \widehat{\phi}](2^{-n}\xi)$ is $2^{n+1}\pi$-periodic, we have

$$\begin{aligned}&\int_{\mathbb{R}} \left|[\widehat{f}(2^n\cdot), \widehat{\phi}](2^{-n}\xi)\right|^2 \left|\widehat{\phi}(2^{-n}\xi)\right|^2 \mathrm{d}\xi \\ &= \sum_{l\in\mathbb{Z}} \int_{2^n([-\pi,\pi]+2l\pi)} \left|[\widehat{f}(2^n\cdot), \widehat{\phi}](2^{-n}\xi)\right|^2 \left|\widehat{\phi}(2^{-n}\xi)\right|^2 \mathrm{d}\xi \\ &= \int_{2^n[-\pi,\pi]} \left|[\widehat{f}(2^n\cdot), \widehat{\phi}](2^{-n}\xi)\right|^2 [\widehat{\phi}, \widehat{\phi}](2^{-n}\xi)\mathrm{d}\xi.\end{aligned}$$

Similarly, we have

$$\begin{aligned}\int_{\mathbb{R}} [\widehat{f}(2^n\cdot), \widehat{\phi}](2^{-n}\xi)\widehat{\phi}(2^{-n}\xi)\overline{\widehat{f}(\xi)}\mathrm{d}\xi &= \int_{2^n[-\pi,\pi]} \left|[\widehat{f}(2^n\cdot), \widehat{\phi}](2^{-n}\xi)\right|^2 \mathrm{d}\xi \\ &= \int_{\mathbb{R}} \overline{[\widehat{f}(2^n\cdot), \widehat{\phi}](2^{-n}\xi)\widehat{\phi}(2^{-n}\xi)}\widehat{f}(\xi)\mathrm{d}\xi.\end{aligned}$$

Now altogether, we have

$$\begin{aligned}\|\widehat{\mathcal{P}_n f} - \widehat{f}\|_{L_2(\mathbb{R})}^2 &= \int_{2^n[-\pi,\pi]} \left|[\widehat{f}(2^n\cdot), \widehat{\phi}](2^{-n}\xi)\right|^2 \left([\widehat{\phi}, \widehat{\phi}](2^{-n}\xi) - 1\right) \mathrm{d}\xi \\ &+ \|\widehat{f}\|_{L_2(\mathbb{R})}^2 - \int_{2^n[-\pi,\pi]} \left|[\widehat{f}(2^n\cdot), \widehat{\phi}](2^{-n}\xi)\right|^2 \mathrm{d}\xi.\end{aligned}$$

Now suppose $\mathrm{supp}\widehat{f} \subset 2^n[-\pi, \pi]$. Then

$$\sum_{k\in\mathbb{Z}} \widehat{f}(\xi + 2^{n+1}k\pi)\overline{\widehat{\phi}(2^{-n}\xi + 2k\pi)} = \widehat{f}(\xi)\overline{\widehat{\phi}(2^{-n}\xi)}$$

on $2^n[-\pi, \pi]$. Thus,

$$\begin{aligned}\|\widehat{\mathcal{P}_n f} - \widehat{f}\|_{L_2(\mathbb{R})}^2 &= \int_{2^n[-\pi,\pi]} \left|\widehat{f}(\xi)\widehat{\phi}(2^{-n}\xi)\right|^2 \left([\widehat{\phi}, \widehat{\phi}](2^{-n}\xi) - 1\right) \mathrm{d}\xi \\ &+ \int_{2^n[-\pi,\pi]} |\widehat{f}(\xi)|^2(1 - |\widehat{\phi}(2^{-n}\xi)|^2)\mathrm{d}\xi \\ &= \int_{2^n[-\pi,\pi]} \left|\widehat{f}(\xi)\widehat{\phi}(2^{-n}\xi)\right|^2 \left([\widehat{\phi}, \widehat{\phi}](2^{-n}\xi) - |\widehat{\phi}(2^{-n}\xi)|^2\right) \mathrm{d}\xi \\ &+ \int_{2^n[-\pi,\pi]} |\widehat{f}(\xi)|^2 \left(|\widehat{\phi}(2^{-n}\xi)|^2 - 1\right)^2 \mathrm{d}\xi.\end{aligned}$$

Since $[\widehat{\phi},\widehat{\phi}]$ is essentially bounded, then when n is large enough, we have

$$\begin{aligned}\|\widehat{\mathcal{P}_n f}-\widehat{f}\|^2_{L_2(\mathbb{R})} &\le C\int_{2^n[-\pi,\pi]}|\widehat{f}(\xi)|^2\left|[\widehat{\phi},\widehat{\phi}](2^{-n}\xi)-|\widehat{\phi}(2^{-n}\xi)|^2\right|\mathrm{d}\xi\\ &\quad+\int_{2^n[-\pi,\pi]}|\widehat{f}(\xi)|^2\left(|\widehat{\phi}(2^{-n}\xi)|^2-1\right)^2\mathrm{d}\xi\\ &= C\int_{2^n[-\pi,\pi]}(1+|\xi|)^{2\tilde{m}}|\widehat{f}(\xi)|^2\frac{\left|[\widehat{\phi},\widehat{\phi}](2^{-n}\xi)-|\widehat{\phi}(2^{-n}\xi)|^2\right|}{(1+|\xi|)^{2\tilde{m}}}\mathrm{d}\xi\\ &\quad+\int_{2^n[-\pi,\pi]}(1+|\xi|)^{2\tilde{m}}|\widehat{f}(\xi)|^2\frac{\left(|\widehat{\phi}(2^{-n}\xi)|^2-1\right)^2}{(1+|\xi|)^{2\tilde{m}}}\mathrm{d}\xi.\end{aligned}$$

Therefore, $\mathcal{P}_n f$ provides approximation order $\tilde{m}$ for all $f\in W_2^{\tilde{m}}(\mathbb{R})$ with $\mathrm{supp}\widehat{f}\subset 2^n[-\pi,\pi]$ whenever we have

$$\left\|\frac{[\widehat{\phi},\widehat{\phi}](2^{-n}\xi)-|\widehat{\phi}(2^{-n}\xi)|^2}{(1+|\xi|)^{2\tilde{m}}}\right\|_{L_\infty(2^n[-\pi,\pi])}\le C2^{-2n\tilde{m}}$$

and

$$\left\|\frac{\left(|\widehat{\phi}(2^{-n}\xi)|^2-1\right)^2}{(1+|\xi|)^{2\tilde{m}}}\right\|_{L_\infty(2^n[-\pi,\pi])}\le C2^{-2n\tilde{m}},$$

for n large enough. The above two inequalities are equivalent to

$$\left\|\frac{[\widehat{\phi},\widehat{\phi}](\xi)-|\widehat{\phi}(\xi)|^2}{(2^{-n}+|\xi|)^{2\tilde{m}}}\right\|_{L_\infty([-\pi,\pi])}\le C$$

and

$$\left\|\frac{\left(|\widehat{\phi}(\xi)|^2-1\right)^2}{(2^{-n}+|\xi|)^{2\tilde{m}}}\right\|_{L_\infty([-\pi,\pi])}\le C,$$

which are all satisfied when $\tilde{m}=m_1=\min\{m,2m_0\}$.

So far we have established that for $f\in W_2^{m_1}(\mathbb{R})$ with $\mathrm{supp}\widehat{f}\subset 2^n[-\pi,\pi]$, we have

$$\|\mathcal{P}_n f-f\|_{L_2(\mathbb{R})}\le C2^{-nm_1}.$$

Now consider general $f\in W_2^{m_1}(\mathbb{R})$ and define $\widehat{g}:=\chi_{2^n[-\pi,\pi]}\widehat{f}$. Then we have

$$\begin{aligned}\|\mathcal{P}_n f-f\|_{L_2(\mathbb{R})}-\|\mathcal{P}_n g-g\|_{L_2(\mathbb{R})} &\le \|\mathcal{P}_n f-f-\mathcal{P}_n g+g\|_{L_2(\mathbb{R})}\\ &\le \|\mathcal{P}_n(f-g)\|_{L_2(\mathbb{R})}+\|f-g\|_{L_2(\mathbb{R})}\\ &\le C\|\widehat{f}-\widehat{g}\|_{L_2(\mathbb{R})},\end{aligned}$$

where the last inequality follows from the boundedness of operator $\mathcal{P}_n$ because $[\widehat{\phi},\widehat{\phi}]$ is essentially bounded (part (1) of Proposition 1.3). Since we have already established that $\|\mathcal{P}_n g-g\|_{L_2(\mathbb{R})}\le C2^{-nm_1}$, all we need to show is that $\|\widehat{f}-\widehat{g}\|_{L_2(\mathbb{R})}\le$

$C2^{-nm_1}$ for every n. Indeed, we have

$$\begin{aligned}\|\widehat{f}-\widehat{g}\|^2_{L_2(\mathbb{R})} &= \int_{\mathbb{R}} \left(1-\chi_{2^n[-\pi,\pi]}(\xi)\right)|\widehat{f}(\xi)|^2 \mathrm{d}\xi \\ &= 2^n \int_{\mathbb{R}} \left(1-\chi_{[-\pi,\pi]}(\xi)\right)|\widehat{f}(2^n\xi)|^2 \mathrm{d}\xi \\ &= 2^n \int_{\mathbb{R}\setminus[-\pi,\pi]} |\widehat{f}(2^n\xi)|^2 \mathrm{d}\xi \\ &= 2^n \int_{\mathbb{R}\setminus[-\pi,\pi]} \frac{(1+|2^n\xi|)^{2m_1}|\widehat{f}(2^n\xi)|^2}{(1+|2^n\xi|)^{2m_1}} \mathrm{d}\xi.\end{aligned}$$

Letting $\nu := (1+|\cdot|)^{m_1}\widehat{f}$, we then have

$$\begin{aligned}\|\widehat{f}-\widehat{g}\|^2_{L_2(\mathbb{R})} &= 2^{n(1-2m_1)} \int_{\mathbb{R}\setminus[-\pi,\pi]} \frac{|\nu(2^n\xi)|^2}{(2^{-n}+|\xi|)^{2m_1}} \mathrm{d}\xi \\ &\leq C_1 2^{n(1-2m_1)} \int_{\mathbb{R}\setminus[-\pi,\pi]} |\nu(2^n\xi)|^2 \mathrm{d}\xi \\ &= C_1 2^{-2nm_1} \int_{2^n(\mathbb{R}\setminus[-\pi,\pi])} |\nu(\xi)|^2 \mathrm{d}\xi \\ &\leq C_2 2^{-2nm_1}.\end{aligned}$$

This concludes the proof of the theorem. ∎

The approximation order of the quasi-interpolatory operator $\mathcal{P}_n$ may not reach the maximal approximation order that space V_n provides. Recall that for a given shift-invariant subspace V of $L_2(\mathbb{R})$ and any function $f \in L_2(\mathbb{R})$, the *approximation error* of V to function f is defined as

$$E(f,V) := \min\{\|f-g\|_{L_2(\mathbb{R})} : g \in V\}.$$

We say that the shift-invariant subspace $V(\phi)$ *provides approximation order* m (or we say ϕ *provides approximation order* m) if, for every $f \in \mathcal{W}_2^m(\mathbb{R})$,

$$E(f,V_n) = O(2^{-nm}). \tag{1.17}$$

If $\{V_n\}_n$ generates an MRA, we say that the MRA provides approximation order m if $V(\phi)$ (or ϕ) provides an approximation order m.

When the shifts of ϕ form a Bessel system, we have $\mathcal{P}_n f \in V_n$. Therefore, the approximation order of $\mathcal{P}_n f$ can never exceed that of $V(\phi)$. The two approximation orders coincide when, for example, $\mathcal{P}_n$ is an orthogonal projector.

It was shown in [**69**] that the approximation order of $V(\phi)$ is completely determined by the following function

$$\Lambda_\phi := \left(1-\frac{|\widehat{\phi}|^2}{[\widehat{\phi},\widehat{\phi}]}\right)^{1/2}, \quad \text{on } [-\pi,\pi]. \tag{1.18}$$

As shown in the following theorem, the behavior of the function Λ_ϕ at the origin is crucial. The proof of the theorem is omitted here and we refer interested readers to the original paper [**69**] for details.

THEOREM 1.6 ([**69**]). *The shift-invariant subspace $V(\phi)$ (or ϕ) provides approximation order $m > 0$ if and only if the function Λ_ϕ defined on the set $([-\pi, \pi])$ as*

$$\Lambda_\phi := \left(1 - \frac{|\widehat{\phi}|^2}{[\widehat{\phi}, \widehat{\phi}]}\right)^{1/2}, \tag{1.19}$$

satisfies $|\cdot|^m \Lambda_\phi \in L_\infty([-\pi, \pi])$.

The condition (1.19) is closely related to the Strang-Fix conditions of ϕ. Recall that a function ϕ satisfies the Strang-Fix (SF) conditions of order $m \in \mathbb{N}_+$ if

$$\widehat{\phi}(0) \neq 0, \quad \widehat{\phi}^{(j)}(2\pi k) = 0, \quad j = 0, 1, 2, ..., m-1, \quad k \in \mathbb{Z}\backslash\{0\}. \tag{1.20}$$

It is shown in [**69**] that under certain conditions on ϕ (e.g. if ϕ is compactly supported and $\widehat{\phi}(0) \neq 0$), (1.19) is equivalent to the SF conditions, meaning $V(\phi)$ (or ϕ) provides approximation order m if and only if ϕ satisfies SF conditions of order m.

When $\widehat{\phi}$ is smooth enough, e.g. when ϕ is compactly supported and $\widehat{\phi}(0) \neq 0$, assumption (2) of Theorem 1.5 is equivalent to the Strang-Fix (SF) conditions. In this case, we can easily see that assumption (2) of Theorem 1.5 ensures that the approximation order of $V(\phi)$ is m. If, in addition, m_0 in assumption (3) of Theorem 1.5 satisfies the inequality $2m_0 \geq m$, then Theorem 1.5 asserts that the operator $\mathcal{P}_n$ attains the approximation order of the space V_n.

If ϕ is a compactly supported refinable function with a finitely supported mask, h_0, satisfying $\widehat{h}_0(0) = 1$, it satisfies the SF conditions of order m. Hence ϕ provides an approximation order m whenever $\widehat{h}_0$ has a zero of order m at π. On the other hand, Proposition 2.1 in the next lecture will show that assumption (3) of Theorem 1.5, i.e. $1 - |\widehat{\phi}|^2 = O(|\cdot|^{2m_0})$, is equivalent to the condition $1 - |\widehat{h}_0|^2 = O(|\cdot|^{2m_0})$. Therefore, the approximation order of $V(\phi)$, with ϕ satisfying some additional mild conditions, is entirely determined by the refinement mask h_0 of ϕ. In fact, as we will see in the following two lectures, not only the approximation order of $V(\phi)$, but also the regularity of ϕ can be easily characterized by h_0. Furthermore, thanks to the MRA structure and the unitary extension principle [**158**], the constructions of tight frames and the approximation properties of the tight frames can also be easily obtained from corresponding properties of h_0.

LECTURE 2

MRA-based tight wavelet frames

This lecture is devoted to MRA-based tight wavelet frames. We first introduce the general notion of tight wavelet frames in space $L_2(\mathbb{R})$ with some other basic concepts and notations.

For any function $f \in L_2(\mathbb{R})$, the dyadic dilation operator $\mathcal{D}$ is defined by $\mathcal{D}f(x) := \sqrt{2}f(2x)$ and the translation operator T is defined by $T_t f(x) := f(x-t)$ for $t \in \mathbb{R}$. Given $j \in \mathbb{Z}$, we have $T_t\mathcal{D}^j = \mathcal{D}^j T_{2^j t}$.

For given $\Psi := \{\psi_1, \ldots, \psi_r\} \subset L_2(\mathbb{R})$, define the *wavelet system* (or *affine system*) as

$$X(\Psi) := \{\psi_{\ell,n,k} : 1 \le \ell \le r;\ n, k \in \mathbb{Z}\}, \tag{2.1}$$

where $\psi_{\ell,n,k} = \mathcal{D}^n T_k \psi_\ell = 2^{n/2}\psi_\ell(2^n \cdot -k)$. The system $X(\Psi) \subset L_2(\mathbb{R})$ is called a *tight wavelet frame* of $L_2(\mathbb{R})$ if

$$\|f\|^2_{L_2(\mathbb{R})} = \sum_{g \in X(\Psi)} |\langle f, g\rangle|^2,$$

holds for all $f \in L_2(\mathbb{R})$, where $\langle \cdot, \cdot \rangle$ is the inner product in $L_2(\mathbb{R})$ and $\|\cdot\|_{L_2(\mathbb{R})} = \sqrt{\langle \cdot, \cdot \rangle}$. This is equivalent to

$$f = \sum_{g \in X(\Psi)} \langle f, g \rangle\, g,$$

for all $f \in L_2(\mathbb{R})$. Here $\{\langle f, g\rangle\}_{g \in X(\Psi)}$ is called *canonical tight frame coefficients.*

It is clear that an orthonormal basis is a tight frame. When $X(\Psi)$ forms an orthonormal basis of $L_2(\mathbb{R})$, then $X(\Psi)$ is called an orthonormal wavelet basis. When $X(\Psi)$ forms a tight frame of $L_2(\mathbb{R})$, then $X(\Psi)$ is called a tight wavelet (affine) frame. We note that in some literature, the definition of tight frame we present is called the tight frame with bound one or Parseval frame.

More generally, the system $X(\Psi) \subset L_2(\mathbb{R})$ is called a *wavelet frame* of $L_2(\mathbb{R})$ if there are constants $0 < C_1 \le C_2 < \infty$ such that

$$C_1\|f\|^2_{L_2(\mathbb{R})} \le \sum_{g \in X(\Psi)} |\langle f, g\rangle|^2 \le C_2\|f\|^2_{L_2(\mathbb{R})}, \tag{2.2}$$

holds for all $f \in L_2(\mathbb{R})$. We call $X(\Psi)$ a Bessel system if only the right inequality holds.

Recall that for a given system $X(\Psi)$, the corresponding *frame operator* is defined by

$$Sf = \sum_{g \in X(\Psi)} \langle f, g\rangle\, g,\ f \in L_2(\mathbb{R}).$$

It is clear that $X(\Psi)$ is a frame of $L_2(\mathbb{R})$ if and only if S is bounded and has a bounded inverse; $X(\Psi)$ is a tight frame of $L_2(\mathbb{R})$ if and only if S is the identity. For comprehensive characterizations of frames in terms of various operators associated with them, the papers [**156, 162**] and the book [**50**] are good references.

The first interesting question is how to choose Ψ such that $X(\Psi)$ is a tight frame, or more generally a frame for $L_2(\mathbb{R})$. A general characterization of all wavelet frames whose dilation matrix is an integral (via dual Gramian analysis of [**156, 162**]) provided in [**158**] and derived from it a special characterization of tight wavelet frames in [**158**] and bi-frame in [**157**].

The dual Gramian analysis identifies the frame operator corresponding to the wavelet system $X(\Psi)$ as the dual Gramian matrix with each entry being written in terms of the Fourier transform of the generators Ψ. The dual Gramian analysis decomposes the operator S into a collection of simpler operators which are called fibers in [**156, 162**] in Fourier domain. The operator S is bounded and has a bounded inverse if and only if almost every fiber operator is uniformly bounded and has a uniformly bounded inverse. S is the identity operator if and only if almost every fiber operator is the identity operator. This leads to the conclusion that the wavelet system $X(\Psi)$ forms a tight frame of $L_2(\mathbb{R})$ if and only if the dual Gramian corresponding to the wavelet system $X(\Psi)$ is the identity operator almost everywhere. By writing each entry of the dual Gramian explicitly, one obtains the following theorem (see, e.g. Corollary 1.3 of [**158**]). For general results and dual Gramian analysis, interested readers should consult [**158, 156, 162**] for details. Note that the dual Gramian analysis was also applied to the Gabor frame analysis in [**159**] to derive the duality principle for the Gabor frames.

THEOREM 2.1 ([**158**]). *The wavelet system $X(\Psi)$ is a tight frame of $L_2(\mathbb{R})$ if and only if the identities*

$$\sum_{\psi\in\Psi}\sum_{k\in\mathbb{Z}}|\widehat{\psi}(2^k\xi)|^2=1;\quad \sum_{\psi\in\Psi}\sum_{k=0}^{\infty}\widehat{\psi}(2^k\xi)\overline{\widehat{\psi}(2^k(\xi+(2j+1)2\pi))}=0\quad j\in\mathbb{Z} \tag{2.3}$$

hold for a.e. $\xi\in\mathbb{R}$. Furthermore, $X(\Psi)$ is an orthonormal basis of $L_2(\mathbb{R})$ if and only if (2.3) *holds and $\|\psi\|=1$ for all $\psi\in\Psi$.*

There were many contributions over the last two decades to the study of the Bessel frame and other related properties of wavelet systems. Examples of univariate wavelet frames can be found in [**63**]; necessary and sufficient conditions for mother wavelets to generate frames were discussed (implicitly) in [**61, 134**]. Characterizations of univariate orthonormal bases associated with integer dilation were established independently in [**98**] and [**108**]. The multivariate counterparts of these results for the dyadic dilation appeared in [**93**]. Characterization of bi-frames (tight frame is a special case) in multivariate case for an integer dilation matrix was given in [**101**]. Independent of all these, as discussed before Theorem 2.1, a general characterization of all wavelet frames whose dilation matrix is an integral (via dual Gramian analysis) were provided in [**158**] and derived from it a special characterization of tight wavelet frames in [**158**] and bi-frame in [**157**].

Although Theorem 2.1 gives a complete characterization of the wavelet system $X(\Psi)$ being a tight frame of $L_2(\mathbb{R})$ and it helps to obtain tight wavelet frame systems with band-limited generators, directly applying this theorem provides little help in the construction of wavelet systems with compactly supported generators. On the other hand, because fast decomposition and reconstruction algorithms exist for compactly supported tight wavelet frames constructed from an MRA, these frames are very handy to use in practice. This motivates the study of MRA-based tight wavelet frames in [**158**]. The key is the introduction of extension principles, e.g. the unitary extension principle. The MRA-based band-limited tight wavelet frames are also constructed in [**6**] without invoking extension principles.

1. Extension principles

In this section, we introduce the unitary extension principle (UEP), which leads to explicit constructions of tight wavelet frames based on the multiresolution analysis generated by a refinable function ϕ. In particular, examples of spline tight wavelet frame will be given. The approximation order of tight wavelet frames is studied and the fast algorithm for tight wavelet frame derived from the unitary extension principle is discussed as well. Also, the decomposition and reconstruction algorithms for the MRA-based tight wavelet frames are provided. We briefly introduce other extension principles as well. Although the main interest here is to construct compactly supported tight frames with finitely supported masks, we state the UEP in a more general setting. An even more general form of the UEP (e.g. multivariate case) can be found in [**65, 158**]).

1.1. Unitary extension principle

Let $\{V_n\}_{n\in\mathbb{Z}}$ be the MRA generated by the refinable function ϕ with refinement mask h_0. The construction of tight frame systems starts with the construction of $\Psi \subset L_2(\mathbb{R})$. The objective of MRA-based construction of tight wavelet frames is to find $\Psi = \{\psi_1, \ldots, \psi_r\} \subset V_1$ such that $X(\Psi)$ forms a tight frame for $L_2(\mathbb{R})$. Since V_1 is a $\frac{1}{2}$-shift-invariant subspace generated by $\phi(2\cdot)$, finding $\Psi \subset V_1$ is the same as finding h_ℓ such that

$$\psi_\ell(x) = 2\sum_{k\in\mathbb{Z}} h_\ell[k]\phi(2x-k). \tag{2.4}$$

The sequences $h_1, \ldots, h_r$ are called wavelet masks, or the high pass filters of the system; the refinement mask h_0 is also known as the low pass filter. In the Fourier domain, (2.4) can be written as

$$\widehat{\psi_\ell}(2\cdot) = \widehat{h}_\ell\widehat{\phi}, \quad \ell = 1, \ldots, r, \tag{2.5}$$

where $\widehat{h}_1, \ldots, \widehat{h}_r$ are 2π periodic functions. Throughout this lecture, ϕ and masks $\{h_\ell, \ell = 0, 1, \ldots, r\}$ are assumed to satisfy the following mild assumptions:

ASSUMPTIONS 2.1. All MRA-based constructions that are considered in this article are assumed to satisfy the following assumptions:

(1) Each mask of $\{h_\ell : \ell = 0, 1, \ldots, r\}$ is a sequence in $\ell_2(\mathbb{Z})$ and its Fourier series $\widehat{h}_\ell$ is measurable and (essentially) bounded.
(2) The refinement mask h_0 of refinable function $\phi \in L_2(\mathbb{R})$ satisfies
$$|\widehat{h}_0(\xi) - 1| \le C|\xi|.$$
(3) The function $[\hat{\phi}, \hat{\phi}] := \sum_{k\in 2\pi\mathbb{Z}} |\hat{\phi}(\cdot + k)|^2$ is essentially bounded, which is equivalent to (1.7) (see [**70**]), i.e. the shifts of ϕ form a Bessel system.

Under the assumption in item 2, the proof of Theorem 3.1 reveals the following property of ϕ

$$\lim_{\xi\to 0} \widehat{\phi}(\xi) = 1.$$

Our analysis in the previous lecture shows that the sequence of subspaces $\{V_n\}_n$ defined in (1.1) forms an MRA. When ϕ is a compactly supported refinable function generated by a finitely supported refinement mask with $\widehat{\phi}(0) = 1$, all conditions in Assumption 2.1 are satisfied. We note that item 2 in Assumption 2.1 is in fact not necessary and were not required in the original proof of UEP [**158**]. We include them here for the simplicity of the proof.

For the UEP in the most general setting with mild assumptions, interested readers should consult [**65, 158**] for the details. The univariate version of the UEP of [**158**] can be stated as following.

THEOREM 2.2 (Unitary Extension Principle, (UEP) [**158**]). *Let $\phi \in L_2(\mathbb{R})$ be a refinable function with refinement mask h_0 and $\{h_1, \ldots, h_r\}$ be a set of sequences. Assume that the refinable function ϕ and masks $\{h_0, h_1, \ldots, h_r\}$ satisfy Assumption 2.1. Then the system $X(\Psi)$ where $\Psi = \{\psi_1, \ldots, \psi_r\}$ defined in (2.4) forms a tight frame in $L_2(\mathbb{R})$ provided the equalities*

$$\sum_{\ell=0}^{r} |\widehat{h}_\ell(\xi)|^2 = 1 \quad \textit{and} \quad \sum_{\ell=0}^{r} \widehat{h}_\ell(\xi)\overline{\widehat{h}_\ell(\xi+\pi)} = 0 \tag{2.6}$$

hold for almost all $\xi \in \sigma(V_0)$, where

$$\sigma(V_0) := \{\xi \in \mathbb{R} : [\hat{\phi}, \hat{\phi}](\xi) \neq 0\}. \tag{2.7}$$

Furthermore, assuming $r = 1$ and $\|\phi\|_{L_2(\mathbb{R})} = 1$, then $X(\Psi)$ is an orthonormal wavelet bases of $L_2(\mathbb{R})$.

First, we note that in order to use the UEP, the mask h_0 must necessarily satisfy

$$|\widehat{h}_0(\xi)|^2 + |\widehat{h}_0(\xi+\pi)|^2 \leq 1. \tag{2.8}$$

Indeed, the UEP condition (2.6) means that the following matrix has the two rows orthonormal to each other

$$\begin{pmatrix} \widehat{h}_0(\xi), & \widehat{h}_1(\xi), & \cdots, & \widehat{h}_r(\xi) \\ \widehat{h}_0(\xi+\pi), & \widehat{h}_1(\xi+\pi), & \cdots, & \widehat{h}_r(\xi+\pi), \end{pmatrix} \quad \text{i.e.} \quad \xi \in \mathbb{R}.$$

Then one can extend the above matrix to a $(r+1)\times(r+1)$ unitary matrix. The fact that the first column of the extended matrix has norm one implies the inequality (2.8).

Conditions in (2.6) can be written in terms of sequences $\{h_0, h_1, \ldots, h_r\}$. The first condition becomes

$$\sum_{\ell=0}^{r}\sum_{k\in\mathbb{Z}} \overline{h_\ell[k]}h_\ell[k-p] = \delta_{p,0}, \quad p \in \mathbb{Z}, \tag{2.9}$$

where $\delta_{p,0} = 1$ when $p = 0$ and 0 otherwise, and the second condition can be written as

$$\sum_{\ell=0}^{r}\sum_{k\in\mathbb{Z}} (-1)^{k-p}\overline{h_\ell[k]}h_\ell[k-p] = 0, \quad p \in \mathbb{Z}. \tag{2.10}$$

Before proving Theorem 2.2, we first prove the following Lemma 2.1. We will prove Lemma 2.1 in the Fourier domain which is similar to the proof of [**65**, Lemma 2.4]. Also, we demonstrate another proof of Lemma 2.1 in the time domain with some additional assumptions. The reason that we are showing the proof of identity

(2.12) in the time domain is because both the identity itself and the proof of it are the foundation of the fast decomposition and reconstruction algorithms for MRA-based wavelet frame systems.

For a given ϕ and its corresponding MRA $\{V_n\}_n$, the quasi-interpolatory operator is defined as (same as (1.13))

$$\mathcal{P}_n : f \mapsto \sum_{k\in\mathbb{Z}} \langle f, \phi_{n,k}\rangle \phi_{n,k}, \tag{2.11}$$

for an arbitrary $f \in L_2(\mathbb{R})$.

LEMMA 2.1. *Let $\phi \in L_2(\mathbb{R})$ be a refinable function with mask h_0. Let*

$$\{h_0, h_1, \ldots, h_r\}$$

be a set of sequences satisfying (2.6) *(or equivalently* (2.9) *and* (2.10)*) and Assumption* 2.1. *Then*

$$\mathcal{P}_n f = \mathcal{P}_{n-1} f + \sum_{\ell=1}^{r}\sum_{k\in\mathbb{Z}} \langle f, \psi_{\ell,n-1,k}\rangle \psi_{\ell,n-1,k}. \tag{2.12}$$

PROOF. To simplify notations, let ϕ be defined as ψ_0, and

$$\mathcal{P}_{n,\ell} f := \sum_{k\in\mathbb{Z}} \langle f, \psi_{\ell,n,k}\rangle \psi_{\ell,n,k},$$

for $\ell = 0, 1, \ldots, r$. Note that $\mathcal{P}_{n,0} = \mathcal{P}_n$. By Assumption 2.1, the system $\{\psi_{\ell,n,k}; k \in \mathbb{Z}\}$ for each ℓ and n, forms a Bessel system. Hence the operator $\mathcal{P}_{n,\ell} : L_2(\mathbb{R}) \mapsto L_2(\mathbb{R})$ is well-defined. Therefore, (2.12) is equivalent to

$$\widehat{\mathcal{P}_n f} = \sum_{\ell=0}^{r} \widehat{\mathcal{P}_{n-1,\ell} f}. \tag{2.13}$$

Since $\mathcal{P}_{n,\ell} = \mathcal{D}^n \mathcal{P}_{0,\ell} \mathcal{D}^{-n}$, we only need to show (2.13) for $n = 1$, i.e.

$$\widehat{\mathcal{P}_1 f} = \sum_{\ell=0}^{r} \widehat{\mathcal{P}_{0,\ell} f}. \tag{2.14}$$

By Proposition 1.3 item (3), (2.14) is equivalent to

$$[\widehat{f}(2\cdot), \widehat{\phi}](\tfrac{\cdot}{2})\widehat{\phi}(\tfrac{\cdot}{2}) = \sum_{\ell=0}^{r} [\widehat{f}, \widehat{\psi}_\ell]\widehat{\psi}_\ell. \tag{2.15}$$

Indeed, since $\widehat{\psi}_\ell = \widehat{h}_\ell(\frac{\cdot}{2})\widehat{\phi}(\frac{\cdot}{2})$, we thus have

$$\begin{aligned}
\sum_{\ell=0}^{r}[\widehat{f},\widehat{\psi}_\ell]\widehat{\psi}_\ell &= \sum_{\ell=0}^{r}[\widehat{f},\widehat{h}_\ell(\tfrac{\cdot}{2})\widehat{\phi}(\tfrac{\cdot}{2})]\widehat{h}_\ell(\tfrac{\cdot}{2})\widehat{\phi}(\tfrac{\cdot}{2})\\
&= \sum_{\ell=0}^{r}\left(\overline{\widehat{h}_\ell(\tfrac{\cdot}{2})}[\widehat{f}(2\cdot),\widehat{\phi}](\tfrac{\cdot}{2}) + \overline{\widehat{h}_\ell(\tfrac{\cdot}{2}+\pi)}[\widehat{f}(2\cdot),\widehat{\phi}](\tfrac{\cdot}{2}+\pi)\right)\widehat{h}_\ell(\tfrac{\cdot}{2})\widehat{\phi}(\tfrac{\cdot}{2})\\
&= \sum_{\ell=0}^{r}|\widehat{h}_\ell(\tfrac{\cdot}{2})|^2[\widehat{f}(2\cdot),\widehat{\phi}](\tfrac{\cdot}{2})\widehat{\phi}(\tfrac{\cdot}{2})\\
&\qquad + \sum_{\ell=0}^{r}\widehat{h}_\ell(\tfrac{\cdot}{2})\overline{\widehat{h}_\ell(\tfrac{\cdot}{2}+\pi)}[\widehat{f}(2\cdot),\widehat{\phi}](\tfrac{\cdot}{2}+\pi)\widehat{\phi}(\tfrac{\cdot}{2})\\
&= [\widehat{f}(2\cdot),\widehat{\phi}](\tfrac{\cdot}{2})\widehat{\phi}(\tfrac{\cdot}{2}).
\end{aligned}$$

The last identity above follows from (2.6) for $\frac{\xi}{2}\in\sigma(V_0)$ and the fact that $\widehat{\phi}(\frac{\xi}{2})=0$ when $\frac{\xi}{2}\notin\sigma(V_0)$. This concludes the proof of the lemma. ■

REMARK 2.1. When Lemma 2.1 is proven in spacial domain, it reveals some details of the fast framelet decomposition and reconstruction algorithms given in Section 1.4. For simplicity, we assume that the refinable function and all framelets are compactly supported, and that their corresponding masks are finitely supported, i.e. we assume that ψ_ℓ, for $\ell = 0,1,\ldots,r$, are compactly supported and the masks h_ℓ are finite sequences; these assumptions are commonly satisfied in applications.

Since the system $\{\psi_{\ell,n,k}; k\in\mathbb{Z}\}$ for any $n\in\mathbb{Z}$, forms a Bessel system, the operators $\mathcal{P}_{n,\ell}: L_2(\mathbb{R})\to L_2(\mathbb{R})$, for each ℓ, is bounded. Therefore, we only need to prove (2.12) for all f that are compactly supported.

By the definition of refinability of ϕ and (2.4), we have for $\ell = 0,1,\ldots,r$,

$$\psi_\ell(x) = 2\sum_{k\in\mathbb{Z}} h_\ell[k]\phi(2x-k) = 2\sum_{k\in\mathbb{Z}} h_\ell[k]\psi_0(2x-k).$$

Thus we have

$$\begin{aligned}
\psi_{\ell,n-1,k} &= 2^{(n-1)/2}\psi_\ell(2^{n-1}\cdot - k)\\
&= 2^{(n-1)/2+1}\sum_{k'\in\mathbb{Z}} h_\ell[k']\psi_0(2^n\cdot -2k-k')\\
&= 2^{1/2}\sum_{k'\in\mathbb{Z}} h_\ell[k']2^{n/2}\psi_0(2^n\cdot -2k-k')\\
&= 2^{1/2}\sum_{k'\in\mathbb{Z}} h_\ell[k']\psi_{0,n,2k+k'}.
\end{aligned}$$

Therefore, we have

$$\begin{aligned}
&\sum_{\ell=0}^{r}\sum_{k\in\mathbb{Z}}\langle f,\psi_{\ell,n-1,k}\rangle\psi_{\ell,n-1,k}\\
&\qquad=2\sum_{\ell=0}^{r}\sum_{k\in\mathbb{Z}}\left(\sum_{k'\in\mathbb{Z}}\overline{h_\ell[k']}\langle f,\psi_{0,n,2k+k'}\rangle\right)\left(\sum_{k''\in\mathbb{Z}}h_\ell[k'']\psi_{0,n,2k+k''}\right)\\
&\qquad=2\sum_{\ell=0}^{r}\sum_{k\in\mathbb{Z}}\left(\sum_{j\in\mathbb{Z}}\overline{h_\ell[j-2k]}\langle f,\psi_{0,n,j}\rangle\right)\left(\sum_{j'\in\mathbb{Z}}h_\ell[j'-2k]\psi_{0,n,j'}\right)\\
&\qquad=2\sum_{j\in\mathbb{Z}}\sum_{j'\in\mathbb{Z}}\left(\sum_{\ell=0}^{r}\sum_{k\in\mathbb{Z}}\overline{h_\ell[j-2k]}h_\ell[j'-2k]\right)\langle f,\psi_{0,n,j}\rangle\psi_{0,n,j'},
\end{aligned}$$

where the switching of summations order is valid since all summations involved are finite sum.

Now we focus on proving that

$$\sum_{\ell=0}^{r}\sum_{k\in\mathbb{Z}}\overline{h_\ell[j-2k]}h_\ell[j'-2k]=\frac{1}{2}\delta_{j,j'}. \tag{2.16}$$

Suppose $j'-j$ is even, i.e. there exists $p\in\mathbb{Z}$ such that $j'=j+2p$. Then we have

$$\begin{aligned}
\sum_{\ell=0}^{r}\sum_{k\in\mathbb{Z}}\overline{h_\ell[j-2k]}h_\ell[j'-2k]&=\sum_{\ell=0}^{r}\sum_{k\in\mathbb{Z}}\overline{h_\ell[2k+j]}h_\ell[2k+j']\\
&=\sum_{\ell=0}^{r}\sum_{k\in\mathbb{Z}}\overline{h_\ell[2k+j]}h_\ell[2k+j+2p]\\
&=\begin{cases}\sum_{\ell=0}^{r}\sum_{k\in\mathbb{Z}}\overline{h_\ell[2k]}h_\ell[2k+2p], & j\text{ even},\\ \sum_{\ell=0}^{r}\sum_{k\in\mathbb{Z}}\overline{h_\ell[2k+1]}h_\ell[2k+2p+1], & j\text{ odd}.\end{cases}\\
&=:\begin{cases}c_1[p], & j\text{ even},\\ c_2[p], & j\text{ odd}.\end{cases}
\end{aligned}$$

Now replacing p by $-2p$ and breaking the summations (2.9) and (2.10) into even and odd parts, we have

$$\begin{aligned}
c_1[p]+c_2[p]&=\delta_{-2p,0}=\delta_{p,0},\\
c_1[p]-c_2[p]&=0,
\end{aligned}$$

which leads to

$$c_1[p]=c_2[p]=\frac{1}{2}\delta_{p,0}.$$

Since $j'-j=2p$, we have $\delta_{p,0}=\delta_{j,j'}$ which concludes that (2.16) is satisfied when $j'-j$ is even.

Suppose $j'-j$ is odd, which means there exists $p\in\mathbb{Z}$ such that $j'=j+2p+1$. Then we have

$$\begin{aligned}\sum_{\ell=0}^{r}\sum_{k\in\mathbb{Z}}\overline{h_\ell[j-2k]}h_\ell[j'-2k] &= \sum_{\ell=0}^{r}\sum_{k\in\mathbb{Z}}\overline{h_\ell[2k+j]}h_\ell[2k+j+2p+1]\\ &= \begin{cases}\sum_{\ell=0}^{r}\sum_{k\in\mathbb{Z}}\overline{h_\ell[2k]}h_\ell[2k+2p+1], & j \text{ even},\\ \sum_{\ell=0}^{r}\sum_{k\in\mathbb{Z}}\overline{h_\ell[2k+1]}h_\ell[2k+2p+2], & j \text{ odd}\end{cases}\\ &=: \begin{cases} c_1[p], & j \text{ even},\\ c_2[p], & j \text{ odd}.\end{cases}\end{aligned}$$

Now replacing p by $-2p-1$ and breaking the summations (2.9) and (2.10) into even and odd parts, we have

$$\begin{aligned} c_1[p]+c_2[p]&=0,\\ -c_1[p]+c_2[p]&=0,\end{aligned}$$

which implies that $c_1[p]=c_2[p]=0$ for all $p\in\mathbb{Z}$. Therefore, we have

$$\sum_{\ell=0}^{r}\sum_{k\in\mathbb{Z}}\overline{h_\ell[j-2k]}h_\ell[j'-2k]=0=\frac{1}{2}\delta_{j,j'}.$$

This concludes the proof of (2.16) and hence the identity (2.12) is proved.

Note that identity (2.12) essentially provides a decomposition and reconstruction formula for any function $f\in L_2(\mathbb{R})$ under the system $X(\Psi)$. This formula and the MRA structure generated by the underlying refinable function ϕ are used to prove that $X(\Psi)$ is a tight frame for $L_2(\mathbb{R})$.

The following two lemmata are necessary in order to prove the UEP. Lemma 2.2 below is in fact a direct consequence of $\cap_n V_n=\{0\}$ and the weakly compactness of the unit ball in $L_2(\mathbb{R})$. However, we provide a direct proof which is similar to the proof of $\cap_n V_n=\{0\}$ (Theorem 1.3).

LEMMA 2.2. *Let refinable function $\phi\in L_2(\mathbb{R})$ satisfying Assumption* 2.1 *and operator $\mathcal{P}_n$ be defined by* (2.11)*. Then, for all $f\in L_2(\mathbb{R})$,*

$$\lim_{n\to-\infty}\mathcal{P}_n f=0.$$

PROOF. Part (1) of Proposition 1.3 gives us

$$\begin{aligned}\|\mathcal{P}_n f\|^2_{L_2(\mathbb{R})} &\le C\sum_{k\in\mathbb{Z}}|\langle f,\phi_{n,k}\rangle|^2\\ &= C\sum_{k\in\mathbb{Z}}\Big(\int_{\mathbb{R}}|f(x)2^{n/2}\phi(2^n x-k)|dx\Big)^2.\end{aligned}$$

If f is supported in the interval $[-R,R]$, then by the Cauchy-Schwartz inequality we have,

$$\|\mathcal{P}_n f\|^2_{L_2(\mathbb{R})}\le C\|f\|^2_{L_2(\mathbb{R})}\sum_{k\in\mathbb{Z}}\int_{\{2^n[-R,R]-k\}}|\phi(x)|^2dx.$$

Now if $n<0$ and $|n|$ is sufficiently large, we have that

$$\{2^n[-R,R]-k_1\}\cap\{2^n[-R,R]-k_2\}=\emptyset,\quad k_1\ne k_2.$$

Therefore,

$$\|\mathcal{P}_n f\|^2_{L_2(\mathbb{R})} \leq C\|f\|^2_{L_2(\mathbb{R})} \int_{E_n} |\phi(x)|^2 dx = C\|f\|^2_{L_2(\mathbb{R})} \int_{\mathbb{R}} \chi_{E_n} |\phi(x)|^2 dx,$$

where

$$E_n = \cup_{k\in\mathbb{Z}}(2^n[-R,R]-k).$$

Since when $n \to -\infty$, $\chi_{E_n}|\phi(x)|^2 \to 0$ for almost every $x \in \mathbb{R}$, and $\chi_{E_n}|\phi(x)|^2 \leq |\phi(x)|^2 \in L_1(\mathbb{R})$, then by the dominated convergence theorem we have

$$\lim_{n\to-\infty} \|f\|^2_{L_2(\mathbb{R})} \int_{\mathbb{R}} \chi_{E_n}|\phi(x)|^2 dx = \|f\|^2_{L_2(\mathbb{R})} \int_{\mathbb{R}} \lim_{n\to-\infty} \chi_{E_n}|\phi(x)|^2 dx = 0.$$

Thus

$$\lim_{n\to-\infty} \|\mathcal{P}_n f\|^2_{L_2(\mathbb{R})} = 0,$$

for all compactly supported $f \in L_2(\mathbb{R})$. By a standard density argument and noting that $\mathcal{P}_n$ is bounded for each n, we can conclude that $\lim_{n\to-\infty} \mathcal{P}_n f = 0$ for all $f \in L_2(\mathbb{R})$. ∎

LEMMA 2.3. *Let refinable function $\phi \in L_2(\mathbb{R})$ satisfy Assumption* 2.1 *and* (2.8) *and operator $\mathcal{P}_n$ be defined by* (2.11). *Then for all $f \in L_2(\mathbb{R})$,*

$$\lim_{n\to\infty} \mathcal{P}_n f = f.$$

PROOF. We will prove this lemma by showing that for any $f \in L_2(\mathbb{R})$

$$\lim_{n\to\infty} \langle \mathcal{P}_n f, f\rangle = \|f\|^2_{L_2(\mathbb{R})}. \tag{2.17}$$

and for $n \in \mathbb{Z}$,

$$\|\mathcal{P}_n\|_{L_2(\mathbb{R})} \leq 1. \tag{2.18}$$

Once (2.17) and (2.18) are established, we will have

$$\begin{aligned} 0 \leq \lim_{n\to\infty} \|\mathcal{P}_n f - f\|^2_{L_2(\mathbb{R})} &= \lim_{n\to\infty} \|\mathcal{P}_n f\|^2_{L_2(\mathbb{R})} - 2\langle \mathcal{P}_n f, f\rangle + \|f\|^2_{L_2(\mathbb{R})} \\ &= \lim_{n\to\infty} \|\mathcal{P}_n f\|^2_{L_2(\mathbb{R})} - \|f\|^2_{L_2(\mathbb{R})} \leq 0, \end{aligned}$$

which will conclude the proof of this lemma.

Now, we show (2.17), whose proof is similar to [**50**, Lemma 11.2.3]. Following the routine arguments in approximation theory, we only need to prove (2.17) for all $f \in L_2(\mathbb{R})$ with $\widehat{f}$ continuous and compactly supported. By a similar proof of part (2) of Proposition 1.3, one can show that $[D^n\widehat{f}, \widehat{\phi}]$ is the Fourier series of $\{\langle f, \phi_{n,k}\rangle : k \in \mathbb{Z}\}$ for each $n \in \mathbb{Z}$. Since $\widehat{f}$ is continuous and compactly supported, the infinite sum

$$[D^n\widehat{f}, \widehat{\phi}] = \sum_{k\in\mathbb{Z}} (D^n\widehat{f})(\xi + 2k\pi)\overline{\widehat{\phi}(\xi + 2k\pi)}$$

is in fact finite. Then, the boundedness of $\widehat{f}$ implies that $[D^n\widehat{f}, \widehat{\phi}] \in L_2([-\pi,\pi])$. Now by Parseval's identity, we have

$$\langle \mathcal{P}_n f, f\rangle = \frac{1}{2\pi}\|[D^n\widehat{f}, \widehat{\phi}]\|^2_{L_2([-\pi,\pi])} = \frac{1}{2\pi}\int_{-\pi}^{\pi} \left|\sum_{k\in\mathbb{Z}} (D^n\widehat{f})(\xi + 2k\pi)\overline{\widehat{\phi}(\xi + 2k\pi)}\right|^2 d\xi.$$

Since $\widehat{f}$ is compactly supported, when n is large enough, we have

$$\langle \mathcal{P}_n f, f\rangle = \frac{1}{2\pi}\int_{-\pi}^{\pi}\left|(D^n\widehat{f})(\xi)\widehat{\phi}(\xi)\right|^2 \mathrm{d}\xi.$$

Since item 2 of Assumption 2.1 implies that $\lim_{\xi\to 0}\widehat{\phi}(\xi) = 1$. Then, for any $\varepsilon > 0$, there exists $r \in (0,\pi)$ such that $1-\varepsilon \le |\widehat{\phi}(\xi)|^2 \le 1+\varepsilon$ for $\xi \in [-r,r]$. Furthermore, when n large enough, $D^n\widehat{f}$ has support in $[-r,r]$. Then we have,

$$\langle \mathcal{P}_n f, f\rangle = \frac{1}{2\pi}\int_{-r}^{r}\left|(D^n\widehat{f})(\xi)\widehat{\phi}(\xi)\right|^2 \mathrm{d}\xi$$

and thus

$$\frac{1-\varepsilon}{2\pi}\|D^n\widehat{f}\|^2_{L_2(\mathbb{R})} \le \langle \mathcal{P}_n f, f\rangle \le \frac{1+\varepsilon}{2\pi}\|D^n\widehat{f}\|^2_{L_2(\mathbb{R})}.$$

Now, (2.17) follows by noting that $\frac{1}{2\pi}\|D^n\widehat{f}\|^2_{L_2(\mathbb{R})} = \|f\|^2_{L_2(\mathbb{R})}$.

From part (3) of Proposition 1.3, it is easy to see that it suffices to prove that $[\widehat{\phi},\widehat{\phi}] \le 1$ a.e. on $[-\pi,\pi]$. Define a sequence of functions $\{\phi_n\}_n$ as (known as the cascade algorithm)

$$\widehat{\phi}_n(\xi) := \widehat{h}_0(\frac{\xi}{2})\widehat{\phi}_{n-1}(\frac{\xi}{2}) = \prod_{j=1}^{n}\widehat{h}_0(2^{-j}\xi)\widehat{\phi}_0(2^{-n}\xi),$$

with $\widehat{\phi}_0 := \chi_{[-\pi,\pi]}$. It is clear that $[\widehat{\phi}_0,\widehat{\phi}_0] \le 1$, by induction, we have

$$\begin{aligned}
[\widehat{\phi}_n,\widehat{\phi}_n](\xi) &= \sum_{k\in\mathbb{Z}}|\widehat{h}_0(\frac{\xi}{2}+k\pi)|^2|\widehat{\phi}_{n-1}(\frac{\xi}{2}+k\pi)|^2\\
&= \sum_{k\in\mathbb{Z}}|\widehat{h}_0(\frac{\xi}{2}+2k\pi)|^2|\widehat{\phi}_{n-1}(\frac{\xi}{2}+2k\pi)|^2\\
&+ \sum_{k\in\mathbb{Z}}|\widehat{h}_0(\frac{\xi}{2}+2k\pi+\pi)|^2|\widehat{\phi}_{n-1}(\frac{\xi}{2}+2k\pi+\pi)|^2\\
&= |\widehat{h}_0(\frac{\xi}{2})|^2\sum_{k\in\mathbb{Z}}|\widehat{\phi}_{n-1}(\frac{\xi}{2}+2k\pi)|^2\\
&+ |\widehat{h}_0(\frac{\xi}{2}+\pi)|^2\sum_{k\in\mathbb{Z}}|\widehat{\phi}_{n-1}(\frac{\xi}{2}+2k\pi+\pi)|^2\\
&\le |\widehat{h}_0(\xi/2)|^2 + |\widehat{h}_0(\xi/2+\pi)|^2\\
&\le 1,\ a.e.\,\xi\in\mathbb{R}.
\end{aligned}$$

The last inequality follows from the fact that $|\widehat{h}_0(\xi/2)|^2 + |\widehat{h}_0(\xi/2+\pi)|^2 \le 1$. By item 2 of Assumption 2.1, we have that $\widehat{\phi}_n$ converges pointwise to $\widehat{\phi}$. (The proof of this argument is the same as that of Theorem 3.1. Although we assume in Theorem 3.1 that h_0 is finitely supported, only item 2 of Assumption 2.1 is necessary to show pointwise convergence of $\widehat{\phi}_n$ to $\widehat{\phi}$.) By Fatou's lemma, we have

$$[\widehat{\phi},\widehat{\phi}] \le \liminf_{n\to\infty}[\widehat{\phi}_n,\widehat{\phi}_n] \le 1,\ a.e.\,\xi\in\mathbb{R}$$

This concludes the proof of (2.18). ∎

PROOF OF THEOREM 2.2. By (2.12), we have

$$\mathcal{P}_n f = \mathcal{P}_{n-1} f + \sum_{\ell=1}^{r} \sum_{k\in\mathbb{Z}} \langle f, \psi_{\ell,n-1,k}\rangle \psi_{\ell,n-1,k}$$

Applying (2.12) inductively, one has that

$$\mathcal{P}_n f = \mathcal{P}_{n'} f + \sum_{\ell=1}^{r} \sum_{j=n'}^{n-1} \sum_{k\in\mathbb{Z}} \langle f, \psi_{\ell,j,k}\rangle \psi_{\ell,j,k}.$$

Letting $n' \to -\infty$ and employing Lemma 2.2, we have

$$\mathcal{P}_n f = \sum_{\ell=1}^{r} \sum_{j<n} \sum_{k\in\mathbb{Z}} \langle f, \psi_{\ell,j,k}\rangle \psi_{\ell,j,k}. \tag{2.19}$$

Then taking $n \to \infty$ at both sides of (2.19), noting that (2.8) holds by (2.6), and applying Lemma 2.3, we have, for all $f \in L_2(\mathbb{R})$,

$$f = \sum_{\ell=1}^{r} \sum_{n\in\mathbb{Z}} \sum_{k\in\mathbb{Z}} \langle f, \psi_{\ell,n,k}\rangle \psi_{\ell,n,k}.$$

This proves that $X(\Psi)$ is a tight frame for $L_2(\mathbb{R})$.

Finally, we show that if $r = 1$ and $\|\phi\|_{L_2(\mathbb{R})} = 1$ (or equivalently $\|\widehat{\phi}\|^2_{L_2(\mathbb{R})} = 2\pi$), then $X(\Psi)$ forms an orthonormal basis for $L_2(\mathbb{R})$. Using (2.6), $\widehat{\phi}(2\cdot) = \widehat{h}_0\widehat{\phi}$ and $\widehat{\psi}(2\cdot) = \widehat{h}_1\widehat{\phi}$, we have

$$\begin{aligned}
\|\widehat{\psi}\|^2_{L_2(\mathbb{R})} = 2\int |\widehat{\psi}(2\xi)|^2 d\xi &= 2\int |\widehat{h}_1(\xi)|^2 |\widehat{\phi}(\xi)|^2 d\xi \\
&= 2\int (1 - |\widehat{h}_0(\xi)|^2) |\widehat{\phi}(\xi)|^2 d\xi \\
&= 4\pi - 2\int |\widehat{\phi}(2\xi)|^2 d\xi \\
&= 2\pi.
\end{aligned}$$

This proves that $\|\psi\|_{L_2(\mathbb{R})} = 1$ and hence $\|\psi_{n,k}\|_{L_2(\mathbb{R})} = 1$ for all $n, k \in \mathbb{Z}$. Since $X(\psi)$ is a tight frame for $L_2(\mathbb{R})$. Then for any atom $h \in X(\psi)$, we have

$$\|h\|^2_{L_2(\mathbb{R})} = \sum_{g\in X(\psi)} |\langle h, g\rangle|^2 = \sum_{g\in X(\psi),\ g\neq h} |\langle h, g\rangle|^2 + \|h\|^4_{L_2(\mathbb{R})},$$

which implies that

$$\sum_{g\in X(\psi),\ g\neq h} |\langle h, g\rangle|^2 = 0, \qquad \text{for all } h \in X(\psi).$$

Therefore, $\langle h, g\rangle = 0$ whenever $h \neq g$ and hence $X(\psi)$ forms an orthonormal basis for $L_2(\mathbb{R})$. ∎

In practice, it is common to decompose a given function down to a certain level instead of down to negative infinity. The following corollary states that if a function is decomposed to a finite level L, the system $\{\phi_{L,k}, \psi_{\ell,n,k} : 1 \leq \ell \leq r, n \geq L, k \in \mathbb{Z}\}$ forms a tight frame of $L_2(\mathbb{R})$. The proof of the corollary follows directly from (2.19) and that $X(\Psi)$ is a tight frame of $L_2(\mathbb{R})$.

COROLLARY 2.1. *Let $\Psi = \{\psi_\ell : 1 \le \ell \le r\}$ be the set of tight framelets constructed from the UEP with ϕ the corresponding refinable function. Then for any given $L \in \mathbb{Z}$, the system*

$$X(\phi, \Psi; L) := \{\phi_{L,k}, \psi_{\ell,n,k} : 1 \le \ell \le r, n \ge L, k \in \mathbb{Z}\}$$

forms a tight frame of $L_2(\mathbb{R})$, i.e. for any $f \in L_2(\mathbb{R})$,

$$f = \sum_{k\in\mathbb{Z}} \langle f, \phi_{L,k}\rangle \phi_{L,k} + \sum_{\ell=1}^{r} \sum_{n\ge L} \sum_{k\in\mathbb{Z}} \langle f, \psi_{\ell,n,k}\rangle \psi_{\ell,n,k}.$$

In general, we refer to the elements in Ψ as *mother wavelets*. If the generators in Ψ are constructed from a MRA, we call them *framelets*. For the special case $r = 1$, the above theorem is given in [**124**]. What makes constructing tight framelets painless is the ability to choose the number of generators r in the UEP. For example, one can easily construct tight framelets from splines. In fact, [**158**] gives a systematic construction of tight wavelet frame systems from B-splines by using the UEP. In the next section, we begin with a simple construction of tight frame from B-splines by using the UEP. We present more constructions of framelets from pseudo-splines using UEP in Lecture 3.

1.2. B-spline tight frames

Consider (centered) B-splines of order m. The corresponding refinement mask $\widehat{h}_0$ is given as $\widehat{h}_0(\xi) = e^{-ij\frac{\xi}{2}} \cos^m(\xi/2)$ with $j = 0$ when m is even and $j = 1$ when m is odd. We define m wavelet masks as

$$\widehat{h}_\ell(\xi) := -i^\ell e^{-ij\frac{\xi}{2}} \sqrt{\binom{m}{\ell}} \sin^\ell(\xi/2) \cos^{m-\ell}(\xi/2), \quad \ell = 1, 2, \ldots, m. \tag{2.20}$$

Clearly, all assumptions in Assumption 2.1 are satisfied. Furthermore, we have

$$\sum_{\ell=0}^{m} |\widehat{h}_\ell(\xi)|^2 = (\cos^2(\xi/2) + \sin^2(\xi/2))^m = 1,$$

and

$$\sum_{\ell=0}^{m} \widehat{h}_\ell(\xi)\overline{\widehat{h}_\ell(\xi+\pi)} = e^{\frac{\pi}{2}ij}(\sin(\xi/2)\cos(\xi/2))^m(1-1)^m = 0.$$

Therefore, the m wavelets defined by

$$\widehat{\psi}_\ell := -i^\ell e^{-ij\frac{\xi}{2}} \sqrt{\binom{m}{\ell}} \frac{\cos^{m-\ell}(\xi/4)\sin^{m+\ell}(\xi/4)}{(\xi/4)^m}$$

generate a tight frame for $L_2(\mathbb{R})$. Note that each framelet ψ_ℓ is a real valued (anti)symmetric function supported in $[-(m+j)/2, (m+j)/2]$. The cases for $m = 2$ and $m = 4$ are given in the following two examples.

EXAMPLE 2.1. Let $h_0 = [\frac{1}{4}, \frac{1}{2}, \frac{1}{4}]$ be the refinement mask of the piecewise linear B-spline $B_2(x) = \max(1 - |x|, 0)$. Define $h_1 = [-\frac{1}{4}, \frac{1}{2}, -\frac{1}{4}]$ and $h_2 = [\frac{\sqrt{2}}{4}, 0, -\frac{\sqrt{2}}{4}]$. Then h_0, h_1 and h_2 satisfy (2.9) and (2.10). Hence, the system $X(\Psi)$ where $\Psi = \{\psi_1, \psi_2\}$ defined in (2.4) by using h_1, h_2 and ϕ is a tight frame of $L_2(\mathbb{R})$ (see Figure 1).

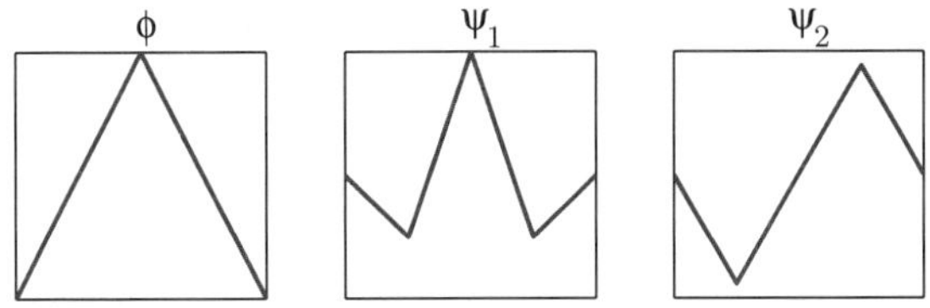

FIGURE 1. Piecewise linear refinable spline and corresponding framelets.

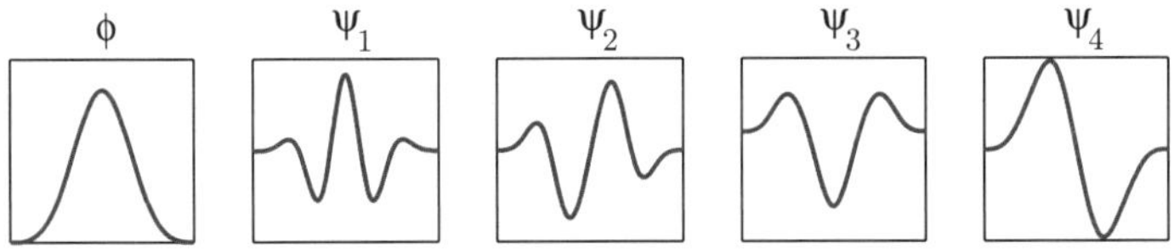

FIGURE 2. Piecewise cubic refinable B-spline and corresponding framelets.

EXAMPLE 2.2. Let $h_0 = [\frac{1}{16}, \frac{1}{4}, \frac{3}{8}, \frac{1}{4}, \frac{1}{16}]$ be the refinement mask of piecewise cubic B-spline B_4. Define h_1, h_2, h_3, h_4 as follows:

$$h_1 = [\tfrac{1}{16}, -\tfrac{1}{4}, \tfrac{3}{8}, -\tfrac{1}{4}, \tfrac{1}{16}], \quad h_2 = [-\tfrac{1}{8}, \tfrac{1}{4}, 0, -\tfrac{1}{4}, \tfrac{1}{8}],$$
$$h_3 = [\tfrac{\sqrt{6}}{16}, 0, -\tfrac{\sqrt{6}}{8}, 0, \tfrac{\sqrt{6}}{16}], \quad h_4 = [-\tfrac{1}{8}, -\tfrac{1}{4}, 0, \tfrac{1}{4}, \tfrac{1}{8}].$$

Then h_0, h_1, h_2, h_3, h_4 satisfy (2.9) and (2.6) and hence the system $X(\Psi)$ where $\Psi = \{\psi_1, \psi_2, \psi_3, \psi_4\}$ defined in (2.4) by h_1, h_2, h_3, h_4 and ϕ is a tight frame of $L_2(\mathbb{R})$ (see Figure 2).

An advantage of the tight wavelet frames derived from the UEP is that those systems have fast decomposition and reconstruction algorithms (see [**65**]), just as the orthonormal wavelet bases of [**60**]. Detailed discussions of decomposition and reconstruction algorithms are given in Section 1.4 for the univariate case and Section 3 for multivariate case.

1.3. Approximations

In this section, we focus on approximation properties of tight frame systems $X(\Psi)$ constructed from the UEP, its relations with approximation order of the underlying MRA, and vanishing moments of Ψ.

In Lecture 1 we introduced the concept of approximation order of MRA. Here, we introduce the concepts of vanishing moments and approximation orders of tight frame systems $X(\Psi)$ as follows.

DEFINITION 2.1. Let $X(\Psi)$ be a tight frame system constructed from the UEP with the underlying MRA generated by the refinable function ϕ.

(1) We say that *the tight frame system* $X(\Psi)$ *has vanishing moments of order* m_0 if $\widehat{\psi_\ell}$ has a zero of order m_0 at the origin, i.e. $|\widehat{\psi_\ell}| = O(|\cdot|^{m_0})$, for every $\ell = 1, 2, \ldots, r$.

(2) Define the *truncated operator* of the tight frame system $X(\Psi)$ as

$$Q_n : f \mapsto \sum_{\psi\in\Psi, k\in\mathbb{Z}, j<n} \langle f, \psi_{j,k}\rangle\, \psi_{j,k}. \tag{2.21}$$

We say that *the tight frame system $X(\psi)$ provides approximation order m_1* if, for all $f \in W_2^{m_1}(\mathbb{R})$,

$$\|f - \mathcal{Q}_n f\|_{L_2(\mathbb{R})} = O(2^{-nm_1}).$$

Since Ψ is constructed from the UEP with the underlying MRA generated by a refinable function ϕ, the order of vanishing moments can be characterized by the masks $\{\widehat{h}_\ell : \ell = 0, 1, \ldots, r\}$, or $\widehat{\phi}$. This provides a very convenient way of analyzing the vanishing moments of Ψ because usually we do not have an explicit formula for Ψ. Now, we summarize this characterization in the following proposition.

PROPOSITION 2.1. *Suppose Ψ is constructed from the UEP. Then it has vanishing moments of order m_0 if and only if one of the following three conditions is satisfied:*

(1) $\sum_{\ell=1}^r |\widehat{h}_\ell|^2 = O(|\cdot|^{2m_0})$;
(2) $1 - |\widehat{h}_0|^2 = O(|\cdot|^{2m_0})$;
(3) $1 - |\widehat{\phi}|^2 = O(|\cdot|^{2m_0})$.

PROOF. Since Ψ is constructed from UEP, then by (2.6) we have

$$\sum_{\ell=1}^r |\widehat{h}_\ell(\cdot)|^2 = 1 - |\widehat{h}_0(\cdot)|^2.$$

Furthermore, since $\widehat{\psi}_\ell = \left(\widehat{h}_\ell \widehat{\phi}\right)(\cdot/2)$ and $\widehat{\phi}(0) = 1$, it is clear that $\sum_{\ell=1}^r |\widehat{h}_\ell(\cdot)|^2 = O(|\cdot|^{2m_0})$ if and only if $|\widehat{\psi}_\ell| = O(|\cdot|^{m_0})$ for all $\ell = 1, 2, \ldots, r$, which establishes (1) and (2).

Let us consider (3). By the refinability of ϕ, we have

$$|\widehat{h}_0|^2 |\widehat{\phi}|^2 = |\widehat{\phi}(2\cdot)|^2.$$

Then

$$(1 - |\widehat{h}_0|^2)|\widehat{\phi}|^2 = |\widehat{\phi}|^2 - |\widehat{\phi}(2\cdot)|^2. \tag{2.22}$$

Suppose $1 - |\widehat{\phi}|^2 = O(|\cdot|^{\tilde{m}})$, i.e. $1 - |\widehat{\phi}|^2 = q + o(|\cdot|^{\tilde{m}})$ near the origin, with q some homogeneous polynomial of total degree $\tilde{m}$ (here $\tilde{m} > 0$ because $\widehat{\phi}(0) = 1$). Then near the origin, we have

$$|\widehat{\phi}|^2 - |\widehat{\phi}(2\cdot)|^2 = q(2\cdot) - q(\cdot) + o(|\cdot|^{\tilde{m}}).$$

Since $q(2\cdot) - q(\cdot)$ is a nonzero homogeneous polynomial of total degree $\tilde{m}$, we see that $|\widehat{\phi}|^2 - |\widehat{\phi}(2\cdot)|^2 = O(|\cdot|^{\tilde{m}})$. Since $\widehat{\phi}(0) = 1$ and (2) gives us $1 - |\widehat{h}_0|^2 = O(|\cdot|^{2m_0})$, we have $\tilde{m} = 2m_0$. ∎

Notice from (2.19) in the proof of Theorem 2.2 that if $X(\Psi)$ is constructed from the UEP with corresponding MRA generated by ϕ, we have

$$\mathcal{P}_n f = \mathcal{Q}_n f, \quad \text{for all } f \in L_2(\mathbb{R}), \tag{2.23}$$

where $\mathcal{P}_n$ is the quasi-interpolatory operator defined by ϕ in (2.11). This means that the approximation order of tight frame system $X(\Psi)$ is the same as that of $\mathcal{P}_n$.

We note that when $X(\Psi)$ forms an orthonormal basis for $L_2(\mathbb{R})$, the set $\{\phi_{n,k} : k \in \mathbb{Z}\}$ forms an orthonormal basis of V_n, and hence $\mathcal{Q}_n$ is an orthogonal projection onto V_n. Therefore, the approximation order of $\mathcal{Q}_n$, coincides with the approximation order of the underlying MRA. The same may not be true when $X(\Psi)$ is a tight frame for $L_2(\mathbb{R})$. In general, approximation order of $\mathcal{Q}_n$ cannot exceed the approximation order provided by the underlying MRA.

The following theorem of [**65**] shows how the approximation order of the tight frame system $X(\Psi)$ depends on the refinable functions and framelets. The proof of it is a direct consequence of Theorem 1.5 and Proposition 2.1.

THEOREM 2.3 ([**65**]). *Let $X(\Psi)$ be a tight frame system constructed from the UEP with underlying MRA generated by ϕ. Assume that Ψ has vanishing moments of order m_0 (equivalently, $1-|\widehat{\phi}|^2 = O(|\cdot|^{2m_0})$ by Proposition 2.1) and that the refinable function ϕ satisfies $[\widehat{\phi},\widehat{\phi}]-|\widehat{\phi}|^2 = O(|\cdot|^{2m})$. Then the approximation order of the tight frame system is $m_1 = \min\{m, 2m_0\}$.*

It is easy to verify that for an arbitrary B-spline B_m, the order of the zero of $1-|\widehat{B}_m|^2$ at the origin cannot exceed 2. Then Theorem 2.3 implies that the tight frame system $X(\Psi)$ constructed from B_m via the UEP (the corresponding masks are given by (2.20)) cannot have approximation order more than 2 (in fact, it is exactly 2 for all $m \geq 2$). Furthermore, Proposition 2.1 tells us that there is at least one wavelet among Ψ that has a vanishing moment of order 1. High order of approximation of a tight frame system $X(\Psi)$ gives good approximations to smooth functions. Furthermore, the high orders of vanishing moments of the framelets provide good sparse approximations to piecewise smooth functions. Hence, in order to have a good tight wavelet system, we need to have refinable functions whose Fourier transform are very flat at the origin. This leads to the introduction of pseudo-splines in [**65, 75**]. We revisit this issue with full details in Lecture 3.

Next, we briefly discuss how the order of vanishing moments plays a role in sparse approximation of piecewise smooth functions. The theory of the sparse approximation is based on nonlinear approximation and the characterization of various function spaces by a weighted norm of framelet coefficients. Here, we only provide some hubristic discussions. Interested readers should consult [**9, 10, 106, 134**] for comprehensive studies of this subject.

We start with the following proposition. It roughly says that when a wavelet function ψ has a high order vanishing moment, the majority of the corresponding wavelet coefficients are small. Hence, if all mother wavelets in Ψ have high vanishing moments, then the system $X(\Psi)$ can have a good sparse approximation to piecewise smooth functions. We note that a compactly supported wavelet ψ having order of vanishing moments m is equivalent to

$$\int x^k \psi(x)dx = 0, \quad \text{for all } 0 \leq k \leq m-1.$$

PROPOSITION 2.2. *If $\psi \in L_2(\mathbb{R})$ is a compactly supported function with order of vanishing moments m, then*

$$\langle f, \psi_{n,k}\rangle = O(\frac{1}{m!}2^{-n(m+1/2)})$$

for any $f \in L_2(\mathbb{R}) \cap C^m(\mathbb{R})$, when n is large enough.

PROOF. Without loss of generality, we assume $k = 0$. Then $\mathrm{supp}\psi_{n,0} = 2^{-n}\mathrm{supp}\,\psi \subseteq (-\varepsilon, \varepsilon)$ for some small $\varepsilon > 0$ when n is large enough. So, by

$f \in C^m(\mathbb{R})$,

$f(x) = f(0)+f'(0)x+\cdots+f^{(m-1)}(0)x^{m-1}/(m-1)!+f^{(m)}(\zeta_x)x^m/m!, \quad x \in (-\varepsilon, \varepsilon).$

So, we have

$$\begin{aligned}\left|\int_{\mathbb{R}} f(x)\psi_{n,0}(x)\,dx\right| &= 2^{n/2}\left|\int_{\mathbb{R}} f(0)\psi(2^n x)\,dx + \int_{\mathbb{R}} f'(0)x\psi(2^n x)\,dx + \cdots\right.\\ &\quad \left.+ \int_{\mathbb{R}} f^{(m-1)}(0)\frac{x^{m-1}}{(m-1)!}\psi(2^n x)\,dx + \int_{\mathbb{R}} f^{(m)}(\zeta_x)\frac{x^m}{m!}\psi(2^n x)\,dx\right|\\ &\le \frac{2^{n/2}}{m!}\|f^{(m)}\|_\infty \int_{\mathbb{R}} |x|^m|\psi(2^n x)|\,dx\\ &= \frac{2^{-n(m+1/2)}}{m!}\|f^{(m)}\|_\infty \int_{\mathbb{R}} |x|^m|\psi(x)|\,dx,\end{aligned}$$

which completes the proof. ∎

Since ψ is compactly supported, the above proof can be modified to show that the majority of wavelet coefficients not near singularities can be small when f is a piecewise smooth function. This indicates that if all mother wavelets in Ψ have high orders of vanishing moments and short supports, then the system $X(\Psi)$ can have an ideally sparse approximation for piecewise smooth functions. However, higher orders of vanishing moments usually imply larger supports of Ψ, and shorter supports usually implies lower orders of vanishing moments of Ψ. Therefore, in practice, one should balance accordingly between the two properties as well as other properties like the regularity of Ψ, for each type of application.

1.4. Decomposition and reconstruction algorithms

The decomposition and reconstruction algorithms for MRA-based tight wavelet frames derived from the UEP is essentially the same as those of MRA-based orthonormal wavelets. Here, we assume that all masks used are finitely supported.

Since

$$\mathcal{P}_L f = D^L \mathcal{P}_0 D^{-L} f,$$

without loss of generality, one may use $\mathcal{P}_0 f \in V_0$ to approximate f. If it is necessary, we can always consider the function $f(2^{-L}\cdot)$ instead of f, since the approximation of a function f in space V_L is the same as that of the function $f(2^{-L}\cdot)$ in space V_0.

When a tight wavelet frame is used, the given data is considered to be sampled as a local average $v[k] = \langle f, \phi(\cdot - k)\rangle$, which means that we can use

$$\mathcal{P}_0 f = \sum_{k\in\mathbb{Z}} v[k]\phi(\cdot - k)$$

to approximate the underlying function f. The accuracy of the approximation $\mathcal{P}_0 f$ to f is discussed in Theorem 1.5 which depends on the sampling function ϕ.

Given the sequence $h_\ell = \{h_\ell[k]\}_{k\in\mathbb{Z}}$ for any $\ell = 0, 1, \ldots, r$, we define an infinite matrix H_ℓ which corresponds to h_ℓ as

$$H_\ell := (H_\ell[l, k]) := (\sqrt{2} \cdot \overline{h_\ell[k - 2l]}),$$

where the (l, k)th entry in H_ℓ is fully determined by the $(k-2l)$th entry in h_ℓ. Then for any $v \in \ell_2(\mathbb{Z})$, we have

$$(H_\ell v)[l] = \sqrt{2}\sum_{k\in\mathbb{Z}} \overline{h_\ell[k - 2l]}v[k].$$

Similarly, we can define H_ℓ^*, the adjoint of H_ℓ, as

$$(H_\ell^* v)[k] = \sqrt{2}\sum_{l\in\mathbb{Z}} h_\ell[k-2l]v[l].$$

Let $\downarrow(\cdot)$ denote the downsampling operator which is defined as

$$\downarrow(v)[k] = v[2k], \quad k\in\mathbb{Z};$$

and $\uparrow(\cdot)$ denote the upsampling operator which is defined as

$$(\uparrow v)[k] = \begin{cases} v[k/2], & k \text{ even},\\ 0, & k \text{ odd}.\end{cases}$$

Then we have

$$H_\ell v = \downarrow(\sqrt{2}\cdot\overline{h_\ell[-\cdot]} * v) \quad \text{and} \quad H_\ell^* v = \sqrt{2}h_\ell * (\uparrow v).$$

In the literature of wavelets, it is traditional to use notations based on convolution with upsampling and downsampling.

We collect the coefficients in each level $n<0$ to form an infinite column vector

$$v_{\ell,n} := [\ldots,\langle f,\psi_{\ell,n,k}\rangle,\ldots]^\top, \quad \ell = 0,1,\ldots,r,$$

where $\psi_0 := \phi$. From the proof of Lemma 2.1, we have shown that (identity (2.16))

$$\sum_{\ell=0}^{r}\sum_{k\in\mathbb{Z}}\overline{h_\ell[j-2k]}h_\ell[j'-2k] = \frac{1}{2}\delta_{j,j'},$$

which is equivalent to

$$\sum_{\ell=0}^{r} H_\ell^* H_\ell = I. \tag{2.24}$$

Thus, the decomposition and reconstruction process (2.12) can be written in the matrix form as

$$v_{0,n+1} = \sum_{\ell=0}^{r} H_\ell^* H_\ell v_{0,n+1}.$$

For a multiple level decomposition operator, we define $\mathcal{A}_L$, $L<0$, as a (rectangular) block matrix given as:

$$\mathcal{A}_L := [H_0^{-L}; H_1 H_0^{-L-1};\ldots;H_r H_0^{-L-1};\ldots;\ H_1;\ldots;H_r]^\top. \tag{2.25}$$

Then we have the reconstruction operator $\mathcal{A}_L^*$, the adjoint operator of $\mathcal{A}_L$, defined as

$$\mathcal{A}_L^* = [H_0^{*-L}; H_0^{*-L-1}H_1^*;\ldots;H_0^{*-L-1}H_r^*;\ldots;\ H_1^*;\ldots;H_r^*]^\top. \tag{2.26}$$

Similar to (2.24), we have a multi-level perfect reconstruction formula $\mathcal{A}_L^*\mathcal{A}_L = I$, which is summarized in the following proposition; the proof is just straightforward calculation and is omitted.

PROPOSITION 2.3. *The decomposition operator $\mathcal{A}_L$, as defined in* (2.25) *satisfies $\mathcal{A}_L^*\mathcal{A}_L = I$, where I is the identity operator.*

In practice, signals are of finite lengths. Therefore, we need a proper definition of operations $\mathcal{H}_\ell v$ when v is of finite length. Now suppose $v = \{v[k]\}_{k=0}^{N-1}$. We define the periodic and symmetric extensions of v as

$$\tilde{v}[k] := \begin{cases} v[k \bmod N], & \text{periodic}, \\ v_{\mathrm{f}}[k \bmod 2N], & \text{symmetric}, \end{cases} \tag{2.27}$$

where v_{f} is defined as

$$v_{\mathrm{f}}[k] := \begin{cases} v[k], & k = 0, 1, \ldots, N-1, \\ v[2N-1-k], & k = N, N+1, \ldots, 2N-1. \end{cases}$$

Therefore, whenever v is of finite length, we define

$$\mathcal{H}_\ell v := \mathcal{H}_\ell \tilde{v},$$

with $\tilde{v}$ being one of the extensions in (2.27).

Note that when v is of finite length, the decomposition and reconstruction process can be realized as matrix multiplications (see e.g. [**15, 35**]). When $\tilde{v}$ is a periodic or symmetric extension of v, the operation $\mathcal{H}_\ell v$ can be understood as convolutions of v (followed by downsampling) with periodic or Neumann boundary conditions; where, the former is also known as circular convolution. In practice, one normally uses convolutions to implement the operation $\mathcal{H}_\ell v$ instead of matrix multiplications due to efficiency considerations. A similar argument applies to $\mathcal{H}_\ell^* v$ as well.

Algorithm 2.1 is the univariate fast framelet decomposition and reconstruction algorithm; Figure 3 is a diagram of the algorithm. For simplicity of notation, we denote $\tilde{h}_\ell := \sqrt{2} \cdot \overline{h_\ell[-\cdot]}$ and $\tilde{h}_\ell^* := \sqrt{2} h_\ell$. For a given finite signal $v \in \mathbb{R}^N$, we denote the finite length convolution operation as $\tilde{h}_\ell \circledast v$ (similarly for $\tilde{h}_\ell^* \circledast v$) which is understood as a restriction of $\tilde{h}_\ell * \tilde{v}$ on the set $\{0, 1, \ldots, N-1\}$:

$$\tilde{h}_\ell \circledast v := (\tilde{h}_\ell * \tilde{v})|_{\{0,1,\ldots,N-1\}},$$

where $\tilde{v}$ is either periodic or symmetric extension of v. Under this definition, $\tilde{h}_\ell \circledast v$ is a convolution of $\tilde{h}_\ell$ with the finite signal v using either periodic or Neumann boundary conditions.

ALGORITHM 2.1. *Given signal $v \in \mathbb{R}^N$ with N assumed to be an integer multiple of 2^L, $L \in \mathbb{N}_+$. Denote $v_{0,0} = v$. Then the L-level fast framelet decomposition and reconstruction are given as follows:*

(1) ***Decomposition:*** *For each $j = 1, 2, \ldots, L$:*
   (a) *Obtain low frequency approximation to v at level j:*

$$v_{0,j} = \downarrow (\tilde{h}_0 \circledast v_{0,j-1});$$

   (b) *Obtain framelet coefficients of v at level j:*

$$v_{\ell,j} = \downarrow (\tilde{h}_\ell \circledast v_{0,j-1}), \quad \ell = 1, 2, \ldots, r.$$

(2) ***Reconstruction:*** *For each $j = L, L-1, \ldots, 1$:*

$$v_{0,j-1} = \sum_{\ell=0}^{r} \tilde{h}_\ell^* \circledast (\uparrow v_{\ell,j}).$$

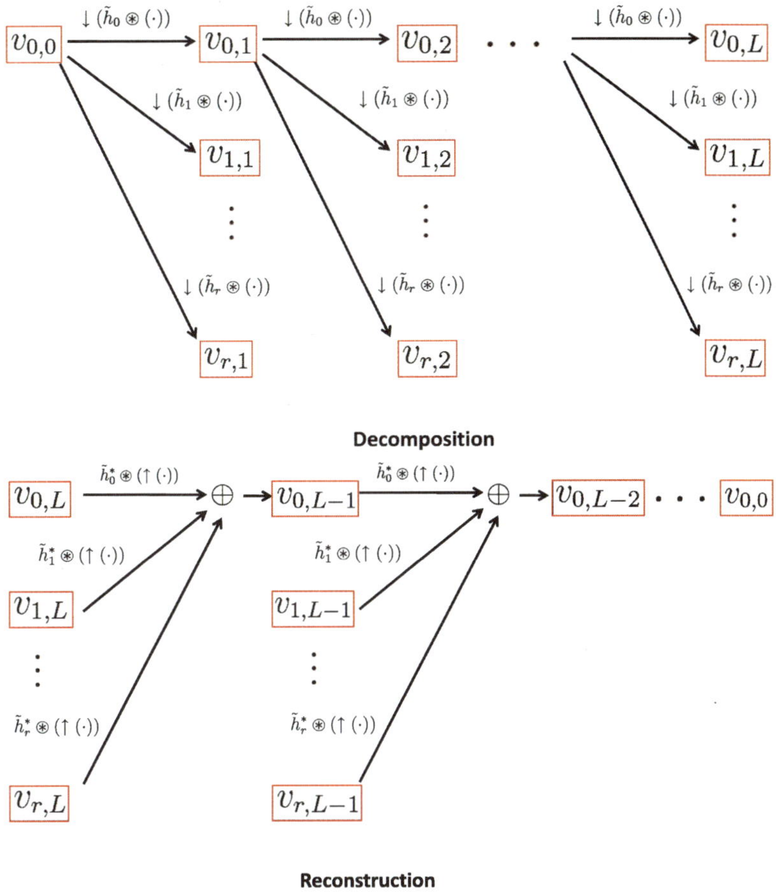

FIGURE 3. The above diagrams illustrate the fast framelet decomposition and reconstruction procedure given by Algorithm 2.1.

1.5. Other extension principles

the unitary extension principle of [**158**] was stated in a more general multivariate setting. In fact, the unitary extension principle is a consequence of a more general theorem on MRA-based tight wavelet frames. The details can be found in [**158**]. Since the publication of [**158**] in 1997, there are many generalizations of the unitary extension principle. Here, we briefly review some of them. The interested reader should consult the references mentioned below for further details.

The first generalization of the unitary extension principle is the oblique extension principle of [**52, 65**]. One of the major motivations to generalize the unitary extension principle is to obtain a spline tight wavelet system with better approximation power. As mentioned before, when the unitary extension principle is applied to construct tight wavelet frames from refinable spline functions, the approximation order of the corresponding truncated wavelet system cannot exceed 2; and there is at least one framelet that has vanishing moments of order 1.

The oblique extension principle can be used to obtain spline tight frame system whose truncated wavelet system has high approximation order and whose generators have high order vanishing moments. The oblique extension principle is stated as following:

PROPOSITION 2.4 (The Oblique Extension Principle (OEP)). *Suppose that the refinable function ϕ and the masks $h_0, h_1, \ldots, h_r$ satisfy Assumptions 2.1; and there exists a 2π-periodic function Θ that is non-negative, essentially bounded, continuous at the origin with $\Theta(0) = 1$. Assume that for $\xi \in \sigma(V_0)$, and $\xi + \pi \in \sigma(V_0)$, the following equalities*

$$\begin{aligned} &|\widehat{h}_0(\xi)|^2\Theta(2\xi) + \textstyle\sum_{\ell=1}^r |\widehat{h}_\ell(\xi)|^2 = \Theta(\xi), \\ &\widehat{h}_0(\xi)\overline{\widehat{h}_0(\xi+\pi)}\Theta(2\xi) + \textstyle\sum_{\ell=1}^r \widehat{h}_\ell(\xi)\overline{\widehat{h}_\ell(\xi+\pi)} = 0. \end{aligned} \tag{2.28}$$

hold. Then the wavelet system $X(\Psi)$ defined by $h_1, \ldots h_r$ is a tight wavelet frame.

PROOF. Setting $\theta := \Theta^{1/2}$, we define a function φ via $\hat{\varphi} := \theta\hat{\phi}$. Since θ is bounded, φ lies in $L_2(\mathbb{R})$. Consider the masks

$$\hat{\hat{h}}_0 := \frac{\theta(2\cdot)\hat{h}_0}{\theta}, \quad \hat{\hat{h}}_i := \frac{\hat{h}_i}{\theta}, \; i = 1, \ldots, r.$$

Applying (2.28), we obtain that $\sum_{i=0}^r |\hat{\hat{h}}_i(\xi)|^2 = 1, a.e.$ on $\sigma(V_0)$, hence $\hat{\hat{h}}_i$, $i = 0, 1, \ldots, r$ is well-defined and bounded, and $\hat{\hat{h}}_0$ is the refinement mask of φ. Moreover, since Θ is continuous at 0 with $\Theta(0) = 1$, we obtain that $\hat{\varphi}$ is continuous at 0 and $\hat{\varphi}(0) = 1$. Now, apply the UEP (Theorem 2.2 and remarks after it) to $\hat{\hat{h}}_i$, and observe that the wavelets obtained from the masks $\hat{\hat{h}}_i$ and φ are the same as the wavelets induced by $\hat{h}_i$ and ϕ. This concludes the proof of the theorem. ■

We note that the unitary extension principle can be viewed as a special case of the oblique extension principle by taking Θ to be 1.

Theorem 2.3 says that when the Fourier transform of the refinable function is not flat at the origin, one can chose a proper Θ which is flat at the origin. The vanishing moments of the resulting framelets have high order of the vanishing moment and the truncated tight wavelet system has a high approximation order. Detailed discussions can be found in [**65**]. This leads to many nice examples of spline tight wavelet frames with high order of vanishing moment and high approximation power in [**52, 64, 65**].

EXAMPLE 2.3. Take $\widehat{h}_0(\xi) = (1 + e^{-i\xi})^4/16$ and

$$\Theta(\xi) = 2452/945 - 1657/840\cos(\xi) + 44/105\cos(2\xi) - 311/7560\cos(3\xi).$$

Let

$$\begin{aligned} \hat{h}_1(\xi) &= t_1\,(1 - e^{-i\xi})^4\Big[1 + 8e^{-i\xi} + e^{-i2\xi}\Big], \\ \hat{h}_2(\xi) &= t_2\,(1 - e^{-i\xi})^4\Big[1 + 8e^{-i\xi} + (\tfrac{7775}{4396}\,t - \tfrac{53854}{1099})e^{-i2\xi} + 8e^{-i3\xi} + e^{-i4\xi}\Big], \\ \hat{h}_3(\xi) &= t_3\,(1 - e^{-i\xi})^4\Big[1 + 8e^{-i\xi} + (21 + \tfrac{t}{8})(e^{-i2\xi} + e^{-i4\xi}) \\ &\qquad\qquad + t\,e^{-i3\xi} + 8e^{-i5\xi} + e^{-i6\xi}\Big], \end{aligned} \tag{2.29}$$

where $t_3 = \sqrt{32655}/20160$, $t = 317784/7775 + 56\sqrt{16323699891}/2418025$, and

$$t_1 = \frac{\sqrt{11113747578360 - 245493856965\,t}}{62697600}, \qquad t_2 = \sqrt{1543080 - 32655\,t}/40320.$$

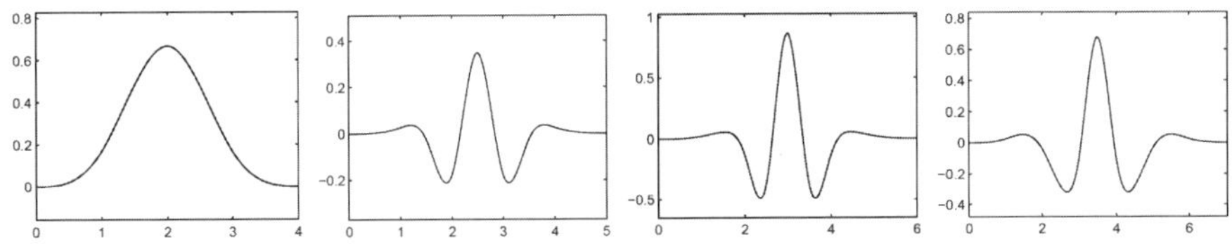

FIGURE 4. Figures from left to right are: refinable function ϕ, and framelets ψ_1, ψ_2 and ψ_3.

The above masks satisfy the OEP conditions, hence they lead to a tight wavelet frame. Here, all the wavelets have vanishing moments of order 4, i.e. $m_0 = 4$. Furthermore, the mother wavelets ψ_1, ψ_2, ψ_3 are symmetric. Note that for the above given ϕ the approximation order of the MRA is $m = 4$. Hence, the approximation order of the tight wavelet system is $4 = \min(m, 2m_0)$. The three filters $\{h_1, h_2, h_3\}$ are of size $7, 9, 11$. The refinable function ϕ and the three framelets ψ_1, ψ_2 and ψ_3 are shown in Figure 4.

Since it is impossible to obtain a tight wavelet frame that provides an arbitrary high approximation order from either the unitary extension principle or the oblique extension principle, a new approach is developed in [**104**]. This new approach of [**104**] starts with a non-stationary multiresolution analysis that has different refinable functions and refinement masks at different levels. A non-stationary version of the unitary extension principle is then established and the corresponding wavelet masks are obtained. For a different level, the corresponding set of wavelet masks is different, since the refinement mask is different. Hence, the wavelets and its masks vary by levels. By a proper choice of pseudo-spline masks, symmetric C^∞ real-valued tight wavelet frames in $L_2(\mathbb{R})$ with compact support and the spectral frame approximation order are obtained in [**104**]. Furthermore, these tight wavelet frames are used to characterize Sobolev spaces with arbitrary regularity in [**105**]. We omit the detailed discussions here and interested readers should consult [**104**] and [**105**] for details.

More recently, a concept of adaptive MRA (AMRA) was introduced in [**102**] to derive a fast and flexible decomposition strategy which adapts to data and provides a sparse approximation to the underlying function. This new AMRA is a variant of the classical MRA. For the general AMRA wavelet system, a unitary extension principle for filter design is derived. Then, it is applied to the directional representation system of shearlets, which, in turn, leads to the unitary extension principle for shearlets. This leads to a comprehensive theory for fast decomposition and reconstruction algorithms associated with 2D as well as 3D-shearlet systems encompassing tight shearlet frames with spatially compactly supported generators within such an AMRA structure. Furthermore, within the framework of [**102**] shearlet-like systems associated with parabolic scaling and unimodular matrices optimally close to rotation are studied.

Finally, both the unitary extension principle and the oblique extension principle can be generalized to a bi-frame setting which is called the mixed extension principle. Interested reader should consult [**52, 65, 157**] where the mixed extension principle is given in the multivariate setting with an arbitrary integer dilation matrix.

Furthermore, the mixed extension principle for $L_2(\mathbb{R}^d)$) of [**157**] is generalized for a pair of dual Sobolev spaces $H^s(\mathbb{R}^d)$ and $H^{-s}(\mathbb{R}^d)$ in [**106**]. Here, we briefly discuss the univariate case and encourage the reader to consult [**106**] for details of the multivariate case. The mixed extension principle is given to ensure that a pair of systems $X^s(\phi;\psi_1,\ldots,\psi_r)$ and $X^{-s}(\tilde{\phi};\tilde{\psi}_1,\ldots,\tilde{\psi}_r)$ form a dual wavelet frame pair in the corresponding dual Sobolev spaces $H^s(\mathbb{R})$ and $H^{-s}(\mathbb{R})$. Recall that the system $X^s(\phi,\Psi):=X^s(\phi;\psi_1,\ldots,\psi_r)$ is called a homogenous wavelet system generated by ϕ and $\Psi:=\{\psi_1,\ldots,\psi_r\}$, i.e.,

$$X^s(\phi,\Psi):=\{\phi(\cdot-k):k\in\mathbb{Z}^d\}\cup\{2^{n(d/2-s)}\psi_\ell(2^n\cdot-k):n\in\mathbb{N}_0,\ k\in\mathbb{Z}^d,1\le\ell\le r\}.$$

The key new feature in this general mixed extension principle is that the regularity and vanishing moment are shared by two different systems in the dual pair. In this case, if ϕ, $\psi_1,\ldots,\psi_r$ are in $H^s(\mathbb{R})$ for $s>0$, then the regularity of ϕ, $\psi_1,\ldots,\psi_r$ is required, but $\psi_1,\ldots,\psi_r$ is not required to have any order of vanishing moment. At the same time, the vanishing moments of $\tilde{\psi}_1,\ldots,\tilde{\psi}_r$ are required, while $\tilde{\phi}$, $\tilde{\psi}_1,\ldots,\tilde{\psi}_r$ are in $H^{-s}(\mathbb{R})$ can be tempered distributions instead of in $L_2(\mathbb{R})$. (Note that in order to have a dual pair in $L_2(\mathbb{R})$, both systems in the dual pair must have a certain order of regularity and vanishing moment at the same time.) This implies that it is not necessary for the systems $X^s(\phi;\psi_1,\ldots,\psi_r)$ and $X^{-s}(\tilde{\phi};\tilde{\psi}_1,\ldots,\tilde{\psi}_r)$ to be normalized into a frame of $L_2(\mathbb{R})$. This leads to many simple constructions of frames in an arbitrarily given Sobolev space. For example, let B_m be the B-spline of order m. Then by [**106**], $\{2^{j(1/2-s)}B_m(2^j\cdot-k)\ :\ j\in\mathbb{N}_0,k\in\mathbb{Z}\}$ is a wavelet frame in $H^s(\mathbb{R})$ for any $0<s<m-1/2$. This construction is applied to multivariate box splines to obtain wavelet frames with small supports, which is known to be difficult if the traditional approach is taken. Recall that in the traditional approach, frames in Sobolev spaces are obtained by normalizing a frame in $L_2(\mathbb{R})$ to a frame in Sobolev space. Since it is hard to construct small support wavelet frames in $L_2(\mathbb{R}^d)$ in general, it is difficult to obtain small support frames in Sobolev spaces by this traditional method. Also, this general mixed extension principle also naturally leads to a characterization of the Sobolev norm of a function in terms of the weighted norm of its wavelet coefficient sequence (decomposition sequence) without requiring that dual wavelet frames be in $L_2(\mathbb{R})$. This approach is quite different from others in the literature (see e.g. [**10, 9, 108, 134**]). Furthermore, applying this general mixed extension principle obtains a characterization for a pair of systems $X^s(\phi;\psi_1,\ldots,\psi_r)$ and $X^{-s}(\tilde{\phi};\tilde{\psi}_1,\ldots,\tilde{\psi}_r)$ in Sobolev spaces $H^s(\mathbb{R})$ and $H^{-s}(\mathbb{R})$) that form a pair of dual Riesz bases. This characterization, for example, leads to a proof of the fact that all interpolatory wavelet systems defined in [**76**] generated by an interpolatory refinable function $\phi\in H^s(\mathbb{R})$ with $s>1/2$ are Riesz bases of the Sobolev space $H^s(\mathbb{R})$.

2. Quasi-affine systems and associated algorithms

In the context of signal and image processing, it is usually preferred to use wavelet systems that are shift-invariant. Recall that a system X, that contains countably many elements, is *τ-shift-invariant* with $\tau\in\mathbb{R}$, if for any $k\in\mathbb{Z}$ and $g\in X$, we have $g(\cdot-\tau k)\in X$. In particular, if a system is 1-shift-invariant, we simply call it *shift-invariant*. Notice that the wavelet system (affine system) $X(\Psi)$ defined by (2.1) is not shift-invariant. In order to achieve shift-invariance, we need to over-sample the affine system $X(\Psi)$ below level 0. This over-sampled affine system is called a quasi-affine system, which was first introduced in [**158**]. One implicitly uses quasi-affine systems when the wavelet decomposition algorithm is applied without downsampling [**57**]. This type of wavelet transform is also known as the undecimated wavelet transform. As a consequence, the theory of quasi-affine systems established in [**158**] also provides a theoretic foundation of using the undecimated wavelet transform which was not available in the literature before. The definition of quasi-affine system is given as follows. Note that most part of this section is taken from [**32**].

2.1. Quasi-affine systems

The notion of quasi-affine systems was first introduced and extensively studied in [**158**].

DEFINITION 2.2. Let $\Psi = \{\psi_1, \ldots, \psi_r\}$ be a set of functions. A *quasi-affine system* from level L is defined as

$$X_L^q(\Psi) = \{\psi_{\ell,n,k}^q : 1 \le \ell \le r; n, k \in \mathbb{Z}\},$$

where $\psi_{\ell,n,k}^q$ is defined by

$$\psi_{\ell,n,k}^q := \begin{cases} \mathcal{D}^n T_k \psi_\ell, & n \ge L, \\ 2^{\frac{n-L}{2}} T_{2^{-L}k} \mathcal{D}^n \psi_\ell, & n < L. \end{cases}$$

The quasi-affine system is obtained by over sampling the wavelet frame system starting from level $L-1$ and downward. Hence, the whole quasi-affine system is a 2^{-L}-shift-invariant system. The quasi-affine system from level 0 was first introduced in [**158**] to convert a non-shift invariant affine system to a shift invariant system. Further, it was shown in [**158**, Theorem 5.5] that a wavelet system $X(\Psi)$ (not necessarily an MRA-based wavelet system) is a tight frame of $L_2(\mathbb{R})$ if and only if the corresponding quasi-affine counterpart $X_L^q(\Psi)$ is a tight frame of $L_2(\mathbb{R})$. In fact, this result plays important roles in the proof of Theorem 2.1 by using the dual Gramian analysis. This result enables us to convert a wavelet system into a shift invariant system which makes it possible to use the powerful tool of the dual Gramian analysis of [**156, 162**]. Note that dual Gramian analysis of [**156**] is designed for shift invariant systems.

Here we provide a direct proof of a special case of [**158**, Theorem 5.5], the case for the MRA-based wavelet system, in Theorem 2.4. Precisely speaking, we prove that the quasi-affine system $X_L^q(\Psi)$ is a tight frame for $L_2(\mathbb{R})$ if Ψ is constructed from UEP (Theorem 2.2).

When we consider the MRA-based quasi-affine system $X_L^q(\Psi)$ generated by Ψ, the spaces V_n, $n < L$ in the MRA for the affine system are replaced by $V_n^{q,L}$, $n < L$, for the quasi-affine system. Note that the space V_n is spanned by the functions $\phi_{n,k}$, while the space $V_n^{q,L}$ is spanned by functions $\phi_{n,k}^q$, where $\phi_{n,k}^q$ is defined by

$$\phi_{n,k}^q := \begin{cases} \mathcal{D}^n T_k \phi, & n \ge L, \\ 2^{\frac{n-L}{2}} T_{2^{-L}k} \mathcal{D}^n \phi, & n < L. \end{cases}$$

The spaces $V_n^{q,L}$ for all $n \in \mathbb{Z}$ are 2^{-L}-shift invariant.

Similar to an affine system, we can define the quasi-interpolatory operator $\mathcal{P}_n^{q,L}$ and the truncated operator $\mathcal{Q}_n^{q,L}$, $n \in \mathbb{Z}$, for the quasi-affine system by

$$\mathcal{P}_n^{q,L} : f \mapsto \sum_{k \in \mathbb{Z}} \langle f, \phi_{n,k}^q \rangle \phi_{n,k}^q \tag{2.30}$$

and

$$\mathcal{Q}_n^{q,L} : f \mapsto \sum_{\ell=1}^{r} \sum_{j<n,k\in\mathbb{Z}} \langle f, \psi_{\ell,j,k}^q \rangle \psi_{\ell,j,k}^q. \tag{2.31}$$

Whenever $\{\phi_{n,k}^q : k \in \mathbb{Z}\}$ is a Bessel system, the quasi-interpolatory operator $\mathcal{P}_n^{q,L}$ maps $f \in L_2(\mathbb{R})$ into $V_n^{q,L}$. From the definition of $\phi_{n,k}^q$, we can see that $\mathcal{P}_n^{q,L} = \mathcal{P}_n$ when $n \geq L$, and that these two operators are different only when $n < L$. Moreover, since for an arbitrary $f \in L_2(\mathbb{R})$ and $n < L$,

$$\begin{aligned}\mathcal{P}_n^{q,L} f = \sum_{k\in\mathbb{Z}} \langle f, \phi_{n,k}^q \rangle \phi_{n,k}^q &= \mathcal{D}^L \sum_{k\in\mathbb{Z}} \langle \mathcal{D}^{-L} f, 2^{\frac{n-L-0}{2}} T_k \mathcal{D}^{n-L}\phi \rangle 2^{\frac{n-L-0}{2}} T_k \mathcal{D}^{n-L}\phi \\ &= \mathcal{D}^L \mathcal{P}_{n-L}^{q,0} D^{-L} f,\end{aligned}$$

one only needs to understand the case $L = 0$. In this case we simplify our notation by setting

$$\mathcal{P}_n^q := \mathcal{P}_n^{q,0}, \quad \mathcal{Q}_n^q := \mathcal{Q}_n^{q,0} \tag{2.32}$$

for the quasi-interpolatory operators and $V_n^q := V_n^{q,0}$, for $n \in \mathbb{Z}$. From now on, we only give the properties for $\mathcal{P}_n^q$, the corresponding spaces V_n^q, and the associated quasi-affine system $X^q(\Psi) := X_0^q(\Psi)$. The corresponding results for the over sampling rate of $2^{-L}\mathbb{Z}$ can be obtained in a similar manner.

Also, for operator $\mathcal{P}_n^q$, $n \in \mathbb{Z}$, we have a decomposition and reconstruction formula similar to (2.12). Similar to the presentation of affine systems, we present proof of Lemma 2.4 in the Fourier domain under Assumption 2.1, and in the time domain with some additional assumptions.

LEMMA 2.4. *Let $X(\Psi)$, where the framelets $\Psi = \{\psi_1, \ldots, \psi_r\}$, be the affine tight frame system obtained from the UEP with corresponding refinable function ϕ and the set of refinement masks $\{h_0, h_1, \ldots, h_r\}$. Let $X^q(\Psi)$ be the quasi-affine system derived from $X(\Psi)$. Assume that Assumption 2.1 is satisfied. Then we have, for all $f \in L_2(\mathbb{R})$,*

$$\mathcal{P}_n^q f = \mathcal{P}_{n-1}^q f + \sum_{\ell=1}^{r} \sum_{k\in\mathbb{Z}} \langle f, \psi_{\ell,n-1,k}^q \rangle \psi_{\ell,n-1,k}^q. \tag{2.33}$$

PROOF. When $n \geq 0$, we have $\phi_{n,k}^q = \phi_{n,k}$ and $\psi_{n,k}^q = \psi_{n,k}$, which imply that

$$\mathcal{P}_n f = \sum_{k\in\mathbb{Z}} \langle f, \phi_{n,k} \rangle \phi_{n,k} = \sum_{k\in\mathbb{Z}} \langle f, \phi_{n,k}^q \rangle \phi_{n,k}^q = \mathcal{P}_n^q f,$$

and

$$\sum_{\ell=1}^{r} \sum_{k\in\mathbb{Z}} \langle f, \psi_{\ell,n,k}^q \rangle \psi_{\ell,n,k}^q = \sum_{\ell=1}^{r} \sum_{k\in\mathbb{Z}} \langle f, \psi_{\ell,n,k} \rangle \psi_{\ell,n,k}.$$

Then (2.12) implies that

$$\begin{aligned}\mathcal{P}_n^q f = \mathcal{P}_n f &= \mathcal{P}_{n-1} f + \sum_{\ell=1}^{r} \sum_{k\in\mathbb{Z}} \langle f, \psi_{\ell,n-1,k} \rangle \psi_{\ell,j,k} \\ &= \mathcal{P}_{n-1}^q f + \sum_{\ell=1}^{r} \sum_{k\in\mathbb{Z}} \langle f, \psi_{\ell,n-1,k}^q \rangle \psi_{\ell,n-1,k}^q,\end{aligned}$$

i.e. the identity (2.33) holds when $n \geq 1$. Next, we show (2.33) holds for $n \leq 0$. We denote ϕ as ψ_0.

First, we note that the system $\{\psi^q_{\ell,n,k} : k \in \mathbb{Z}\}$ for any given $\ell = 0, 1, \ldots, r$ and $n \leq 0$ is a Bessel system. Indeed, by Assumption 2.1, we know that $\{\psi_{\ell,n,k} : k \in \mathbb{Z}\}$ is a Bessel system, i.e. there exists constant $0 < C < \infty$ such that for all $f \in L_2(\mathbb{R})$,

$$\sum_{k\in\mathbb{Z}} |\langle f, \psi_{\ell,n,k}\rangle|^2 \leq C\|f\|^2_{L_2(\mathbb{R})}.$$

Notice that

$$\begin{aligned}\{\psi^q_{\ell,n,k} : k \in \mathbb{Z}\} &= \{2^n\psi_\ell(2^n x - 2^n k) : k \in \mathbb{Z}\} \\ &= \bigcup_{l\in\mathbb{Z}/2^{-n}\mathbb{Z}} \{2^n\psi_\ell(2^n(\cdot - l) - k) : k \in \mathbb{Z}\}.\end{aligned}$$

By a simple change of variables, one can easily check that the system $\{2^n\psi_\ell(2^n(\cdot - l) - k) : k \in \mathbb{Z}\}$ is also a Bessel system with bound $2^{n/2}\sqrt{C}$ for every $l \in \mathbb{Z}/2^{-n}\mathbb{Z}$. Therefore, we have

$$\sum_{k\in\mathbb{Z}} |\langle f, \psi^q_{\ell,n,k}\rangle|^2 \leq C\|f\|^2_{L_2(\mathbb{R})}.$$

Denote

$$\mathcal{P}^q_{n,\ell} f := \sum_{k\in\mathbb{Z}} \langle f, \psi^q_{\ell,n,k}\rangle {\psi_{\ell,n,k}}^q,$$

for $\ell = 0, 1, \ldots, r$. Note that $\mathcal{P}^q_{n,0} = \mathcal{P}^q_n$. Since, the system $\{\psi^q_{\ell,n,k}; k \in \mathbb{Z}\}$ for each ℓ and n, forms a Bessel system, the operator $\mathcal{P}^q_{n,\ell} : L_2(\mathbb{R}) \to L_2(\mathbb{R})$ is well-defined. Therefore, to prove (2.33) for $n \leq 0$, it is equivalent to show

$$\widehat{\mathcal{P}^q_n f} = \sum_{\ell=0}^{r} \widehat{\mathcal{P}^q_{n-1,\ell} f}. \tag{2.34}$$

Letting $I_n := 2^n(\mathbb{Z}/2^{-n}\mathbb{Z})$, we observe that

$$\begin{aligned}\mathcal{P}^q_{n,\ell} f &= \sum_{k\in\mathbb{Z}} \langle f, \psi^q_{\ell,n,k}\rangle \psi^q_{\ell,n,k} \\ &= 2^n \sum_{\alpha\in I_n}\sum_{k\in\mathbb{Z}} \langle f, (T_\alpha\psi_\ell)_{n,k}\rangle (T_\alpha\psi_\ell)_{n,k}.\end{aligned}$$

Then by Proposition 1.3 item (3), we have

$$\begin{aligned}\widehat{\mathcal{P}^q_{n,\ell} f}(\xi) &= 2^n \sum_{\alpha\in I_n} [\widehat{f}(2^n\cdot), \widehat{T_\alpha\psi_\ell}](2^{-n}\xi)\widehat{T_\alpha\psi_\ell}(2^{-n}\xi) \\ &= 2^n \sum_{\alpha\in I_n}\sum_{k\in\mathbb{Z}} \widehat{f}(\xi + 2^{n+1}k\pi)\overline{\widehat{T_\alpha\psi_\ell}(\frac{\xi}{2^n} + 2k\pi)}\widehat{T_\alpha\psi_\ell}(2^{-n}\xi) \\ &= 2^n \sum_{\alpha\in I_n}\sum_{k\in\mathbb{Z}} \widehat{f}(\xi + 2^{n+1}k\pi)e^{i\alpha(2^{-n}\xi+2k\pi)}\overline{\widehat{\psi_\ell}(\frac{\xi}{2^n} + 2k\pi)}e^{-i\alpha2^{-n}\xi}\widehat{\psi_\ell}(2^{-n}\xi) \\ &= 2^n \sum_{\alpha\in I_n}\sum_{k\in\mathbb{Z}} \widehat{f}(\xi + 2^{n+1}k\pi)\overline{\widehat{\psi_\ell}(\frac{\xi}{2^n} + 2k\pi)}e^{i\alpha2k\pi}\widehat{\psi_\ell}(2^{-n}\xi) \\ &= \sum_{k\in\mathbb{Z}} \widehat{f}(\xi + 2^{n+1}k\pi)\overline{\widehat{\psi_\ell}(\frac{\xi}{2^n} + 2k\pi)}\left(2^n \sum_{\alpha\in I_n} e^{i\alpha2k\pi}\right)\widehat{\psi_\ell}(2^{-n}\xi).\end{aligned}$$

Notice that

$$2^n \sum_{\alpha \in I_n} e^{i\alpha 2k\pi} = 2^n \sum_{l \in \mathbb{Z}/2^{-n}\mathbb{Z}} \left(e^{i2^{n+1}k\pi}\right)^l = \begin{cases} 1, & k \in 2^{-n}\mathbb{Z}, \\ 0, & k \notin 2^{-n}\mathbb{Z}. \end{cases}$$

Therefore

$$\begin{aligned} \widehat{\mathcal{P}^q_{n,\ell} f} &= \sum_{k \in 2^{-n}\mathbb{Z}} \widehat{f}(\xi + 2^{n+1}k\pi)\overline{\widehat{\psi}_\ell(\frac{\xi}{2^n} + 2k\pi)}\widehat{\psi}_\ell(2^{-n}\xi) \\ &= [\widehat{f}, \widehat{\psi}_\ell(\frac{\cdot}{2^n})]\widehat{\psi}_\ell(\frac{\cdot}{2^n}). \end{aligned}$$

Now, identity (2.34) is equivalent to

$$[\widehat{f}, \widehat{\phi}(\frac{\cdot}{2^n})]\widehat{\phi}(\frac{\cdot}{2^n}) = \sum_{\ell=0}^{r} [\widehat{f}, \widehat{\psi}_\ell(\frac{\cdot}{2^{n-1}})]\widehat{\psi}_\ell(\frac{\cdot}{2^{n-1}}), \tag{2.35}$$

for $n \le 0$. Using $\widehat{\psi}_\ell = \widehat{h}_\ell(\frac{\cdot}{2})\widehat{\phi}(\frac{\cdot}{2})$ and the first identity of the UEP condition (2.6), we have

$$\sum_{\ell=0}^{r} [\widehat{f}, \widehat{\psi}_\ell(\frac{\cdot}{2^{n-1}})]\widehat{\psi}_\ell(\frac{\cdot}{2^{n-1}}) = \sum_{\ell=0}^{r} |\widehat{h}_\ell(\frac{\cdot}{2^n})|^2 [\widehat{f}, \widehat{\phi}(\frac{\cdot}{2^n})]\widehat{\phi}(\frac{\cdot}{2^n}) = [\widehat{f}, \widehat{\phi}(\frac{\cdot}{2^n})]\widehat{\phi}(\frac{\cdot}{2^n}).$$

This concludes the proof of (2.35) and hence the lemma follows. ∎

REMARK 2.2. Lemma 2.4 can be proven in the time domain. We prove a simple case here because the proof will reveal some details of the fast framelet decomposition and reconstruction algorithm for quasi-affine systems. For simplicity, we assume that ψ_ℓ, for $\ell = 0, 1, \ldots, r$, are compactly supported and the masks h_ℓ are finite sequences.

We only need to prove (2.33) for $n \le 0$. Since system $\{\psi^q_{\ell,n,k} : k \in \mathbb{Z}\}$ for any given $\ell = 0, 1, \ldots, r$ and $n \le 0$ is a Bessel system, the operators $\mathcal{P}^q_{n,\ell} : L_2(\mathbb{R}) \mapsto L_2(\mathbb{R})$, for each ℓ is bounded. Therefore, we only need to show (2.33) for all compactly supported f.

By the definitions of a refinable equation (1.5) and a framelet (2.4), one obtains for $\ell = 0, 1, \ldots, r$,

$$\psi_\ell = 2 \sum_{k \in \mathbb{Z}} h_\ell[k]\phi(2 \cdot -k).$$

This leads to

$$\begin{aligned} \psi^q_{\ell,n-1,k} &= 2^{n-1} T_k \psi_\ell(2^{n-1} \cdot) = 2^n T_k \left(\sum_{k' \in \mathbb{Z}} h_\ell[k']\psi_0(2^n \cdot -k') \right) \\ &= \sum_{k' \in \mathbb{Z}} h_\ell[k'] 2^n \psi_0(2^n(\cdot - k - 2^{-n}k')) \\ &= \sum_{k' \in 2^{-n}\mathbb{Z}} h_\ell[2^n k'] 2^n \psi_0(2^n(\cdot - k - k')). \end{aligned}$$

We define the dilated sequence $h_{\ell,n}$ by

$$h_{\ell,n}[k] = \begin{cases} h_\ell[2^n k], & k \in 2^{-n}\mathbb{Z}, \\ 0, & k \notin 2^{-n}\mathbb{Z}. \end{cases} \tag{2.36}$$

The sequence $h_{\ell,n}$ is obtained inductively by inserting 0 between every two entries in $h_{\ell,n+1}$, and $h_{\ell,0} = h_\ell$. With the dilated sequence, we have

$$\psi^q_{\ell,n-1,k} = \sum_{k'\in\mathbb{Z}} h_{\ell,n}[k']\psi^q_{0,n,k+k'}.$$

Therefore, we have

$$\begin{aligned}
&\sum_{\ell=0}^{r}\sum_{k\in\mathbb{Z}}\langle f,\psi^q_{\ell,n-1,k}\rangle\psi^q_{\ell,n-1,k}\\
&\quad=\sum_{\ell=0}^{r}\sum_{k\in\mathbb{Z}}\left(\sum_{k'\in\mathbb{Z}}\overline{h_{\ell,n}[k']}\langle f,\psi^q_{0,n,k'+k}\rangle\right)\left(\sum_{k''\in\mathbb{Z}}h_{\ell,n}[k'']\psi^q_{0,n,k''+k}\right)\\
&\quad=\sum_{k'\in\mathbb{Z}}\sum_{k''\in\mathbb{Z}}\left(\sum_{\ell=0}^{r}\sum_{k\in\mathbb{Z}}\overline{h_{\ell,n}[k]}h_{\ell,n}[k+k''-k']\right)\langle f,\psi^q_{0,n,k'}\rangle\psi^q_{0,n,k''},
\end{aligned}$$

where the switching of summation order is valid since all summations involved are finite sums.

Now, we focus on showing

$$\sum_{\ell=0}^{r}\sum_{k\in\mathbb{Z}}\overline{h_{\ell,n}[k]}h_{\ell,n}[k+k''-k'] = \delta_{k'-k'',0}.$$

When $k'-k''\in 2^{-n}\mathbb{Z}$, there exists $p\in\mathbb{Z}$ such that $k'-k''=2^{-n}p$ and we have

$$\begin{aligned}
\sum_{\ell=0}^{r}\sum_{k\in\mathbb{Z}}\overline{h_{\ell,n}[k]}h_{\ell,n}[k+k''-k'] &= \sum_{\ell=0}^{r}\sum_{k\in\mathbb{Z}}\overline{h_{\ell,n}[k]}h_{\ell,n}[k-2^{-n}p]\\
&=\sum_{\ell=0}^{r}\sum_{k\in 2^{-n}\mathbb{Z}}\overline{h_{\ell,n}[k]}h_{\ell,n}[k-2^{-n}p]\\
&=\sum_{\ell=0}^{r}\sum_{k\in\mathbb{Z}}\overline{h_\ell[k]}h_\ell[k-p] = \delta_{p,0}.
\end{aligned}$$

The last identity follows by (2.9). The sum is nonzero if and only if $p=0$, which is equivalent to $k'=k''$. When $k'-k''\notin 2^{-n}\mathbb{Z}$, there exist $p_1,p_2\in\mathbb{Z}$ and $p_2\notin 2^{-n}\mathbb{Z}$ such that $k'-k''=2^{-n}p_1+p_2$. Then we have

$$\begin{aligned}
\sum_{\ell=0}^{r}\sum_{k\in\mathbb{Z}}\overline{h_{\ell,n}[k]}h_{\ell,n}[k+k''-k'] &= \sum_{\ell=0}^{r}\sum_{k\in\mathbb{Z}}\overline{h_{\ell,n}[k]}h_{\ell,n}[k-2^{-n}p_1-p_2]\\
&=\sum_{\ell=0}^{r}\sum_{k\in 2^{-n}\mathbb{Z}}\overline{h_{\ell,n}[k]}h_{\ell,n}[k-2^{-n}p_1-p_2].
\end{aligned}$$

Since $k-2^{-n}p_1-p_2\notin 2^{-n}\mathbb{Z}$ when $k\in 2^{-n}\mathbb{Z}$, we have $h_{\ell,n}[k-2^{-n}p_1-p_2]=0$ for any $k\in 2^{-n}\mathbb{Z}$ and the last identity is equal to 0. In conclusion, for the dilated filters $h_{0,n}, h_{1,n},\dots,h_{r,n}$, we still have a result similar to (2.9)

$$\sum_{\ell=0}^{r}\sum_{k\in\mathbb{Z}}\overline{h_{\ell,n}[k]}h_{\ell,n}[k-p] = \delta_{p,0},\quad n\le 0,\ p\in\mathbb{Z}. \tag{2.37}$$

Thus we have proven (2.33), which leads to

$$\sum_{\ell=0}^{r}\sum_{k\in\mathbb{Z}}\langle f,\psi^q_{\ell,n-1,k}\rangle\psi^q_{\ell,n-1,k}=\sum_{k\in\mathbb{Z}}\langle f,\psi^q_{0,n,k}\rangle\psi^q_{0,n,k}=\mathcal{P}^q_n f.$$

This is the identity we need to prove the case when $n\le 0$. In all, identity (2.33) holds for any $n\in\mathbb{Z}$.

REMARK 2.3. We note that in the proof of identity (2.12) for the affine system, one needs both identities of (2.6); while in the proof of identity (2.33) for $n\le 0$, when the quasi-affine system is used, one only needs the first identity of (2.6).

The next theorem shows that a result similar to (2.23) also holds for quasi-affine systems, and that $X^q(\Psi)$ is a tight frame for $L_2(\mathbb{R})$.

THEOREM 2.4. *Let $X(\Psi)$, where the framelets $\Psi=\{\psi_1,\dots,\psi_r\}$, be the affine tight frame system obtained from the UEP. Then the quasi-affine system $X^q(\Psi)$ forms a tight frame for $L_2(\mathbb{R})$, and $\mathcal{P}^q_n f=\mathcal{Q}^q_n f$ for all $f\in L_2(\mathbb{R})$.*

PROOF. We begin by showing that $\mathcal{P}^q_n f=\mathcal{Q}^q_n f$. Telescoping on (2.33) we have, for any $f\in L_2(\mathbb{R})$ and $n>n'$,

$$\mathcal{P}^q_n f=\mathcal{P}^q_{n'}f+\sum_{\ell=1}^{r}\sum_{j=n'}^{n-1}\sum_{k\in\mathbb{Z}}\langle f,\psi^q_{\ell,j,k}\rangle\psi^q_{\ell,j,k}. \tag{2.38}$$

Thus the proof of $\mathcal{P}^q_n f=\mathcal{Q}^q_n f$ is transferred to the proof of $P^q_{n'}f\to 0$ as $n'\to-\infty$. Since we are interested in the behavior of $P^q_{n'}$ as $n'\to-\infty$, we now assume that $n'<0$.

As shown in the proof of Lemma 2.4, $\{\phi^q_{n',k}:k\in\mathbb{Z}\}$ is a Bessel system for any given $n'\in\mathbb{Z}$. Thus, we have

$$\sum_{k\in\mathbb{Z}}|\langle f,\phi^q_{n',k}\rangle|^2\le C\|f\|^2_{L_2(\mathbb{R})}.$$

Using the above inequality and

$$\mathcal{P}^q_{n'}f\ =\ \sum_{k\in\mathbb{Z}}\langle f,\phi^q_{n',k}\rangle\phi^q_{n',k},$$

we have the norm of $\mathcal{P}^q_{n'}f$ satisfying

$$\|\mathcal{P}^q_{n'}f\|^2_{L_2(\mathbb{R})}\le C\sum_{k\in\mathbb{Z}}|\langle f,\phi^q_{n',k}\rangle|^2. \tag{2.39}$$

Based on routine arguments in approximation theory, we only need to check the value of $\|\mathcal{P}^q_{n'}f\|_{L_2(\mathbb{R})}$ when f is supported on an interval $[-R,R]$ for an arbitrary given $R>0$. Applying the Cauchy-Schwartz inequality to (2.39), we have,

$$\begin{aligned}\|\mathcal{P}^q_{n'}f\|^2_{L_2(\mathbb{R})}&\le C\|f\|^2_{L_2(\mathbb{R})}\sum_{k\in\mathbb{Z}}\int_{[-R,R]}\left|2^{n'}\phi(2^{n'}(x-k))\right|^2\mathrm{d}x\\&=2^{n'}C\|f\|^2_{L_2(\mathbb{R})}\sum_{k\in\mathbb{Z}}\int_{E_{n',k}}|\phi(x)|^2\,\mathrm{d}x,\end{aligned} \tag{2.40}$$

where $E_{n',k}=2^{n'}([-R,R]-k)$. Now, we prove that

$$2^{n'}\sum_{k\in\mathbb{Z}}\int_{E_{n',k}}|\phi(x)|^2\,\mathrm{d}x\to 0,\quad \text{as } n'\to-\infty. \tag{2.41}$$

By the monotone convergence theorem, we have

$$2^{n'}\sum_{k\in\mathbb{Z}}\int_{E_{n',k}}|\phi(x)|^2\,\mathrm{d}x = 2^{n'}\int\sum_{k\in\mathbb{Z}}\chi_{E_{n',k}}|\phi(x)|^2\,\mathrm{d}x.$$

Observe that $\sum_{k\in\mathbb{Z}}\chi_{E_{n',k}} \le C_R$ with the constant C_R only depending on R. Therefore, we have

$$2^{n'}\sum_{k\in\mathbb{Z}}\chi_{E_{n',k}}|\phi(x)|^2 \to 0 \text{ a.e.} \quad \text{and} \quad 2^{n'}\sum_{k\in\mathbb{Z}}\chi_{E_{n',k}}|\phi(x)|^2 \le C_R|\phi(x)|^2.$$

Then, the dominated convergence theorem implies (2.41), and hence (2.38) becomes

$$\mathcal{P}_n^q f = \sum_{\ell=1}^{r}\sum_{j<n}\sum_{k\in\mathbb{Z}}\langle f, \psi_{\ell,j,k}^q\rangle\psi_{\ell,j,k}^q = \mathcal{Q}_n^q f.$$

This completes the proof of $\mathcal{P}_n^q f = \mathcal{Q}_n^q f$.

Since $\mathcal{P}_n^q f = \mathcal{P}_n f$ for $n \ge 0$, we have $P_n^q f = P_n f \to f$ as $n \to \infty$ (by Lemma 2.3). This implies that $X^q(\Psi)$ forms a tight frame of $L_2(\mathbb{R})$ and thus concludes the proof of this theorem. ∎

From the above theorem, the following result on approximation orders of quasi-affine tight frame systems $X^q(\Psi)$ is obvious. Note that the approximation order of $X^q(\Psi)$ is defined similarly as $X(\Psi)$ in Definition 2.1.

PROPOSITION 2.5. *Let $X^q(\Psi)$ be a quasi-affine tight frame system constructed from UEP with the underlying MRA generated by a refinable function ϕ. Assume that Ψ has vanishing moments of order m_0 and ϕ provides approximation order m. Then the approximation order of the quasi-affine tight frame system is $m_1 = \min\{m, 2m_0\}$.*

Similar to the affine counterpart, we normally do not perform full levels of decomposition and reconstruction as given by Theorem 2.4. We normally stop at a certain level instead. The following corollary states that by doing so, we still have a tight frame system which includes the shifts of ϕ at a certain scale.

COROLLARY 2.2. *Let $\Psi = \{\psi_\ell : 1 \le \ell \le r\}$ be the set of tight framelets constructed from the UEP with ϕ the corresponding refinable function. Then for any given $L \in \mathbb{Z}$, the system*

$$X^q(\phi, \Psi; L) := \{\phi_{L,k}^q, \psi_{\ell,n,k}^q : 1 \le \ell \le r, n \ge L, k \in \mathbb{Z}\}$$

forms a tight frame of $L_2(\mathbb{R})$, i.e. for any $f \in L_2(\mathbb{R})$,

$$f = \sum_{k\in\mathbb{Z}}\langle f, \phi_{L,k}^q\rangle\phi_{L,k}^q + \sum_{\ell=1}^{r}\sum_{n\ge L}\sum_{k\in\mathbb{Z}}\langle f, \psi_{\ell,n,k}^q\rangle\psi_{\ell,n,k}^q.$$

In practical calculations, we always assume that the function we have is $\mathcal{P}_0^q f \in V_0^q$, and perform decomposition down to V_{-L}^q for some $L \in \mathbb{Z}^+$. Then the decomposition and reconstruction formula can be written as

$$\mathcal{P}_0^q f = \mathcal{P}_{-L}^q f + \sum_{\ell=1}^{r}\sum_{j=-L}^{-1}\sum_{k\in\mathbb{Z}}\langle f, \psi_{\ell,j,k}^q\rangle\psi_{\ell,j,k}^q.$$

This process corresponds to the so-called undecimated wavelet decomposition and reconstruction. Details of this process are described in the next section.

2.2. Algorithms for quasi-affine tight frames

In this subsection we provide the decomposition and reconstruction algorithms for quasi-affine tight frame systems following a similar discussion as in Section 1.4. Since

$$\mathcal{P}_L^{q,L} f = \mathcal{P}_L f = D^L \mathcal{P}_0 D^{-L} f = D^L \mathcal{P}_0^q D^{-L} f,$$

one may use $\mathcal{P}_0^q f \in V_0^q$ to approximate f without loss of generality. If it is necessary, we can always consider function $f(2^{-L}\cdot)$ instead of f, since the approximation of a function f in space $V_L^{q,L}$ is the same as that of the function $f(2^{-L}\cdot)$ in space V_0^q.

In level $n \le 0$, the dilated filter is denoted by $h_{\ell,n}$, which is defined by (same as (2.36))

$$h_{\ell,n}[k] = \begin{cases} h_\ell[2^n k], & k \in 2^{-n}\mathbb{Z}, \\ 0, & k \in \mathbb{Z} \backslash 2^{-n}\mathbb{Z}. \end{cases} \tag{2.42}$$

Given a sequence $h_{\ell,n} = \{h_{\ell,n}[k]\}_{k\in\mathbb{Z}}$ for any $\ell = 0, 1, \ldots, r$, and $n \le 0$, we define an infinite matrix $\mathsf{H}_{\ell,n}^q$ corresponding to $h_{\ell,n}$ as

$$\mathsf{H}_{\ell,n}^q := (\mathsf{H}_{\ell,n}^q[l,k]) := (\overline{h_{\ell,n}[k-l]}),$$

where the (l,k)th entry in $\mathsf{H}_{\ell,n}^q$ is fully determined by the $(k-l)$th entry in $h_{\ell,n}$. Then for any vector $v \in \ell_2(\mathbb{Z})$, we have

$$(H_{\ell,n}^q v)[l] = \sum_{k\in\mathbb{Z}} \overline{h_{\ell,n}[k-l]}v[k].$$

Note that the matrix multiplication defined above is in fact the convolution of v with filter $\overline{h_{\ell,n}[-\cdot]}$, i.e.

$$H_{\ell,n}^q v = \overline{h_{\ell,n}[-\cdot]} * v.$$

We denote $\mathsf{H}_{\ell,n}^{q*}$ as the adjoint of $\mathsf{H}_{\ell,n}^q$ with

$$(H_{\ell,n}^{q*} v)[k] = \sum_{l\in\mathbb{Z}} h_{\ell,n}[k-l]v[l] = (h_{\ell,n} * v)[k].$$

We collect the coefficients in each level $n < 0$ to form an infinite column vector

$$v_{\ell,n} := [\ldots, \langle f, \psi_{\ell,n,k}^q \rangle, \ldots]^\top, \quad \ell = 0, 1, \ldots, r,$$

where $\psi_0^q := \phi^q$. From the proof of Lemma 2.4, we have shown that (identity (2.37))

$$\sum_{\ell=0}^{r} \sum_{k\in\mathbb{Z}} \overline{h_{\ell,n}[k]} h_{\ell,n}[k-p] = \delta_{p,0}, \quad p \in \mathbb{Z}.$$

which is equivalent to

$$\sum_{\ell=0}^{r} \mathsf{H}_{\ell,n}^{q*} \mathsf{H}_{\ell,n}^q = \mathsf{I}. \tag{2.43}$$

Thus, the decomposition and reconstruction process (2.33) can be written in matrix form as

$$v_{0,n} = \sum_{\ell=0}^{r} \mathsf{H}_{\ell,n}^{q*} \mathsf{H}_{\ell,n}^q v_{0,n} \quad n \le 0.$$

For a multiple level decomposition operator, we define $\mathcal{A}_L^q$, for $L < 0$, as a (rectangular) block matrix given as:
(2.44)

$$\mathcal{A}_L^q := [(\prod_{n=L}^{0} H_{0,n}^q); (H_{1,L}^q \prod_{n=L+1}^{0} H_{0,n}^q); \ldots; (H_{r,L}^q \prod_{n=L+1}^{0} H_{0,n}^q); \ldots; H_{1,0}^q; \ldots; H_{r,0}^q]^\top,$$

where the product of operators is defined as,

$$\prod_{n=L_1}^{L_2} T_n := T_{L_1} T_{L_1+1} \cdots T_{L_2}.$$

Then, we have the reconstruction operator $\mathcal{A}_L^{q*}$, the adjoint operator of $\mathcal{A}_L^q$, defined as
(2.45)

$$\mathcal{A}_L^{q*} = [(\prod_{n=0}^{L} H_{0,n}^{q*}); (\prod_{n=0}^{L+1} H_{0,n}^{q*} H_{1,L}^{q*}); \ldots; (\prod_{n=0}^{L+1} H_{0,n}^{q*} H_{r,L}^{q*}); \ldots; H_{1,0}^{q*}; \ldots; H_{r,0}^{q*}].$$

For the case $L = 0$, we denote

$$\mathcal{A}_0^q := [H_{0,0}^q; H_{1,0}^q; \ldots; H_{r,0}^q]^\top; \quad \mathcal{A}_0^{q*} := [H_{0,0}^{q*}; H_{1,0}^{q*}; \ldots; H_{r,0}^{q*}].$$

The next proposition shows that the decomposition and reconstruction process is perfect, i.e. $\mathcal{A}_L^{q*}\mathcal{A}_L^q = I$.

PROPOSITION 2.6. *The decomposition operator $\mathcal{A}_L^q$, as defined in (2.44) satisfies $\mathcal{A}_L^{q*}\mathcal{A}_L^q = I$, where I is the identity operator.*

Here, we provide the univariate fast framelet decomposition and reconstruction algorithm for quasi-affine tight frame systems in Algorithm 2.2. For simplicity of notation, we denote $\tilde{h}_{\ell,n} := \overline{h_{\ell,n}[-\cdot]}$. For a given finite signal $v \in \mathbb{R}^N$, we denote the finite length convolution operation as $\tilde{h}_{\ell,n} \circledast v$, and understand it to be the restriction of $H_{\ell,n}\tilde{v} = \tilde{h}_{\ell,n} * \tilde{v}$ on the set $\{0, 1, \ldots, N-1\}$:

$$\tilde{h}_{\ell,n} \circledast v := (H_{\ell,n}\tilde{v})|_{\{0,1,\ldots,N-1\}} = (\tilde{h}_{\ell,n} * \tilde{v})|_{\{0,1,\ldots,N-1\}},$$

where $\tilde{v}$ is either periodic or a symmetric extension of v given by (2.27).

ALGORITHM 2.2. *Given signal $v \in \mathbb{R}^N$ with $N \in \mathbb{N}_+$. Denote $v_{0,0} = v$. Then the L-level fast quasi-affine framelet decomposition and reconstruction is given as follows:*

(1) ***Decomposition:*** *For each $j = 1, 2, \ldots, L$:*
   (a) *Obtain low frequency approximation to v at level j:*

$$v_{0,j} = \tilde{h}_{0,1-j} \circledast v_{0,j-1}.$$

   (b) *Obtain framelet coefficients of v at level j:*

$$v_{\ell,j} = \tilde{h}_{\ell,1-j} \circledast v_{0,j-1}, \quad \ell = 1, 2, \ldots, r.$$

(2) ***Reconstruction:*** *For each $j = L, L-1, \ldots, 1$:*

$$v_{0,j-1} = \sum_{\ell=0}^{r} h_{\ell,1-j} \circledast v_{\ell,j}.$$

3. Higher dimension tight frame systems

So far we have focused on univariate tight frame systems. However, almost all theorems and propositions mentioned so far were established in a multivariate setting. For example, the unitary extension principle, the mixed extension principle and the oblique extension principle are stated for a multivariate setting in [**65, 158**]. In fact, examples of multivariate spline tight frames from Box splines by using the unitary extension principle are provided in [**160**]. Here, we present simplified versions of them.

3.1. MRA and tight frame systems for $L_2(\mathbb{R}^d)$

For a given function $\phi \in L_2(\mathbb{R}^d)$, we define the *shift-invariant* subspace $V(\phi) \subset L_2(\mathbb{R}^d)$ generated by ϕ as

$$V(\phi) := \overline{\operatorname{span}\{\phi(\cdot - k),\ k \in \mathbb{Z}^d\}},$$

and denote V_n as the 2^n-dilate of $V(\phi)$, i.e.

$$V_n = \overline{\operatorname{span}\{\phi(2^n \cdot -k),\ k \in \mathbb{Z}^d\}},\ n \in \mathbb{Z}. \tag{2.46}$$

We say that $\{V_n\}_n$ forms a *multiresolution analysis (MRA)* for $L_2(\mathbb{R}^d)$ if the following conditions are satisfied:

$$V_n \subset V_{n+1}; \quad \overline{\cup_n V_n} = L_2(\mathbb{R}^d); \quad \text{and} \quad \cap_n V_n = \{0\}.$$

The function ϕ is refinable if

$$\phi(x) = 2^d \sum_{k\in\mathbb{Z}^d} h_0[k]\phi(2x-k) \tag{2.47}$$

for some $h_0 \in \ell_2(\mathbb{Z}^d)$. In frequency domain, the above equation can be rewritten as

$$\widehat{\phi}(2\cdot) = \widehat{h}_0\widehat{\phi}.$$

The Fourier transform of a function $f \in L_1(\mathbb{R}^d)$ is defined by

$$\widehat{f}(\xi) := \int_{\mathbb{R}^d} f(t)e^{-i\xi\cdot t}dt, \qquad \xi \in \mathbb{R}^d,$$

which can be extended to more general function spaces (e.g. $L_2(\mathbb{R}^d)$) in a natural manner. The Fourier series of a sequence $h_0 \in \ell_2(\mathbb{Z}^d)$ is defined as

$$\widehat{h}_0(\xi) := \sum_{j\in\mathbb{Z}^d} h_0[j]e^{-ij\cdot\xi}, \qquad \xi \in \mathbb{R}^d.$$

The following theorem is the corresponding $\mathbb{R}^d$ version of Theorem 1.4.

THEOREM 2.5 ([**115**]). *Let $\phi \in L_2(\mathbb{R}^d)$, and V_n defined as (2.46). Then,*

(1) $\cap_n V_n = \{0\}$;
(2) *Assume, in addition, that ϕ is refinable. Then,*

$$\overline{\cup_n V_n} = L_2(\mathbb{R}^d)$$

if and only if

$$\cap_n 2^n Z(\hat{\phi})$$

is a set of measure zero, where $Z(\hat{\phi}) := \{\xi \in \mathbb{R}^d : \hat{\phi}(\xi) = 0\}$.

In particular, the sequence of subspaces $\{V_n\}_{n\in\mathbb{Z}}$ generated by a refinable function $\phi \in L_2(\mathbb{R}^d)$ forms an MRA if either of the following conditions is satisfied:

(1) *ϕ is a compactly supported refinable function;*

(2) ϕ *is refinable and* $\widehat{\phi}$ *is continuous at* 0 *with* $\widehat{\phi}(0) \neq 0$.

PROOF. The proof is a direct analog to that of Theorem 1.4. ■

With a refinable function ϕ generating an MRA for $L_2(\mathbb{R}^d)$, we can construct tight frames systems for $L_2(\mathbb{R}^d)$.

DEFINITION 2.3. Let $\Psi = \{\psi_1, \ldots, \psi_r\}$ be a set of functions. A wavelet/affine system generated by Ψ is defined as

$$X(\Psi) = \{\psi_{\ell,n,k} : \ 1 \leq \ell \leq r; n \in \mathbb{Z}; k \in \mathbb{Z}^d\}$$

where $\psi_{\ell,n,k}$ is defined by

$$\psi_{\ell,n,k} := \mathcal{D}^n T_k \psi_\ell := 2^{nd/2} \psi_\ell(2^n \cdot -k).$$

A quasi-affine system from level 0 is defined as

$$X^q(\Psi) = \{\psi^q_{\ell,n,k} : \ 1 \leq \ell \leq r; n \in \mathbb{Z}; k \in \mathbb{Z}^d\},$$

where $\psi^q_{\ell,n,k}$ is defined by

$$\psi^q_{\ell,n,k} := \begin{cases} \mathcal{D}^n T_k \psi_\ell, & n \geq 0, \\ 2^{\frac{nd}{2}} T_{2^{-L}k} \mathcal{D}^n \psi_\ell, & n < 0. \end{cases}$$

The following is the unitary extension principle for $L_2(\mathbb{R}^d)$ which was originally proved in [**158**]. For simplicity, we assume the underlying refinable function ϕ is compactly supported with $\widehat{\phi}(0) = 1$ and the associated masks are finely supported. Under these assumptions, items 1-3 in Assumption 2.1 are satisfied.

THEOREM 2.6 (Unitary Extension Principle for $L_2(\mathbb{R}^d)$ [**158**]). *Let* $\phi \in L_2(\mathbb{R}^d)$ *be the compactly supported refinable function with finitely supported refinement mask* h_0 *satisfying* $\widehat{\phi}(0) = 1$. *Let* $\{h_1, \ldots, h_r\}$ *be a set of finitely supported sequences. Then the wavelet system* $X(\Psi)$ *and its quasi-affine counterpart* $X^q(\Psi)$, *with* $\Psi = \{\psi_1, \ldots, \psi_r\}$ *defined by* $\widehat{\psi_\ell}(2\cdot) = \widehat{h_\ell}\widehat{\phi}$, *forms a tight frame in* $L_2(\mathbb{R}^d)$ *provided the equalities*

$$\sum_{\ell=0}^{r} |\widehat{h}_\ell(\xi)|^2 = 1 \quad \textit{and} \quad \sum_{\ell=0}^{r} \widehat{h}_\ell(\xi)\overline{\widehat{h}_\ell(\xi+\nu)} = 0, \tag{2.48}$$

hold for all $\nu \in \{0, \pi\}^d \setminus \{\mathbf{0}\}$ *and* $\xi \in [-\pi, \pi]^d$. *Furthermore, assuming* $r = 2^d - 1$ *and* $\|\phi\|_{L_2(\mathbb{R}^d)} = 1$, *then* $X(\Psi)$ *is an orthonormal wavelet bases of* $L_2(\mathbb{R}^d)$.

PROOF. The proof is analogous to that of Theorem 2.2 and Theorem 2.4; hence we only sketch it here.

First, one can establish the following equalities using (2.48),

$$\begin{aligned} \mathcal{P}_n f &= \mathcal{P}_{n-1} f + \sum_{\ell=1}^{r} \sum_{k \in \mathbb{Z}^d} \langle f, \psi_{\ell,n-1,k} \rangle \psi_{\ell,n-1,k} \\ &= \mathcal{P}_{n'} f + \sum_{\ell=1}^{r} \sum_{j=n'}^{n-1} \sum_{k \in \mathbb{Z}^d} \langle f, \psi_{\ell,j,k} \rangle \psi_{\ell,j,k}, \end{aligned}$$

and

$$(2.49)\quad \begin{aligned}\mathcal{P}_n^q f &= \mathcal{P}_{n-1}^q f + \sum_{\ell=1}^{r}\sum_{k\in\mathbb{Z}^d}\langle f, \psi_{\ell,n-1,k}^q\rangle\psi_{\ell,n-1,k}^q\\ &= \mathcal{P}_{n'}^q f + \sum_{\ell=1}^{r}\sum_{j=n'}^{n-1}\sum_{k\in\mathbb{Z}^d}\langle f, \psi_{\ell,j,k}^q\rangle\psi_{\ell,j,k}^q.\end{aligned}$$

Finally one can show that

$$\mathcal{P}_{n'}f\to 0 \quad\text{and}\quad \mathcal{P}_{n'}^q f\to 0,\quad n'\to-\infty$$

and

$$\mathcal{P}_n f\to f \quad\text{and}\quad \mathcal{P}_n^q f\to f,\quad n\to\infty$$

which establishes that $X(\Psi)$ and $X^q(\Psi)$ form tight frames for $L_2(\mathbb{R}^d)$. ∎

One way to construct tight frames for $L_2(\mathbb{R}^d)$ is by taking *tensor products* of univariate tight frames. For simplicity of notation, we consider 2D case, i.e. $d=2$. Arguments for higher dimensions are similar. Given a set of univariate masks $\{h_\ell : \ell = 0,1,\ldots,r\}$, define the 2D masks $h_{i,j}[k_1,k_2]$ as

$$(2.50)\qquad h_{i,j}[k_1,k_2] := h_i[k_1]h_j[k_2],\quad 0\le i,j\le r;\ (k_1,k_2)\in\mathbb{Z}^2,$$

or equivalently in Fourier domain

$$\widehat{h}_{i,j}[\xi_1,\xi_2] := \widehat{h}_i[\xi_1]\widehat{h}_j[\xi_2],\quad 0\le i,j\le r;\ (\xi_1,\xi_2)\in\mathbb{R}^2.$$

Then the corresponding 2D refinable function and mother wavelets are defined by

$$\psi_{i,j}(x,y) = \psi_i(x)\psi_j(y),\quad 0\le i,j\le r; (x,y)\in\mathbb{R}^2.$$

Denote the collection of mother wavelets as

$$\Psi_2 := \{\psi_{i,j};\ 0\le i,j\le r;\ (i,j)\ne(0,0)\},$$

which contains $(r+1)^2-1$ elements. (Here, ψ_0 is defined to be the underlying univariate refinable function.) If the univariate masks $\{h_\ell\}$ are constructed from UEP, then it is easy to verify that $\{h_{i,j}\}$ satisfies (2.48) and thus $X(\Psi_2)$ and $X^q(\Psi_2)$ are tight frames for $L_2(\mathbb{R}^2)$.

3.2. Algorithms for tensor product tight frame systems

In this section, we only discuss the fast framelet decomposition and reconstruction algorithm for 2D finite signals. Since we are using tensor product tight frame systems, the 2D fast algorithms can be easily derived based on 1D fast algorithms for both wavelet and quasi-affine systems.

Now, we introduce some notation. Let the set of masks $\{h_\ell : \ell=0,1,\ldots,r\}$ be constructed from univariate UEP (Theorem 2.2). For a given signal $v\in\mathbb{R}^{N_1}\times\mathbb{R}^{N_2}$ with $N_1,N_2\in\mathbb{N}_+$, let $D_{\ell,j}^x$ and $D_{\ell,j}^y$ denote the one level of decomposition with respect to the x-index and y-index of v, which are defined as follows:

$$(D_{\ell,j}^x v)[\cdot,k_2] = \begin{cases}\downarrow(\tilde{h}_\ell \circledast v[\cdot,k_2]), & \text{wavelet system,}\\ \tilde{h}_{\ell,j}\circledast v[\cdot,k_2], & \text{quasi-affine system;}\end{cases}$$

$$(D_{\ell,j}^y v)[k_1,\cdot] = \begin{cases}\downarrow(\tilde{h}_\ell \circledast v[k_1,\cdot]), & \text{wavelet system,}\\ \tilde{h}_{\ell,j}\circledast v[k_1,\cdot], & \text{quasi-affine system,}\end{cases}$$

where $\tilde{h}_\ell = \sqrt{2} \cdot \overline{h_\ell[-\cdot]}$ and $\tilde{h}_{\ell,j} = \overline{h_{\ell,j}[-\cdot]}$ with $h_{\ell,j}$ defined as in (2.42). Similarly, we can define one level of reconstruction $D^{x*}_{\ell,j}$ and $D^{y*}_{\ell,j}$ as

$$(D^{x*}_{\ell,j}v)[\cdot, k_2] = \begin{cases} \tilde{h}^*_\ell \circledast \uparrow (v[\cdot, k_2]), & \text{wavelet system,} \\ h_{\ell,j} \circledast v[\cdot, k_2], & \text{quasi-affine system;} \end{cases}$$

$$(D^{y*}_{\ell,j}v)[k_1, \cdot] = \begin{cases} \tilde{h}^*_\ell \circledast \uparrow (v[k_1, \cdot]), & \text{wavelet system,} \\ h_{\ell,j} \circledast v[k_1, \cdot], & \text{quasi-affine system,} \end{cases}$$

where $\tilde{h}^*_\ell = \sqrt{2}h_\ell$.

Now, we present the following 2D fast framelet decomposition and reconstruction algorithm, which unifies the algorithm for wavelet systems and quasi-affine systems. A diagram illustrating one level of decomposition and reconstruction is given in Figure 5.

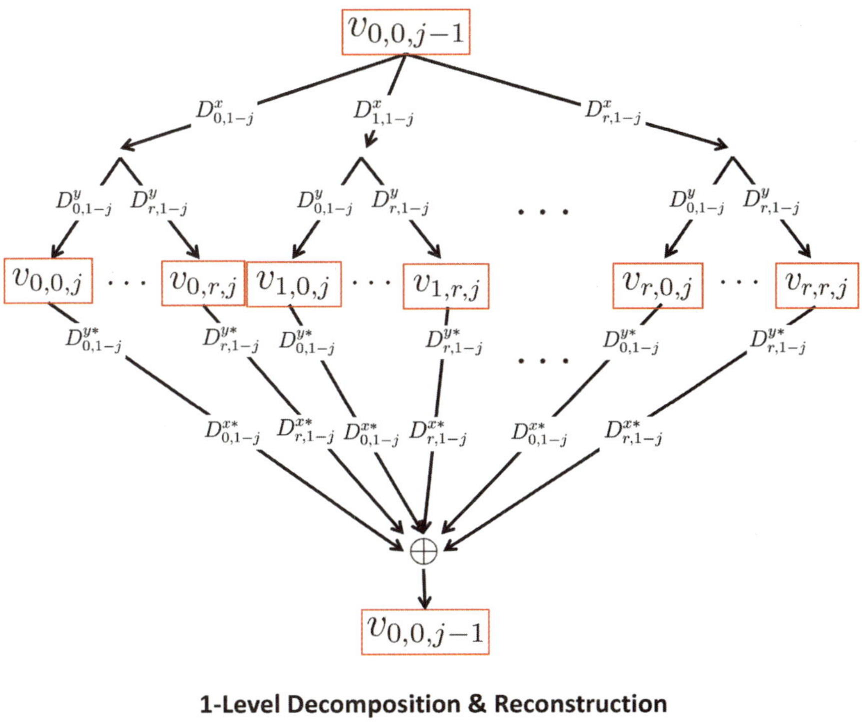

FIGURE 5. The above diagram illustrates one level of the fast framelet decomposition and reconstruction procedure given by Algorithm 2.3.

ALGORITHM 2.3. *Given signal* $v \in \mathbb{R}^{N_1} \times \mathbb{R}^{N_2}$ *with* $N_1, N_2 \in \mathbb{N}_+$. *If the underlying system is affine, then* N_1 *and* N_2 *are assumed to be integer multiples of* 2^L *with* L *the total level of decompositions. Denote* $v_{0,0,0} = v$. *Then the* L*-level fast framelet decomposition and reconstruction is given as follows:*

(1) ***Decomposition:*** *For each $j = 1, 2, \dots, L$:*
   (a) *Obtain low frequency approximation to v at level j:*

$$v_{0,0,j} = D^y_{0,1-j}(D^x_{0,1-j} v_{0,0,j-1});$$

   (b) *Obtain framelet coefficients of v at level j: for each $\ell_1 = 0, 1, \dots, r$ and $\ell_2 = 0, 1, \dots, r$ with $(\ell_1, \ell_2) \neq (0, 0)$,*

$$v_{\ell_1,\ell_2,j} = D^y_{\ell_2,1-j}(D^x_{\ell_1,1-j} v_{0,j-1}).$$

(2) ***Reconstruction:*** *For each $j = L, L-1, \dots, 1$:*

$$v_{0,0,j-1} = \sum_{\ell_1=0}^{r} \sum_{\ell_2=0}^{r} D^{x*}_{\ell_1,1-j}(D^{y*}_{\ell_2,1-j} v_{\ell_1,\ell_2,j}).$$

LECTURE 3

Pseudo-splines and tight frames

A class of refinable and compactly supported functions known as pseudo-splines encompasses a wide variety of refinable functions which provide much flexibility in the construction of wavelets and framelets. Functions such as B-splines, interpolatory, or Daubechies' orthogonal refinable functions (see [**61, 60**]) are special cases of pseudo-splines. Pseudo-splines were first introduced in [**65, 165**] in order to construct tight framelets with satisfactory approximation order for the truncated frame series of the tight wavelet system generated by the unitary extension principle.

The organization of this lecture is as follows. After a general introduction of pseudo-splines in Section 1, a construction of tight framelets is presented in Section 2. As a special example, we can obtain orthonormal wavelet bases for $L_2(\mathbb{R})$ which were originally constructed by Daubechies in [**60**]. Furthermore, the technique used to estimate the regularity of pseudo-splines can be applied to discover that the tight frame systems constructed in [**65**] and [**75**] always have one framelet whose dilations and shifts already form a Riesz basis for $L_2(\mathbb{R})$. This leads to a new understanding about the structure of pseudo-spline tight frame systems.

An optimal regularity analysis of pseudo-splines does not come easily, as has already been illustrated in a regularity estimate of the orthonormal refinable functions, which is a special case of pseudo-splines (see [**54**] and [**61**]). In [**75**], a systematic regularity analysis of both types of pseudo-splines was given. In Section 3, we provide regularity analysis for pseudo-splines. In particular, the exact decay of the Fourier transforms of the pseudo-splines is given which leads to an estimate of their Hölder regularity. The last section contains two technical lemmata which are needed in the proofs of the properties of pseudo-splines. Finally, we note that some key part of this lecture (e.g. regularity analysis) follows from [**75**].

1. Definitions and basics

This section is devoted to the definitions of various pseudo-splines and their basic properties.

1.1. Definitions

Pseudo-splines are defined in terms of their refinement masks. It starts with the simple trigonometric identity $1 = \cos^2(\xi) + \sin^2(\xi)$. For given nonnegative integers l and m with $l \le m-1$, we have

$$1 = \left(\cos^2(\xi/2) + \sin^2(\xi/2)\right)^{m+l}. \tag{3.1}$$

The refinement masks of pseudo-splines are defined by the summation of the first $l+1$ terms of the binomial expansion of (3.1). In particular, the refinement mask of a *pseudo-spline of Type I with order* (m,l) is given by, for $\xi \in [-\pi,\pi]$,

$$|_1\widehat{a}(\xi)|^2 := |_1\widehat{a}_{(m,l)}(\xi)|^2 := \cos^{2m}(\xi/2)\sum_{j=0}^{l}\binom{m+l}{j}\sin^{2j}(\xi/2)\cos^{2(l-j)}(\xi/2) \tag{3.2}$$

and the refinement mask of a *pseudo-spline of Type II with order* (m,l) is given by, for $\xi \in [-\pi,\pi]$,

$$_2\widehat{a}(\xi) := {_2\widehat{a}_{(m,l)}}(\xi) := \cos^{2m}(\xi/2)\sum_{j=0}^{l}\binom{m+l}{j}\sin^{2j}(\xi/2)\cos^{2(l-j)}(\xi/2). \tag{3.3}$$

Define

$$P_{m,l}(y) := \sum_{j=0}^{l}\binom{m+l}{j}y^j(1-y)^{l-j} \tag{3.4}$$

and

$$R_{m,l}(y) := (1-y)^m P_{m,l}(y), \tag{3.5}$$

where m,l are nonnegative integers with $l \le m-1$. Then, it is obvious that

$$_2\widehat{a}(\xi) = R_{m,l}(\sin^2(\xi/2)).$$

Except for some special circumstances, we simply notation by dropping the subscript "(m,l)" in $_1\widehat{a}_{(m,l)}(\xi)$ and $_2\widehat{a}_{(m,l)}(\xi)$ for simplicity. We note that the mask of Type I is obtained by taking the square root of the mask of Type II using the Fejér-Riesz lemma (see e.g. [**61**] and [**152**]), i.e. $_2\widehat{a}(\xi) = |_1\widehat{a}(\xi)|^2$. Type I was introduced and used by [**65**] to construct tight framelets.

Finally, a family of *dual pseudo-splines* was first discovered by [**81**] and the general refinement mask of a dual pseudo-spline of order (m,l) with $0 \le l \le m-1$ is

$$\widehat{b}_{m,l}(\xi) = e^{i\xi/2}\cos^{2m+1}(\xi/2)\sum_{j=0}^{l}\binom{m+1/2+l}{j}\sin^{2j}(\xi/2)\cos^{2(l-j)}(\xi/2). \tag{3.6}$$

Notice that dual pseudo-spline of order $(m,0)$ is the B-spline of degree $2m$ and dual pseudo-spline of order $(m,m-1)$ is the limit function of the dual $2m$-point scheme. The masks of dual pseudo-splines are very similar in format to those of pseudo-splines. In fact, if we let $r = m+\frac{1}{2}$, then (3.6) can be rewritten as

$$\widehat{b}_{r,l}(\xi) = e^{i\xi/2}\cos^{2r}(\xi/2)\sum_{j=0}^{l}\binom{r+l}{j}\sin^{2j}(\xi/2)\cos^{2(l-j)}(\xi/2).$$

Therefore, the analysis of dual pseudo-splines is similar to that of pseudo-splines by using Newton's generalized binomial theorem. Hence, we focus on the analysis of pseudo-splines because the analysis of dual pseudo-splines is analogous. Interested readers should work out the details as exercises.

1.2. Underlying refinable functions

Given refinement masks ${}_k\widehat{a}$, the corresponding pseudo-splines can be formally defined in terms of their Fourier transforms ${}_k\widehat{\phi}(\xi) := \prod_{j=1}^{\infty} {}_k\widehat{a}(2^{-j}\xi)$ for $k = 1, 2$. To make sure that ${}_k\widehat{\phi}$ is well defined, we need to consider the convergence of the right-hand-side which is given below. For finitely supported refinement masks, the following theorem is well known and one can find the proof in e.g. [**61**]. Here, we provide an alternative proof motivated from a proof of [**116**].

PROPOSITION 3.1 ([**116**]). *Suppose that h_0 is a finitely supported refinement mask satisfying $\widehat{h}_0(0) = 1$. Then, the product*

$$\widehat{\phi}(\xi) := \prod_{j=1}^{\infty} \widehat{h}_0(2^{-j}\xi) \tag{3.7}$$

converges uniformly on any compact subset of $\mathbb{R}$. Furthermore, $\widehat{\phi}$ is the Fourier transform of a compactly supported tempered distribution ϕ that satisfies refinement equation (1.6).

PROOF. Since $\widehat{h}_0$ is a trigonometric polynomial with $\widehat{h}_0(0) = 1$, we have

$$|\widehat{h}_0(\xi)| \le 1 + C|\xi| \le e^{C|\xi|}, \tag{3.8}$$

for some $C > 0$. This leads to

$$\prod_{j=1}^{\infty} |\widehat{h}_0(2^{-j}\xi)| \le e^{C|\xi|}.$$

Therefore, the right-hand side of (3.7) converges uniformly on any compact subset K of $\mathbb{R}$. It is clear that $\widehat{\phi}$ satisfies refinement equation (1.6). Note that in the above proof, we only require that the refinement mask satisfy (3.8). The finiteness of the support of the refinement mask is not needed.

Let $c_0 = \max_\xi |\widehat{h}_0(\xi)|$ and $c_1 = \max_{|\xi| \le 2} |\widehat{\phi}(\xi)|$. For any positive j, consider ξ, such that $2^j \le |\xi| < 2^{j+1}$. Since

$$\widehat{\phi}(\xi) = \prod_{k=1}^{j} \widehat{h_0}(2^{-k}\xi)\widehat{\phi}(2^{-j}\xi),$$

we have

$$|\widehat{\phi}(\xi)| \le c_1 c_0^j \le c_1 c_0^{\log_2 |\xi|} = c_1 |\xi|^{\log_2 c_0}. \tag{3.9}$$

Therefore, $\widehat{\phi}$ has at most polynomial growth at infinity. Hence, ϕ is a tempered distribution with $\widehat{\phi}(0) = 1$.

Finally it remains to prove that ϕ is compactly supported. Let $\phi_0 = \delta$ be the Dirac function. Then it is a compactly supported tempered distribution whose Fourier transform is 1. Define the cascade algorithm

$$\widehat{\phi}_{n+1}(\xi) = \widehat{h}_0(\xi/2)\widehat{\phi}_n(\xi/2).$$

Then, the infinity product in the right-hand side of (3.7) converges uniformly on any compact subset $\mathbb{R}$ if and only if ϕ_n does. Furthermore, (3.9) shows that $\widehat{\phi}_n$ has at most polynomial growth at infinity. Assume that the support of the refinement mask is $[-q, q]$. Let K be any compact subset in $\mathbb{R}$ containing $[-q, q]$. Then it can be shown inductively that the support of ϕ_n is in K for all $n \ge 0$. Now, let g be an arbitrary test function whose support is disjoint with K. Then, applying the dominated converge theorem, one has

$$\langle \phi, g\rangle = (2\pi)^{1/2}\langle \widehat{\phi}, \widehat{g}\rangle = (2\pi)^{1/2}\langle \lim_n \widehat{\phi}_n, \widehat{g}\rangle = (2\pi)^{1/2}\lim_n \langle \widehat{\phi}_n, \widehat{g}\rangle = \lim_n \langle \phi_n, g\rangle = 0.$$

Hence, ϕ is compactly supported. ∎

One of the fundamental questions for the study of refinable functions is that given a refinement mask h_0, does there exist a corresponding refinable function $\phi \in L_2(\mathbb{R})$ such that the refinement equation (1.5) is satisfied. This question will be answered by the following result where part of the proof is taken from [**32**] (see [**125**] for a more general analysis). To establish these results, we need to introduce the cascade algorithm. For a given refinement mask h_0, define the *cascade algorithm* as

$$\widehat{f}_n(\xi) = \widehat{h}_0(\frac{\xi}{2})\widehat{f}_{n-1}(\frac{\xi}{2}) = \prod_{j=1}^{n} \widehat{h}_0(2^{-j}\xi)\widehat{f}_0(2^{-n}\xi), \quad n > 0, \tag{3.10}$$

with initial function f_0 satisfying $\widehat{f}_0(\xi) = \chi_{[-\pi,\pi]}(\xi)$. Proposition 3.1 shows that f_n converges uniformly on each compact set.

THEOREM 3.1 ([**32**]). *Suppose h_0 is finitely supported and satisfies the following condition:*

$$\begin{cases} |\widehat{h}_0(\xi)|^2 + |\widehat{h}_0(\xi+\pi)|^2 \leqslant 1, & a.e. \quad \xi \in \mathbb{R}, \\ \widehat{h}_0(0) = 1. \end{cases} \tag{3.11}$$

Then, the limit ϕ of the cascade algorithm (3.10) *is a compactly supported function in $L_2(\mathbb{R})$ satisfying $\widehat{\phi}(0) = 1$.*

PROOF. Since h_0 is finitely supported and $\widehat{h}_0(0) = 1$, then by Proposition 3.1, there exists a unique compactly supported refinable distribution ϕ whose Fourier transform given by (3.7) satisfies $\widehat{\phi}(0) = 1$ (also see [**61, 29**]). Now, consider

$$\begin{aligned} \|\widehat{f}_n\|^2_{L_2(\mathbb{R})} &= \int_{-2^n\pi}^{2^n\pi} \prod_{j=1}^{n} |\widehat{h}_0(2^{-j}\xi)|^2 \mathrm{d}\xi, \\ (\widehat{h}_0 \text{ is } 2\pi\text{-periodic}) &= \int_0^{2^{n+1}\pi} \prod_{j=1}^{n} |\widehat{h}_0(2^{-j}\xi)|^2 \mathrm{d}\xi \\ &= \int_0^{2^n\pi} \prod_{j=1}^{n} |\widehat{h}_0(2^{-j}\xi)|^2 \mathrm{d}\xi + \int_{2^n\pi}^{2^{n+1}\pi} \prod_{j=1}^{n} |\widehat{h}_0(2^{-j}\xi)|^2 \mathrm{d}\xi \\ &= \int_0^{2^n\pi} \prod_{j=1}^{n} |\widehat{h}_0(2^{-j}\xi)|^2 \mathrm{d}\xi + \int_0^{2^n\pi} \prod_{j=1}^{n} |\widehat{h}_0(2^{-j}(\xi+2^n\pi))|^2 \mathrm{d}\xi \\ &= \int_0^{2^n\pi} \prod_{j=1}^{n-1} |\widehat{h}_0(2^{-j}\xi)|^2 \left(|\widehat{h}_0(2^{-n}\xi)|^2 + |\widehat{h}_0(2^{-n}\xi+\pi)|^2\right) \mathrm{d}\xi. \end{aligned}$$

By (3.11), we have

$$\|\widehat{f}_n\|^2_{L_2(\mathbb{R})} \le \int_0^{2^n\pi} \prod_{j=1}^{n-1} |\widehat{h}_0(2^{-j}\xi)|^2 \mathrm{d}\xi = \|\widehat{f}_{n-1}\|^2_{L_2(\mathbb{R})}.$$

Then by induction, we have

$$\|\widehat{f}_n\|^2_{L_2(\mathbb{R})} \le \|\widehat{f}_0\|^2_{L_2(\mathbb{R})} = 2\pi, \quad \text{for all } n \ge 0.$$

Since the sequence $\{\widehat{f}_n\}_n$ converges to $\widehat{\phi}$ pointwise, then by Fatou's lemma

$$\|\widehat{\phi}\|^2_{L_2(\mathbb{R})} \le \liminf_{n\to\infty} \|\widehat{f}_n\|^2_{L_2(\mathbb{R})} \le 2\pi < \infty,$$

and hence $\phi \in L_2(\mathbb{R})$. ∎

By the definition of the refinement masks of the pseudo-splines, the pseudo-splines of both types satisfy (3.11), and hence all pseudo-splines are compactly supported functions in $L_2(\mathbb{R})$. In fact, they are much smoother than merely being in $L_2(\mathbb{R})$ as is shown in Section 3. Furthermore, from Theorem 1.4 in Lecture 1 the sequence of subspaces $\{V_n\}_n$ (defined via (1.1)) generated by any pseudo-spline forms an MRA for $L_2(\mathbb{R})$.

Since most pseudo-splines do not have an analytical form (B-splines as the exception). It is important to know whether the infinite product (3.7) converges in $L_2(\mathbb{R})$, so that one can obtain the approximation of the corresponding refinable function.

PROPOSITION 3.2. *The cascade algorithm defined in* (3.10) *by the refinement mask h_0 of an arbitrary given pseudo-spline with any type and order converges to the corresponding pseudo-spline ϕ in $L_2(\mathbb{R})$.*

PROOF. Since we already have pointwise convergence of the cascade algorithm by Proposition 3.1, all we need to show is that $\|\widehat{\phi}\|^2_{L_2(\mathbb{R})} = \lim_{n\to\infty} \|\widehat{f}_n\|^2_{L_2(\mathbb{R})}$. Indeed, we have

$$0 \le \|\widehat{f}_n - \widehat{\phi}\|^2_{L_2(\mathbb{R})} = \langle \widehat{f}_n - \widehat{\phi}, \widehat{f}_n - \widehat{\phi}\rangle = \|\widehat{f}_n\|^2_{L_2(\mathbb{R})} - 2\mathrm{Re}\langle \widehat{f}_n, \widehat{\phi}\rangle + \|\widehat{\phi}\|^2_{L_2(\mathbb{R})}.$$

Since $\widehat{f}_n$ pointwise converges to $\widehat{\phi}$, one needs to show that $\lim_{n\to\infty}\langle \widehat{f}_n, \widehat{\phi}\rangle = \|\widehat{\phi}\|^2_{L_2(\mathbb{R})}$ and $\|\widehat{\phi}\|^2_{L_2(\mathbb{R})} = \lim_{n\to\infty} \|\widehat{f}_n\|^2_{L_2(\mathbb{R})}$. This leads to

$$0 \le \lim_{n\to\infty} \|\widehat{f}_n - \widehat{\phi}\|^2_{L_2(\mathbb{R})} \le \lim_{n\to\infty} \left(\|\widehat{f}_n\|^2_{L_2(\mathbb{R})} - \|\widehat{\phi}\|^2_{L_2(\mathbb{R})}\right).$$

By the dominated convergence theorem, it suffices to show that

$$|\widehat{f}_n|^2 \le C|\widehat{\phi}|^2 \quad \text{and} \quad |\widehat{f}_n\widehat{\phi}| \le C|\widehat{\phi}|^2. \tag{3.12}$$

Since we have already shown that $\phi \in L_2(\mathbb{R})$, for any given $n \ge 0$, the existence of $C > 0$ satisfying (3.12) is obvious when $|\xi| > 2^n\pi$ since $\widehat{f}_n(\xi) = 0$. Now, for $|\xi| \le 2^n\pi$, since ϕ is refinable, it follows that

$$|\widehat{\phi}(\xi)|^2 = \left|\prod_{j=1}^{\infty} \widehat{h}_0(2^{-j}\xi)\right|^2 = \left|\prod_{j=1}^{n} \widehat{h}_0(2^{-j}\xi) \prod_{j=n+1}^{\infty} \widehat{h}_0(2^{-j}\xi)\right|^2 = |\widehat{f}_n(\xi)|^2 |\widehat{\phi}(2^{-n}\xi)|^2.$$

By (3.16) one can easily check that the inequality

$$|\widehat{\phi}(\xi)|^2 \ge 1/C > 0, \quad \xi \in [-\pi, \pi]$$

holds for some $C > 0$ for all pseudo-spines. Hence, (3.12) holds for all pseudo-splines. ∎

1.3. Approximation

Recall the definition of the refinement mask of a pseudo-spline of Type II with order (m,l)

$$_2\widehat{a}(\xi)=\cos^{2m}(\xi/2)\sum_{j=0}^{l}\binom{m+l}{j}\sin^{2j}(\xi/2)\cos^{2(l-j)}(\xi/2).$$

One can see that a pseudo-spline of Type II with order (m,l) is a convolution of a B-spline of order $2m$ with a distribution. The mask of the distribution is $P_{m,l}(\sin^2(\xi/2))$, where $P_{m,l}(y)$ is given by (3.4). Since Lemma 3.3 says that $P_{m,l}(y)=\sum_{j=0}^{l}\binom{m-1+j}{j}y^j$, one has $P_{m,l}(\sin^2(\xi/2))\geq 1$ for all ξ. Hence, pseudo-spline of Type II with order (m,l) can be viewed as a convolution of a B-spline with a distribution whose Fourier transform is at least 1 at any point ξ. Therefore, the pseudo-spline $_2\phi$ of Type II with order (m,l) satisfies the SF condition of order $2m$. By a similar discussion with some special treatment of the case when m is odd, one can conclude that the pseudo-spline $_1\phi$ of Type I with order (m,l) satisfies the SF condition of order m.

Consider the masks of $_1\widehat{a}$ of the pseudo-splines of type I of order (m,l). Since $1-|_1\widehat{a}|^2$ has a factor $\sin^{2(l+1)}(\xi/2)$, we have $1-|_1\widehat{a}|^2=O(|\cdot|^{2l+2})$. Then by Theorem 1.5 and Proposition 2.1, the operator $\mathcal{P}_nf$ defined by (1.13) corresponding to the pseudo-spline $_1\phi$ of order (m,l) provides approximation order $\min\{m,2l+2\}$ for sufficiently smooth functions. Hence, Theorem 2.3 implies that any tight frame system constructed by the UEP using pseudo-splines of type I of order (m,l) provides approximation order $\min\{m,2l+2\}$.

The following theorem shows that, if we start from pseudo-splines of Type II with order (m,l) in Construction 3.3, the corresponding tight frame system $X(\Psi)$ provides approximation order $2l+2$.

THEOREM 3.2. *Let m and l be nonnegative integers satisfying $l\leq m-1$. Let $_k\phi$, $k=1,2$ be the pseudo-spline of Type I and II with order (m,l) and $_k\widehat{a}$, $k=1,2$ be its refinement mask. Let $\mathcal{P}_n$ be defined by* (1.13) *for a given refinable function $_k\phi$ Then, $\mathcal{P}_n$ provides approximation order* $\min\{m,2l+2\}$ *for Type I and* $2l+2$ *for type II. Consequently, the corresponding truncated operator $\mathcal{Q}_n$ defined in* (2.21) *for any tight wavelet frame system constructed from a refinable function $_k\phi$, $k=1,2$ of order (m,l) via the unitary extension principle provides approximation order* $\min\{m,2l+2\}$ *for Type I and* $2l+2$ *for Type II.*

PROOF. Notice that $_2\widehat{a}=|_1\widehat{a}|^2$ and $1-{_2\widehat{a}}^2=(1+{_2\widehat{a}})(1-{_2\widehat{a}})$. Since $1+{_2\widehat{a}}(0)\neq 0$, the order of zeros of $1-{_2\widehat{a}}^2$ at the origin is the same as that of $1-{_2\widehat{a}}$, which is exactly $2l+2$ based on the discussion above. This, together with Theorem 1.5 and Proposition 2.1, concludes the proof of the theorem. ∎

For fixed m, since the value of the mask $|_k\widehat{a}(\xi)|$, for $k=1,2$ and $\xi\in\mathbb{R}$, increases with l (by (1) of Lemma 3.3 in Section 4), and the length of the mask $_ka$ also increases with l, we conclude that the decay rate of the Fourier transform of a pseudo-spline decreases with l (see Section 3 for details) and the support of the corresponding pseudo-spline increases with l. In particular, for fixed m, the pseudo-spline with order $(m,0)$ has the highest order of smoothness with the shortest support, the pseudo-spline with order $(m,m-1)$ has the lowest order of smoothness with the largest support in the family. When we move from B-splines to orthogonal or interpolatory refinable functions, we sacrifice the smoothness and short support of the B-splines to gain some other desirable properties, such as orthogonality or interpolatory property. What do we get for the pseudo-splines of other orders? When we move from B-splines to pseudo-splines, we gain the approximation power of the truncated tight frame systems derived from them and the vanishing moments of wavelets.

As mentioned in Lecture 1, if ϕ satisfies the SF condition of order m, then ϕ (or $V(\phi)$, or the underlying MRA) provides approximation order m. Therefore, even though the $(V_n)_{n\in\mathbb{Z}}$ may be generated by a different pseudo-spline of order (m,l) with m fixed, the corresponding spaces $(V_n)_{n\in\mathbb{Z}}$ provide the same approximation order. However, in many applications we are more interested in the performance of the tight frame system constructed using UEP. Therefore, we normally use the truncated operator $\mathcal{Q}_n$ defined in (2.21) or, equivalently, the quasi-interpolatory operator $\mathcal{P}_n : L_2(\mathbb{R}) \mapsto V_n$ defined in (2.11) to approximate f. Theorem 3.2 tells us that for B-splines, the approximation order of the corresponding $\mathcal{Q}_n$ (hence $\mathcal{P}_n$) can never exceed 2 even if a high order B-spline is used. On the other hand, for pseudo-spline of either type with order (m,l), $0 \le l \le m-1$, the approximation order of $\mathcal{Q}_n$ (hence $\mathcal{P}_n$) is $\min\{m, 2l+2\}$ for Type I and $2l+2$ for Type II. This indicates that the tight frame system derived from a pseudo-spline normally gives better approximation order when the truncated series is used to approximate the underlying functions than that derived from B-splines. For fixed m, the choice of l depends entirely on applications. According to the practical problems in hand, one must properly balance the approximation order, the length of support of the wavelet, and regularity.

1.4. Stability and orthogonality

Recall that ϕ is *stable*, if $\{\phi(\cdot - k)\}_{k\in\mathbb{Z}}$ forms a Riesz basis for $V(\phi)$, i.e. there exist $0 < C_1, C_2 < \infty$, such that for any sequence $b \in \ell_2(\mathbb{Z})$,

$$C_1\|b\|_{\ell_2(\mathbb{Z})} \le \Big\| \sum_{k\in\mathbb{Z}} b[k]\phi(\cdot - k)\Big\|_{L_2(\mathbb{R})} \le C_2\|b\|_{\ell_2(\mathbb{Z})}. \tag{3.13}$$

In particular if only the upper bound of (3.13) is satisfied, we say that $\{\phi(\cdot - k)\}_{k\in\mathbb{Z}}$ forms a *Bessel* sequence (as we have already discussed in Lecture 1). On the other hand, if $C_1 = C_2 = 1$, then we say that ϕ is *orthonormal*, i.e. $\{\phi(\cdot - k)\}_{k\in\mathbb{Z}}$ forms an orthonormal basis for $V(\phi)$. Recall that ϕ is orthonormal if and only if $\langle\phi, \phi(\cdot - k)\rangle = \delta_{k,0}$ for all $k \in \mathbb{Z}$.

The stability of function $\phi \in L_2(\mathbb{R})$ can also be characterized by its bracket product which is well known in the literature (see e.g. [**70, 61, 114, 156**]). Now, we summarize and prove this characterization in the following lemma.

LEMMA 3.1. *For any given $\phi \in L_2(\mathbb{R})$, then ϕ is stable, i.e.* (3.13) *is satisfied, if and only if $[\widehat{\phi}, \widehat{\phi}]$ satisfies*

$$0 < C_1 \le [\widehat{\phi}, \widehat{\phi}](\xi) \le C_2; \tag{3.14}$$

holds for almost every $\xi \in \mathbb{R}$ with

$$\operatorname{ess\,inf}\, [\widehat{\phi}, \widehat{\phi}] = C_1 \quad \textit{and} \quad \operatorname{ess\,sup}\, [\widehat{\phi}, \widehat{\phi}] = C_2.$$

As a consequence, ϕ and its shifts form an orthonormal basis of $V(\phi)$ if and only if $[\widehat{\phi}, \widehat{\phi}] = 1$ holds almost everywhere. Furthermore, assume that ϕ is compactly supported, and then the upper bound of (3.14) *or equivalently* (3.13) *is satisfied.*

PROOF. For simplicity, we denote $\tilde{\phi} := [\widehat{\phi}, \widehat{\phi}]^{\frac{1}{2}}$. Then for all finite sequences $b \in \ell_2(\mathbb{Z})$, we have the following identity (see proof of part (1) of Proposition 1.3)

$$2\pi \left\| \sum_{k\in\mathbb{Z}} b[k]\phi(\cdot - k) \right\|^2_{L_2(\mathbb{R})} = \|\widehat{b}\tilde{\phi}\|^2_{L_2([-\pi,\pi])}. \tag{3.15}$$

Suppose $\tilde{\phi}$ satisfies (3.14). Then the upper bound of (3.14) implies that (3.15) is satisfied for all $b \in \ell_2(\mathbb{Z})$. Therefore,

$$C_1\|\widehat{b}\|^2_{L_2([-\pi,\pi])} \le \|\widehat{b}\tilde{\phi}\|^2_{L_2([-\pi,\pi])} \le C_2\|\widehat{b}\|^2_{L_2([-\pi,\pi])}$$

with

$$C_1 = \text{ess inf } [\widehat{\phi}, \widehat{\phi}] \quad \text{and} \quad C_2 = \text{ess sup } [\widehat{\phi}, \widehat{\phi}],$$

and hence (3.13) follows.

Now suppose the upper bound of (3.13) is satisfied. If $\tilde{\phi}$ is not essentially bounded, then there exists a sequence of domains $\Omega_n \subset [-\pi, \pi]$ with positive measure such that $\tilde{\phi} > n$ on Ω_n. Let c^n be a sequence in $\ell_2(\mathbb{Z})$ satisfying $\widehat{c^n} = \chi_{\Omega_n}$ as $n \to \infty$. Then by (3.15), we have

$$2\pi \left\| \sum_{k\in\mathbb{Z}} c^n[k]\phi(\cdot - k) \right\|^2_{L_2(\mathbb{R})} = \|\widehat{c^n}\tilde{\phi}\|^2_{L_2([-\pi,\pi])} \ge n\|\widehat{c^n}\|^2_{L_2([-\pi,\pi])} = 2\pi n\|c^n\|^2_{\ell_2(\mathbb{Z})}.$$

This contradicts with the upper bound of (3.13). Therefore, $\tilde{\phi}$ must be essentially bounded and the upper bound of (3.14) follows. Notice that $\tilde{\phi}$ is essentially bounded away from 0 if and only if $1/\tilde{\phi}$ is essentially bounded. Then using a similar argument we can show that the lower bound of (3.14) follows from that of (3.13).

Now, we show that whenever ϕ is compactly supported, the upper bound of (3.14), or equivalently (3.13), is satisfied automatically. Indeed, part (2) of Proposition 1.3 implies that $\tilde{\phi}$ is the Fourier series of $\{\langle \phi, \phi(\cdot - k)\rangle : k \in \mathbb{Z}\}$ which is a finite sequence since ϕ is compactly supported. Then $\tilde{\phi}$ is in fact a trigonometric polynomial which is continuous and hence bounded on $[-\pi, \pi]$. ∎

Consider the compactly supported centered B-splines

$$\widehat{B}_m(\xi) = e^{-ij\frac{\xi}{2}} \left(\frac{\sin(\xi/2)}{\xi/2} \right)^m.$$

Since it is obvious that there is $C_1 > 0$, such that $|\widehat{B}_m(\xi)| > \sqrt{C_1}$ for all $\xi \in [-\pi, \pi]$, we have

$$[\widehat{B}_m, \widehat{B}_m](\xi) = |\widehat{B}_m(\xi)|^2 + \sum_{k\in\mathbb{Z}\backslash 0} |\widehat{B}_m(\xi + 2k\pi)|^2 \ge |\widehat{B}_m(\xi)|^2 \ge C_1.$$

Hence all B-splines are stable. Next, we prove that all pseudo-splines are stable as given by [**74**].

PROPOSITION 3.3 ([**74**]). *All pseudo-splines for both types with arbitrary orders are stable.*

PROOF. Here, we only prove stability for pseudo-splines of type II. The arguments for pseudo-splines of type I are similar.

Let $\phi_{(m,l)}$ be pseudo-spline of type II with order (m,l) and $\widehat{a}_{(m,l)}$ be its refinement mask. Since pseudo-splines are compactly supported functions in $L_2(\mathbb{R})$,

Lemma 3.1 tells us that we only need to show that there is a constant $C_1 > 0$ such that $C_1 \leq [\widehat{\phi}_{(m,l)}, \widehat{\phi}_{(m,l)}](\xi)$ holds for every $\xi \in \mathbb{R}$.

Since for each fixed $m \geq 1$ and for every $0 \leq l \leq m-1$, we have

$$P_{m,l}(\sin^2(\xi/2)) \geq 1,$$

and the following inequality

$$\cos^{2m}(\xi/2) \leq \widehat{a}_{(m,l)}(\xi)$$

holds for all $\xi \in \mathbb{R}$. Therefore, by (3.7), we have for all $\xi \in \mathbb{R}$,

$$|\widehat{B}_{2m}(\xi)| \leq |\widehat{\phi}_{(m,l)}(\xi)|. \tag{3.16}$$

Since B_{2m} is stable, there is a constant $C_1 > 0$ such that $C_1 \leq [\widehat{\phi}_{(m,l)}, \widehat{\phi}_{(m,l)}](\xi)$ holds for every $\xi \in \mathbb{R}$. Altogether, we have that $\phi_{(m,l)}$ is stable. ∎

Let ϕ_m be the pseudo-splines of type I with order $(m, m-1)$. Next, we show that ϕ_m is orthonormal, which means ϕ_m satisfies (3.13) with $C_1 = C_2 = 1$. By Lemma 3.1, we know that ϕ_m is orthonormal if and only if

$$[\widehat{\phi}_m, \widehat{\phi}_m] = 1 \tag{3.17}$$

almost everywhere. We now have the following proposition.

PROPOSITION 3.4. *Pseudo-splines of type I with order $(m, m-1)$ are orthonormal refinable functions.*

PROOF. All we need to show is that $[\widehat{\phi}_m, \widehat{\phi}_m] = 1$ a.e. We first note that for any $k \in \mathbb{Z}$,

$$\int_{-\pi}^{\pi} [\widehat{\phi}_m, \widehat{\phi}_m](\xi) e^{ik\xi} \mathrm{d}\xi = \int_{\mathbb{R}} |\widehat{\phi}_m(\xi)|^2 e^{ik\xi} \mathrm{d}\xi.$$

Define f_0 such that $\widehat{f_0}(\xi) = \chi_{[-\pi,\pi]}(\xi)$. Using the identity (3.27) and following the same arguments as in the proof of Theorem 3.4, we have

$$\int_{\mathbb{R}} |\widehat{f_n}(\xi)|^2 e^{ik\xi} \mathrm{d}\xi = \int_{\mathbb{R}} |\widehat{f_0}(\xi)|^2 e^{ik\xi} \mathrm{d}\xi = \int_{-\pi}^{\pi} e^{ik\xi} \mathrm{d}\xi = 2\pi\delta_{k,0}.$$

Then by the dominated convergence theorem (similarly as the proof of Theorem 3.4), we have

$$\int_{\mathbb{R}} |\widehat{\phi}_m(\xi)|^2 e^{ik\xi} \mathrm{d}\xi = 2\pi\delta_{k,0},$$

which means

$$\int_{-\pi}^{\pi} [\widehat{\phi}_m, \widehat{\phi}_m](\xi) e^{ik\xi} \mathrm{d}\xi = 2\pi\delta_{k,0}.$$

This proves that $[\widehat{\phi}_m, \widehat{\phi}_m] = 1$ a.e. ∎

Following arguments similar to those used in proving the previous proposition, we can show that all pseudo-splines of type II with order $(m, m-1)$ are indeed interpolatory as given by the following proposition. Recall that a continuous function $\phi \in L_2(\mathbb{R})$ is said to be interpolatory if $\phi(k) = \delta_{k,0}$ for $k \in \mathbb{Z}$ (see e.g. [**79**]).

One of the key steps in the proof of the following proposition is to show that $|\widehat{\phi}(\xi)| \leq C(1+|\xi|)^{-1-\varepsilon}$. In fact, the Fourier transform of pseudo-splines of type II with order $(m, m-1)$ has a faster decay than $O((1+|\xi|)^{-1-\varepsilon})$, which means they are actually smoother than merely continuous. We provide an optimal decay analysis of all pseudo-splines in Theorem 3.8.

PROPOSITION 3.5. *All pseudo-splines of type II with order $(m, m-1)$ are interpolatory.*

PROOF. Let ϕ be pseudo-spline of type II with order $(m, m-1)$ and $\widehat{a}$ be its refinement mask. First, we show that $|\widehat{\phi}(\xi)| \leq C(1+|\xi|)^{-1-\varepsilon}$ for some $\varepsilon > 0$. Such decay makes sure that $\widehat{\phi} \in L_1(\mathbb{R})$, which ensures the continuity of ϕ and thus $\phi(k)$ is well-defined for all $k \in \mathbb{Z}$.

Now, we show the decay of $\widehat{\phi}$. Recall from (3.3) that

$$\widehat{a}(\xi) = \cos^{2m}(\xi/2) \sum_{j=0}^{m-1} \binom{2m-1}{j} \sin^{2j}(\xi/2) \cos^{2(m-j-1)}(\xi/2).$$

Define

$$\mathcal{L}(\xi/2) := \sum_{j=0}^{m-1} \binom{2m-1}{j} \sin^{2j}(\xi/2) \cos^{2(m-j-1)}(\xi/2).$$

Then

$$\begin{aligned} |\widehat{\phi}(\xi)| &= \prod_{j=1}^{\infty} \left| \left(\frac{1+e^{-i(2^{-j}\xi)}}{2} \right) \right|^{2m} \prod_{j=1}^{\infty} \mathcal{L}(2^{-j}\xi) \\ &= \left| \left(\frac{1-e^{-i\xi}}{i\xi} \right) \right|^{2m} \prod_{j=1}^{\infty} \mathcal{L}(2^{-j}\xi) \\ &\leq C(1+|\xi|)^{-2m} \prod_{j=1}^{\infty} \mathcal{L}(2^{-j}\xi). \end{aligned} \tag{3.18}$$

By item 1 of Lemma 3.3, it is easy to see that $\sup_\xi \mathcal{L}(\xi) = 2^{2m-2}$. We now provide a decay estimate of $\prod_{j=1}^{\infty} \mathcal{L}(2^{-j}\xi)$. Since $\mathcal{L}$ is a trigonometric polynomial with $\mathcal{L}(0) = 1$, we have $\mathcal{L}(\xi) \leq 1 + C|\xi|$. Therefore,

$$\sup_{|\xi| \leq 1} \prod_{j=1}^{\infty} \mathcal{L}(2^{-j}\xi) \leq \sup_{|\xi| \leq 1} \prod_{j=1}^{\infty} e^{C 2^{-j}|\xi|} \leq e^C.$$

For a given $|\xi| \geq 1$, there exists $J \geq 1$ such that $2^{J-1} \leq |\xi| \leq 2^J$. Therefore,

$$\begin{aligned} \prod_{j=1}^{\infty} \mathcal{L}(2^{-j}\xi) &= \prod_{j=1}^{J} \mathcal{L}(2^{-j}\xi) \prod_{j=1}^{\infty} \mathcal{L}(2^{-j-J}\xi) \\ &\leq 2^{J(2m-2)} e^C \\ &\leq C(1+|\xi|)^{2m-2}. \end{aligned} \tag{3.19}$$

Combining inequalities (3.18) and (3.19), we have the desired decay estimate of $\widehat{\phi}$.

Now, proving ϕ is interpolatory is equivalent to showing that

$$\sum_{k \in \mathbb{Z}} \phi(k) e^{-ik\xi} = 1 \tag{3.20}$$

almost everywhere. Define $\phi^o(\xi) := \sum_{k \in \mathbb{Z}} \widehat{\phi}(\xi + 2\pi k)$. Since $\widehat{\phi} \in L_1(\mathbb{R})$, it is easy to show that $\phi^o \in L_1([-\pi, \pi])$. Considering the Fourier coefficients of ϕ^o, we have

$$\widehat{\phi^o}[k] = \int_{-\pi}^{\pi} \phi^o(\xi) e^{ik\xi} \mathrm{d}\xi = \int_{\mathbb{R}} \widehat{\phi}(\xi) e^{ik\xi} \mathrm{d}\xi = 2\pi \phi(k).$$

Since ϕ is compactly supported, we know that $\{\phi[k] : k \in \mathbb{Z}\}$ is a finite sequence. Then $\sum_{k\in\mathbb{Z}} |\widehat{\phi^o}[k]| < \infty$, which means ϕ^o has a Fourier series with absolute convergence. Therefore,

$$\sum_{k\in\mathbb{Z}} \widehat{\phi}(\xi + 2\pi k) = \phi^o(\xi) = \sum_{k\in\mathbb{Z}} \phi(k)e^{-ik\xi}. \tag{3.21}$$

Identity (3.21) is known as the *Poisson summation formula.* Combining (3.21) with (3.20) we have that ϕ is interpolatory if and only if

$$\sum_{k\in\mathbb{Z}} \widehat{\phi}(\xi + 2\pi k) = 1$$

almost everywhere. Noting that $\widehat{a}+\widehat{a}(\cdot+\pi) = 1$, the rest of the proof follows similar to that of Proposition 3.4 and we leave the details to the readers as an exercise. ∎

Note that pseudo-splines with order $(m, 0)$ for both types are in fact B-splines. Pseudo-splines of Type I with order $(m, m-1)$ are the refinable functions with orthonormal shifts (called orthogonal refinable functions) given in [**60**]. The pseudo-splines of Type II with order $(m, m-1)$ are the interpolatory refinable functions (which were first introduced in [**79**] and a systematic construction was given in [**60**]). The other pseudo-splines fill the gap between the B-splines and orthogonal or interpolatory refinable functions.

Similarly, we have the stability of all dual pseudo-splines.

PROPOSITION 3.6 ([**72**]). *All dual pseudo-splines are stable.*

2. Wavelets from pseudo-splines

For a given pseudo-spline, there are many ways to construct pseudo-spline tight wavelet frames from it. Here, we provide a general construction of pseudo-spline tight wavelet frames.

2.1. Framelets

In this Section, we give a construction of a tight wavelet frame system from pseudo-splines. The construction is from [**75**] which is motivated from [**51**] and one of the constructions of [**65**]. The construction can be applied to any refinable function whose mask is a trigonometric polynomial and satisfies (3.11).

CONSTRUCTION 3.3. *Let $\phi \in L_2(\mathbb{R})$ be a compactly supported refinable function with its trigonometric polynomial refinement mask $\widehat{h}_0$ satisfying $\widehat{h}_0(0) = 1$ and (3.11). Let*

$$T = 1 - |\widehat{h}_0|^2 - |\widehat{h}_0(\cdot+\pi)|^2 \quad \text{and} \quad \mathcal{A} := \frac{\sqrt{T}}{2},$$

where $\sqrt{T}$ is obtained via the Fejér-Riesz lemma. Define

$$\widehat{h}_1(\xi) := e^{-i\xi}\overline{\widehat{h}_0(\xi+\pi)}, \quad \widehat{h}_2(\xi) := \mathcal{A}(\xi) + e^{-i\xi}\mathcal{A}(-\xi)$$

and

$$\widehat{h}_3(\xi) := e^{-i\xi}\overline{\widehat{h}_2(\xi+\pi)}.$$

Let $\Psi := \{\psi_1,\ \psi_2,\ \psi_3\}$, where

$$\widehat{\psi}_\ell(\xi) = \widehat{h}_\ell(\xi/2)\widehat{\phi}(\xi/2), \qquad \ell = 1, 2, 3.$$

Then $X(\Psi)$ is a tight frame for $L_2(\mathbb{R})$. Moreover, assume that ϕ, hence its refinement mask $\widehat{h}_0$, is symmetric about the origin. Then, ψ_1 is symmetric about $\frac{1}{2}$, ψ_2 is symmetric about $\frac{1}{4}$ and ψ_3 is antisymmetric about $\frac{1}{4}$.

PROOF. By Theorem 2.2, in order to verify that $X(\Psi)$ is a tight frame for $L_2(\mathbb{R})$, one needs to show that the masks $\{\widehat{h}_0, \widehat{h}_1, \widehat{h}_2, \widehat{h}_3\}$ satisfy (2.6). Using the definitions of $\widehat{h}_1$ and $\widehat{h}_3$

$$\widehat{h}_1 = e^{-i\xi}\overline{\widehat{h}_0(\cdot+\pi)} \quad \text{and} \quad \widehat{h}_3 = e^{-i\xi}\overline{\widehat{h}_2(\cdot+\pi)},$$

we have

$$\sum_{\ell=0}^{3} \widehat{h}_\ell \overline{\widehat{h}_\ell(\cdot+\pi)} = \widehat{h}_0\overline{\widehat{h}_0(\cdot+\pi)} - \widehat{h}_0\overline{\widehat{h}_0(\cdot+\pi)} + \widehat{h}_2\overline{\widehat{h}_2(\cdot+\pi)} - \widehat{h}_2\overline{\widehat{h}_2(\cdot+\pi)} = 0.$$

Next, we show that

$$\sum_{\ell=0}^{3} |\widehat{h}_\ell|^2 = 1. \tag{3.22}$$

Since

$$|\widehat{h}_0|^2 + |\widehat{h}_1|^2 = |\widehat{h}_0|^2 + |\widehat{h}_0(\cdot+\pi)|^2,$$

it remains to show that

$$|\widehat{h}_2|^2 + |\widehat{h}_3|^2 = 1 - |\widehat{h}_0|^2 - |\widehat{h}_0(\cdot+\pi)|^2 = T.$$

Since h_0 has real coefficients, we have $\widehat{h}_0(-\xi) = \overline{\widehat{h}_0(\xi)}$, and thus $T(\xi) = T(-\xi)$. Furthermore, $T(\xi)$ is nonnegative and π-periodic. Then by the Fejér-Riesz lemma (and spectral factorization [**61**, Lemma 6.1.3]), the function $\mathcal{A}(\xi)$ is also π-periodic and has real Fourier coefficients. In summary, we have

$$\mathcal{A}(\xi) = \mathcal{A}(\xi+\pi), \quad \text{and} \quad |\mathcal{A}(\xi)|^2 = |\mathcal{A}(-\xi)|^2, \quad \text{for all } \xi \in \mathbb{R}. \tag{3.23}$$

Since

$$\widehat{h}_2(\xi) = \mathcal{A}(\xi) + e^{-i\xi}\mathcal{A}(-\xi) \quad \text{and} \quad \widehat{h}_3(\xi) = e^{-i\xi}\overline{\widehat{h}_2(\cdot+\pi)} = e^{-i\xi}\mathcal{A}(-\xi) - \mathcal{A}(\xi),$$

applying (3.23), one obtains

$$\begin{aligned} |\widehat{h}_2(\xi)|^2 &= \left(\mathcal{A}(\xi) + e^{-i\xi}\mathcal{A}(-\xi)\right)\left(\overline{\mathcal{A}(\xi)} + e^{i\xi}\overline{\mathcal{A}(-\xi)}\right) \\ &= |\mathcal{A}(\xi)|^2 + |\mathcal{A}(-\xi)|^2 + e^{i\xi}\mathcal{A}(\xi)\overline{\mathcal{A}(-\xi)} + e^{-i\xi}\mathcal{A}(-\xi)\overline{\mathcal{A}(\xi)} \\ &= 2|\mathcal{A}(\xi)|^2 + e^{i\xi}\mathcal{A}(\xi)\overline{\mathcal{A}(-\xi)} + e^{-i\xi}\mathcal{A}(-\xi)\overline{\mathcal{A}(\xi)} \end{aligned}$$

and

$$\begin{aligned} |\widehat{h}_3(\xi)|^2 &= \left(e^{-i\xi}\mathcal{A}(-\xi) - \mathcal{A}(\xi)\right)\left(e^{i\xi}\overline{\mathcal{A}(-\xi)} - \overline{\mathcal{A}(\xi)}\right) \\ &= |\mathcal{A}(\xi)|^2 + |\mathcal{A}(-\xi)|^2 - e^{i\xi}\mathcal{A}(\xi)\overline{\mathcal{A}(-\xi)} - e^{-i\xi}\mathcal{A}(-\xi)\overline{\mathcal{A}(\xi)} \\ &= 2|\mathcal{A}(\xi)|^2 - e^{i\xi}\mathcal{A}(\xi)\overline{\mathcal{A}(-\xi)} - e^{-i\xi}\mathcal{A}(-\xi)\overline{\mathcal{A}(\xi)}. \end{aligned}$$

Hence,

$$|\widehat{h}_2(\xi)|^2 + |\widehat{h}_3(\xi)|^2 = 4|\mathcal{A}(\xi)|^2 = T(\xi),$$

which gives (3.22) and thus concludes that the masks $\{\widehat{h}_0, \widehat{h}_1, \widehat{h}_2, \widehat{h}_3\}$ satisfy (2.6). Therefore, $X(\Psi)$ is indeed a tight frame for $L_2(\mathbb{R})$ by the unitary extension principle.

Now we show that ψ_1 is symmetric about $\frac{1}{2}$ while ψ_2 is symmetric about $\frac{1}{4}$ and ψ_3 is antisymmetric about $\frac{1}{4}$. It is well known that a function $f \in L_2(\mathbb{R})$, is symmetric about the point $\gamma_1 \in \mathbb{R}$ if and only if

$$f(x) = f(2\gamma_1 - x) \quad \text{a.e.},$$

which is equivalent to

$$\widehat{f}(\xi) = e^{-i2\gamma_1\xi}\widehat{f}(-\xi) \quad \text{a.e.} \tag{3.24}$$

Similarly, a function $f \in L_2(\mathbb{R})$ is antisymmetric about the point $\gamma_2 \in \mathbb{R}$ if and only if

$$f(x) = -f(2\gamma_2 - x) \quad \text{a.e.},$$

which is equivalent to

$$\widehat{f}(\xi) = -e^{-i2\gamma_2\xi}\widehat{f}(-\xi) \quad \text{a.e.} \tag{3.25}$$

By the definition of $\widehat{h}_1$ and the fact that $\widehat{h}_0$ is symmetric about the origin and 2π-periodic, one obtains

$$\widehat{h}_1(\xi) = e^{-i\xi}\overline{\widehat{h}_0(\xi+\pi)} = e^{-2i\xi}\big(e^{i\xi}\overline{\widehat{h}_0(-\xi+\pi)}\big) = e^{-2i\xi}\widehat{h}_1(-\xi).$$

Since ϕ is symmetric about the origin, then by (3.24) one obtains

$$\widehat{\phi}(\xi) = \widehat{\phi}(-\xi), \quad \text{for all } \xi \in \mathbb{R}. \tag{3.26}$$

Therefore,

$$\widehat{\psi_1}(\xi) = \widehat{h}_1(\xi/2)\widehat{\phi}(\xi/2) = e^{-i\xi}\widehat{h}_1(-\xi/2)\widehat{\phi}(-\xi/2) = e^{-i\xi}\widehat{\psi_1}(-\xi),$$

which, by (3.24), means that ψ_1 is symmetric about $\frac{1}{2}$. Similarly by the definition of $\widehat{h}_2$, one obtains

$$\widehat{h}_2(\xi) = \mathcal{A}(\xi) + e^{-i\xi}\mathcal{A}(-\xi) = e^{-i\xi}\big(\mathcal{A}(-\xi) + e^{i\xi}\mathcal{A}(\xi)\big) = e^{-i\xi}\widehat{h}_2(-\xi).$$

Applying (3.26) and the definition of $\widehat{\psi_2}$, one obtains,

$$\widehat{\psi_2}(\xi) = \widehat{h}_2(\xi/2)\widehat{\phi}(\xi/2) = e^{-i\frac{\xi}{2}}\widehat{h}_2(-\xi/2)\widehat{\phi}(-\xi/2) = e^{-i\frac{\xi}{2}}\widehat{\psi_2}(-\xi),$$

which, by (3.24), means that ψ_2 is symmetric about $\frac{1}{4}$. Similarly, we can show that ψ_3 is antisymmetric about $\frac{1}{4}$. ∎

REMARK 3.1. Theorem 3.6 proves that the first framelet ψ_1 constructed from dual pseudo-splines generates a Riesz basis for $L_2(\mathbb{R})$. Furthermore, when pseudo-splines of type I with order $(m, m-1)$ are used in Construction 3.3, the set Ψ contains only one element ψ which coincides with Daubechies' orthonormal wavelets of order m [**61, 60**] (see the next subsection for more details).

Next, we give one example of (anti)symmetric tight framelets constructed from Construction 3.3 using pseudo-splines of Type II with order $(3,1)$.

EXAMPLE 3.1. Let $\widehat{h}_0$ to be the mask of the pseudo-spline of Type II with order $(3,1)$ i.e.

$$\widehat{h}_0(\xi) = \cos^6(\xi/2)\big(1 + 3\sin^2(\xi/2)\big).$$

We define

$$\widehat{h}_1(\xi) := e^{-i\xi}\overline{\widehat{h}_0(\xi+\pi)} = e^{-i\xi}\sin^6(\xi/2)\big(1 + 3\cos^2(\xi/2)\big),$$

$$\widehat{h}_2(\xi) := \mathcal{A}(\xi) + e^{-i\xi}\mathcal{A}(-\xi) \quad \text{and} \quad \widehat{h}_3(\xi) := e^{-i\xi}\mathcal{A}(-\xi) - \mathcal{A}(\xi),$$

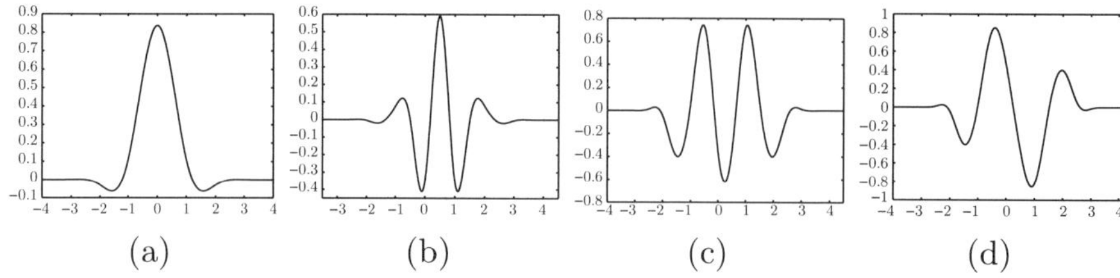

FIGURE 1. (a) is the pseudo-spline of Type II with order (3, 1) and (b)-(d) are the corresponding (anti)symmetric tight framelets.

where

$$\mathcal{A} = \frac{1}{2}\Big(0.00123930398199e^{-4i\xi} + 0.00139868605052e^{-2i\xi} - 0.22813823298962 + 0.44712319189971e^{2i\xi} - 0.22162294894260e^{4i\xi}\Big).$$

The graphs of Ψ are given by (b)-(d) in Figure 1. The tight frame system has approximation order 4.

Since Construction 3.3 is generic and applies on any refinement mask, we can apply this construction on dual pseudo-splines. Notice that dual pseudo-splines are symmetric about $-1/2$. Then, we have the following simple results on the symmetry of the framelets constructed via Construction 3.3. Furthermore, arguments similar to those used in proving Theorem 3.6 can be used to prove that the first framelet ψ_1 constructed from dual pseudo-splines generates a Riesz basis for $L_2(\mathbb{R})$.

PROPOSITION 3.7. *Let $\Psi = \{\psi_1, \psi_2, \psi_3\}$ be the tight framelets constructed from Construction* 3.3 *from dual pseudo-splines. Then, ψ_1 is antisymmetric about $\frac{1}{2}$, ψ_2 is symmetric about* 0 *and ψ_3 is antisymmetric about* 0.

2.2. Orthonormal wavelets

In this subsection, we show that the tight frame systems constructed from Construction 3.3 using pseudo-splines of type I with order $(m, m-1)$ are orthonormal bases for $L_2(\mathbb{R})$. In fact, they are the orthonormal bases originally constructed by Daubechies in [**60**]. For simplicity, we denote pseudo-splines of type I with order $(m, m-1)$ simply as ϕ_m, and its mask as a_m. In fact, one can see easily that the masks, a_m, for all m are exactly the same as the refinement masks of the orthonormal refinable functions in [**60**] by item 1 of Lemma 3.3.

By definition of pseudo-splines, we have

$$|\widehat{a}_m(\xi)|^2 + |\widehat{a}_m(\xi+\pi)|^2 = 1, \quad \text{and} \quad \widehat{a}_m(0) = 1. \tag{3.27}$$

Under the condition (3.27), Construction 3.3 produces only one framelet ψ with mask

$$\widehat{h}_1(\xi) = e^{-i\xi}\overline{\widehat{a}_m(\xi+\pi)}.$$

Now, we show that the tight frame system $X(\psi)$ is in fact an orthonormal basis for $L_2(\mathbb{R})$. Based on the unitary extension principle (Theorem 2.2), all we need to show is that $\|\phi_m\|_2 = 1$ for any given $m \geq 1$.

THEOREM 3.4. *Given ϕ_m, pseudo-spline of type I with order $(m, m-1)$, let ψ be the single framelet constructed from Construction* 3.3 *using ϕ_m. Then $X(\psi)$ forms a compactly supported orthonormal wavelet basis for $L_2(\mathbb{R})$.*

PROOF. As we discussed above, all we need to show is that $\|\phi_m\|_2 = 1$ or equivalently, $\|\widehat{\phi}_m\|_2^2 = 2\pi$.

Consider the cascade algorithm defined by (same as (3.10))

$$\widehat{f}_n(\xi) = \widehat{a}_m(\frac{\xi}{2})\widehat{f}_{n-1}(\frac{\xi}{2}) = \prod_{j=1}^{n} \widehat{a}_m(2^{-j}\xi)\widehat{f}_0(2^{-n}\xi), \quad n > 0,$$

with initial function f_0 satisfying $\widehat{f}_0(\xi) = \chi_{[-\pi,\pi]}(\xi)$. By Proposition 3.1, $\widehat{f}_n \to \widehat{\phi}_m$ pointwise as $n \to \infty$.

Following a proof similar to that of Theorem 3.1, we have

$$\begin{aligned}
\|\widehat{f}_n\|_{L_2(\mathbb{R})}^2 &= \int_{-2^n\pi}^{2^n\pi} \prod_{j=1}^{n} |\widehat{a}_m(2^{-j}\xi)|^2 d\xi \\
&= \int_0^{2^n\pi} \prod_{j=1}^{n-1} |\widehat{a}_m(2^{-j}\xi)|^2 \left(|\widehat{a}_m(2^{-n}\xi)|^2 + |\widehat{a}_m(2^{-n}\xi + \pi)|^2\right) d\xi, \\
\text{(by (3.27))} \quad &= \int_0^{2^n\pi} \prod_{j=1}^{n-1} |\widehat{a}_m(2^{-j}\xi)|^2 d\xi \\
&= \|\widehat{f}_{n-1}\|_{L_2(\mathbb{R})}^2 = \cdots = \|\widehat{f}_0\|_{L_2(\mathbb{R})}^2 = 2\pi.
\end{aligned}$$

Note that

$$|\widehat{\phi}_m(\xi)|^2 \geq |B_{2m}(\xi)| = \left(\frac{\sin(\xi/2)}{\xi/2}\right)^{2m} \geq C > 0, \quad \xi \in [-\pi, \pi].$$

Then arguments similar to those in Proposition 3.2 imply that

$$\|\widehat{\phi}_m\|_{L_2(\mathbb{R})}^2 = \lim_{n\to\infty} \|\widehat{f}_n\|_{L_2(\mathbb{R})}^2 = 2\pi,$$

which concludes the proof of this theorem. ∎

2.3. Riesz wavelets from pseudo-splines

In this section, we focus on the structure of the tight frame systems constructed from pseudo-splines by applying the unitary extension principle [**158**]. We show that in all pseudo-spline tight frame systems constructed using Construction 3.3, there is one wavelet whose dilations and shifts already form a Riesz basis for $L_2(\mathbb{R})$.

A system $X(\psi)$ is a *Riesz basis* if there exist $1 < C_1 \leq C_2 < \infty$ such that, for all sequences $c \in \ell_2(\mathbb{Z}^2)$,

$$C_1\|c\|_{\ell_2(\mathbb{Z}^2)} \leq \left\| \sum_{(n,k)\in\mathbb{Z}^2} c[n,k]\psi_{n,k} \right\|_{L_2(\mathbb{R})} \leq C_2\|c\|_{\ell_2(\mathbb{Z}^2)}$$

holds and the span of $\{\psi_{n,k} : n, k \in \mathbb{Z}\}$ is dense in $L_2(\mathbb{R})$. The function ψ is called *Riesz wavelet* if $X(\psi)$ forms a Riesz basis for $L_2(\mathbb{R})$ which is also called the *Riesz wavelet system*. When only the right hand side of the above inequalities holds, the wavelet system $X(\psi)$ is a Bessel system. Note that the definition of a Bessel system used here is equivalent to the earlier definition which uses the right hand side of the inequality in the definition of the frame in (2.2) (see e.g. [**50, 156**]).

For a given stable refinable function $\phi \in L_2(\mathbb{R})$, the key step in the MRA-based construction of the Riesz wavelet ψ is to select some desirable wavelet mask h_1 with

$$\widehat{\psi}(2\cdot) = \widehat{h}_1\widehat{\phi}.$$

When $\{\phi(\cdot - k) : k \in \mathbb{Z}\}$ forms an orthonormal basis for $V_0(\phi)$, e.g. ϕ is a pseudo-spline of Type I with order $(m, m-1)$, and if we choose h_1 as

$$h_1[k] = (-1)^{k-1}\overline{h_0[1-k]}, \quad k \in \mathbb{Z},$$

or equivalently

$$\widehat{h}_1(\xi) = e^{-i\xi}\overline{\widehat{h}_0(\xi + \pi)}. \tag{3.28}$$

then Theorem 2.2 implies that the corresponding wavelet system $X(\psi)$ is an orthonormal basis for $L_2(\mathbb{R})$. It was shown in [**103**] that if ϕ is a B-spline, then the wavelet system $X(\psi)$ with the corresponding wavelet mask h_1 given by (3.28) is a Riesz basis for $L_2(\mathbb{R})$. The first goal of this section is to show that if ϕ is any pseudo-spline with mask h_0, then the wavelet system $X(\psi)$ with the corresponding wavelet mask h_1 given by (3.28) is a Riesz basis for $L_2(\mathbb{R})$. To prove this, we use the following theorem which is a special case of [**103**, Theorem 2.1]. Note that when both refinement masks are finitely supported the result is similar to those already obtained in [**55, 56, 53**].

THEOREM 3.5. *Let h_0 be a finitely supported refinement mask of a refinable function $\phi \in L_2(\mathbb{R})$ with $\widehat{h}_0(0) = 1$ and $\widehat{h}_0(\pi) = 0$, such that $\widehat{h}_0$ can be factorized into the form*

$$|\widehat{h}_0(\xi)| = \left|\left(\frac{1+e^{-i\xi}}{2}\right)^n \mathcal{L}(\xi)\right| = \cos^n(\xi/2)|\mathcal{L}(\xi)|, \quad \xi \in [-\pi, \pi], \tag{3.29}$$

where $\mathcal{L}$ is the Fourier series of a finitely supported sequence with $\mathcal{L}(\pi) \neq 0$. Suppose that

$$|\widehat{h}_0(\xi)|^2 + |\widehat{h}_0(\xi+\pi)|^2 \neq 0, \quad \xi \in [-\pi, \pi].$$

Define

$$\widehat{\psi}(2\xi) := e^{-i\xi}\overline{\widehat{h}_0(\xi+\pi)}\widehat{\phi}(\xi)$$

and

$$\tilde{\mathcal{L}}(\xi) := \frac{\mathcal{L}(\xi)}{|\widehat{h}_0(\xi)|^2 + |\widehat{h}_0(\xi+\pi)|^2}. \tag{3.30}$$

Assume that

$$\rho_{\mathcal{L}} := \|\mathcal{L}(\xi)\|_{L_\infty(\mathbb{R})} < 2^{n-\frac{1}{2}} \quad \text{and} \quad \rho_{\tilde{\mathcal{L}}} := \|\tilde{\mathcal{L}}(\xi)\|_{L_\infty(\mathbb{R})} < 2^{n-\frac{1}{2}}, \tag{3.31}$$

Then $X(\psi)$ is a Riesz basis for $L_2(\mathbb{R})$.

Recall that the refinement masks of pseudo-splines of Type I and II are

$$|{}_1\widehat{a}(\xi)| := \cos^m(\xi/2)\left(\sum_{j=0}^{l}\binom{m+l}{j}\sin^{2j}(\xi/2)\cos^{2(l-j)}(\xi/2)\right)^{\frac{1}{2}}$$

and

$${}_2\widehat{a}(\xi) := \cos^{2m}(\xi/2)\sum_{j=0}^{l}\binom{m+l}{j}\sin^{2j}(\xi/2)\cos^{2(l-j)}(\xi/2).$$

Using the polynomial $P_{m,l}(y)$ given by (3.4) with $y = \sin^2(\xi/2)$, we can rewrite ${}_1\widehat{a}$ and ${}_2\widehat{a}$ as

$$|{}_1\widehat{a}| = \Big((1-y)^m P_{m,l}(y)\Big)^{\frac{1}{2}}, \qquad {}_2\widehat{a} = (1-y)^m P_{m,l}(y). \tag{3.32}$$

Hence, the corresponding $\mathcal{L}$ function in (3.29) for pseudo-splines of Type I and II are

$$|_1\mathcal{L}| = (P_{m,l}(y))^{\frac{1}{2}}, \qquad |_2\mathcal{L}| = P_{m,l}(y). \tag{3.33}$$

Furthermore, using $R_{m,l}(y)$ defined by (3.5), we have

$$|_1\widehat{a}(\xi)|^2 + |_1\widehat{a}(\xi+\pi)|^2 = R_{m,l}(y) + R_{m,l}(1-y)$$

and

$$|_2\widehat{a}(\xi)|^2 + |_2\widehat{a}(\xi+\pi)|^2 = R^2_{m,l}(y) + R^2_{m,l}(1-y),$$

with $y = \sin^2(\xi/2)$. Hence,

$$|_1\tilde{\mathcal{L}}| = \frac{(P_{m,l}(y))^{\frac{1}{2}}}{R_{m,l}(y) + R_{m,l}(1-y)} \quad \text{and} \quad |_2\tilde{\mathcal{L}}| = \frac{P_{m,l}(y)}{R^2_{m,l}(y) + R^2_{m,l}(1-y)}. \tag{3.34}$$

Theorem 3.5 tells us that the key step is to estimate the upper bounds of $|\mathcal{L}(\xi)|$ and $|\tilde{\mathcal{L}}(\xi)|$. The estimation of $\|_1\tilde{\mathcal{L}}\|_{L_\infty(\mathbb{R})}$ and $\|_2\tilde{\mathcal{L}}\|_{L_\infty(\mathbb{R})}$ are based on the following result:

PROPOSITION 3.8. *Let m and l be given nonnegative integers with $l \le m-1$ and $|_1\tilde{\mathcal{L}}|$ and $|_2\tilde{\mathcal{L}}|$ be defined in (3.34). Then,*

(1) $\|_1\tilde{\mathcal{L}}\|_{L_\infty(\mathbb{R})} = \sup_{y\in[0,1]} \frac{(P_{m,l}(y))^{\frac{1}{2}}}{R_{m,l}(y)+R_{m,l}(1-y)} < 2^{m-\frac{1}{2}}$.
(2) $\|_2\tilde{\mathcal{L}}\|_{L_\infty(\mathbb{R})} = \sup_{y\in[0,1]} \frac{P_{m,l}(y)}{R^2_{m,l}(y)+R^2_{m,l}(1-y)} < 2^{2m-\frac{1}{2}}$.

PROOF. Item 1 of Lemma 3.3 gives

$$P_{m,l}(y) = \sum_{j=0}^{l} \binom{m+l}{j} y^j(1-y)^{l-j} = \sum_{j=0}^{l} \binom{m-1+j}{j} y^j, \tag{3.35}$$

with $y \in [0,1]$. Hence both $(P_{m,l}(y))^{\frac{1}{2}}$ and $P_{m,l}(y)$ attain their maximum at $y=1$ and the maximum values are:

$$(P_{m,l}(1))^{\frac{1}{2}} = \binom{m+l}{l}^{\frac{1}{2}} \quad \text{and} \quad P_{m,l}(1) = \binom{m+l}{l}.$$

By item 3 of Lemma 3.3, one obtains

$$\begin{aligned}
\|_1\tilde{\mathcal{L}}\|_{L_\infty(\mathbb{R})} &= \sup_{y\in[0,1]} \frac{(P_{m,l}(y))^{\frac{1}{2}}}{R_{m,l}(y)+R_{m,l}(1-y)} \\
&\le \binom{m+l}{l}^{\frac{1}{2}} \max_{y\in[0,1]} \frac{1}{R_{m,l}(y)+R_{m,l}(1-y)} \\
&\le \frac{2^{m+l-1}\binom{m+l}{l}^{\frac{1}{2}}}{\sum_{j=0}^{l}\binom{m+l}{j}}.
\end{aligned}$$

Applying item 3 of Lemma 3.2, i.e.

$$\frac{2^l\binom{m+l}{l}^{\frac{1}{2}}}{\sum_{j=0}^{l}\binom{m+l}{j}} \le 1, \tag{3.36}$$

one obtains

$$\|_1\tilde{\mathcal{L}}\|_{L_\infty(\mathbb{R})} \le 2^{m-1} < 2^{m-\frac{1}{2}}.$$

The proof of (2) is similar to that of (1). Indeed, by item 4 of Lemma 3.3 and applying (3.36) again, we have

$$\begin{aligned}
\|{}_2\tilde{\mathcal{L}}\|_{L_\infty(\mathbb{R})} &= \sup_{y\in[0,1]} \frac{P_{m,l}(y)}{R^2_{m,l}(y)+R^2_{m,l}(1-y)} \\
&\le \binom{m+l}{l} \max_{y\in[0,1]} \frac{1}{R^2_{m,l}(y)+R^2_{m,l}(1-y)} \\
&= \frac{2^{2m+2l-1}\binom{m+l}{l}}{\left(\sum_{j=0}^{l}\binom{m+l}{j}\right)^2} \le 2^{2m-1} < 2^{2m-\frac{1}{2}}.
\end{aligned}$$

■

THEOREM 3.6. *Let ${}_k\phi$, $k=1,2$ be the pseudo-spline of Type I and II with order (m,l). The refinement masks ${}_ka$, $k=1,2$, are given in (3.2) and (3.3). Define*

$$\widehat{{}_k\psi}(2\xi) := e^{-i\xi}\overline{\widehat{{}_ka}(\xi+\pi)}\widehat{{}_k\phi}(\xi), \quad k=1,2, \tag{3.37}$$

then $X({}_k\psi)$ forms a Riesz basis for $L_2(\mathbb{R})$.

PROOF. To apply Theorem 3.5, we first note that by items 3 and 4 of Lemma 3.3,

$$|\widehat{{}_1a}(\xi)|^2 + |\widehat{{}_1a}(\xi+\pi)|^2 = R_{m,l}(\sin^2(\xi/2)) + R_{m,l}(\cos^2(\xi/2)) \neq 0$$

and

$$|\widehat{{}_2a}(\xi)|^2 + |\widehat{{}_2a}(\xi+\pi)|^2 = R^2_{m,l}(\sin^2(\xi/2)) + R^2_{m,l}(\cos^2(\xi/2)) \neq 0,$$

for all $\xi\in[-\pi,\pi]$.

Since Proposition 3.8 showed that

$$\rho_{{}_1\tilde{\mathcal{L}}} = \|{}_1\tilde{\mathcal{L}}\|_{L_\infty(\mathbb{R})} < 2^{m-\frac{1}{2}} \quad \text{and} \quad \rho_{{}_2\tilde{\mathcal{L}}} = \|{}_2\tilde{\mathcal{L}}\|_{L_\infty(\mathbb{R})} < 2^{2m-\frac{1}{2}},$$

we only need to check whether

$$\rho_{{}_1\mathcal{L}} = \|{}_1\mathcal{L}\|_{L_\infty(\mathbb{R})} < 2^{m-\frac{1}{2}}, \quad \rho_{{}_2\mathcal{L}} = \|{}_2\mathcal{L}\|_{L_\infty(\mathbb{R})} < 2^{2m-\frac{1}{2}}.$$

Indeed, we have

$$|\widehat{{}_ka}(\xi)|^2 + |\widehat{{}_ka}(\xi+\pi)|^2 \le 1 \quad \text{for all } \xi\in\mathbb{R}.$$

Hence,

$$|{}_k\mathcal{L}(\xi)| \le |{}_k\tilde{\mathcal{L}}(\xi)| \quad \text{for all } \xi\in\mathbb{R}.$$

This concludes the proof. ■

REMARK 3.2. The Riesz wavelet constructed in the above theorem has the same length of support and at least the same order of smoothness as that of the corresponding pseudo-spline. The order of its vanishing moments is the same as the order of the B-spline factor of the pseudo-spline. But, in general, its dual Riesz wavelet system is not compactly supported. However, this is not a problem for some applications. In applications like image compression, the short Riesz wavelet system can be applied to obtain a fast reconstruction algorithm, while decomposition is obtained by solving a linear system of equations (see [**121**]).

In Construction 3.3 as well as in the three constructions of [**65**] for pseudo-splines of Type I where the number of wavelets is either two or three (see [**65**] Section 3.1 for details), we observe that one of the wavelets ψ_1 is defined by

$$\widehat{\psi_1} := e^{-i\xi}\overline{\widehat{h_0}(\xi+\pi)}\widehat{\phi}(\xi/2),$$

where $\widehat{h}_0$ is the refinement mask of a pseudo-spline. Then, from Theorem 3.6, $X(\psi_1)$ forms a Riesz basis for $L_2(\mathbb{R})$. This means that each pseudo-spline tight frame system constructed in **[65]** or by Construction 3.3, already has a subsystem form a Riesz basis for $L_2(\mathbb{R})$. Finally, we note that if dual pseudo-splines are used in Construction 3.3 or any of the constructions in **[65]**, we will have the same conclusion by a similar argument as for pseudo-splines.

3. Regularity of pseudo-splines

This section is on theoretical analysis of pseudo-splines. Those who are interested in the basics of MRA-base wavelet frames and applications may skip this section. Also, we note that the material of this section is mainly taken from **[75]**.

For $\alpha = n+\beta$, $n \in \mathbb{N}$, $0 \leq \beta < 1$, the Hölder space C^α (see e.g. **[61]**) is defined to be the set of functions which are n times continuously differentiable and whose n^{th} derivative $f^{(n)}$ satisfies the condition,

$$|f^{(n)}(x+h) - f^{(n)}(x)| \leq C|h|^\beta, \ \forall x, h.$$

It is well known (see **[61]**) that if

$$\int_{\mathbb{R}} |\widehat{f}(\xi)|(1+|\xi|)^\alpha < \infty,$$

then $f \in C^\alpha$. In particular, if $|\widehat{f}(\xi)| \leq C(1+|\xi|)^{-1-\alpha-\varepsilon}$, then $f \in C^\alpha$.

The main idea here is to obtain a lower bound of the regularity of pseudo-splines with order (m,l) by estimating the decay of the Fourier transform of them. It turns out that this lower bound coincides with the upper bound when m goes to infinity, as will be shown in Theorem 3.9. It is well known that the exact Sobolev regularity of a given refinable function can be obtained via its mask by applying the transfer operator (see e.g. **[61, 161]** and references in there). Although the Sobolev exponent of a given refinable function can be computed exactly through finding the spectrum of the transfer operator derived from the corresponding refinement mask, this approach does not lead itself to a systematic computation of Sobolev exponents for a class of refinable functions, such as pseudo-splines. This is simply because different refinable functions lead to different transfer operators. This is the main reason we give a systematic estimate of the decay of the Fourier transform of pseudo-splines instead.

Since for any compactly supported refinable function ϕ in $L_2(\mathbb{R})$ with $\widehat{\phi}(0) = 1$, the refinement mask h_0 must satisfy $\widehat{h}_0(0) = 1$ and $\widehat{h}_0(\pi) = 0$ (see e.g. **[61]** or **[116]**), then $\widehat{h}_0(\xi)$ can be factored as

$$\widehat{h}_0(\xi) = \Big(\frac{1+e^{-i\xi}}{2}\Big)^n \mathcal{L}(\xi),$$

where n is the maximum multiplicity of zeros of $\widehat{h}_0$ at π and $\mathcal{L}(\xi)$ is a trigonometric polynomial with $\mathcal{L}(0) = 1$. Hence, we have

$$\widehat{\phi}(\xi) = \prod_{j=1}^\infty \widehat{a}(2^{-j}\xi) = \prod_{j=1}^\infty \Big(\frac{1+e^{-i2^{-j}\xi}}{2}\Big)^n \prod_{j=1}^\infty \mathcal{L}(2^{-j}\xi) = \Big(\frac{1-e^{-i\xi}}{i\xi}\Big)^n \prod_{j=1}^\infty \mathcal{L}(2^{-j}\xi).$$

This shows that any compactly supported refinable function in $L_2(\mathbb{R})$ is the convolution of a B-spline of some order, say n, with a distribution (see **[155]**). Indeed, a

B-spline of order n can be rewritten as

$$\widehat{B}_n = \Big(\frac{1-e^{-i\xi}}{i\xi}\Big)^n.$$

It is a piecewise polynomial of degree $n-1$ in $C^{n-1-\varepsilon}(\mathbb{R})$, supported on $[0,n]$, and has refinement mask $\left(\frac{1+e^{-i\xi}}{2}\right)^n$. Since $\mathcal{L}(\xi)$ is bounded, $\mathcal{L}(\xi)$ is actually the refinement mask of a refinable distribution (see e.g. [**61**]). Therefore, ϕ is the convolution of the B-spline B_n with the distribution. The regularity of ϕ comes from the B-spline factor while the distribution factor takes away the regularity. But the distribution component provides some desirable properties for ϕ, such as interpolatory properties, orthogonality of its shifts and approximation order of certain quasi-interpolants.

The decay of $|\widehat{\phi}|$ can be characterized by $|\widehat{h}_0|$ as stated in the following theorem. The proof of this theorem can be found in [**61**]. Note that in the following theorem, we write $|\widehat{h}_0|$ in the form of

$$|\widehat{h}_0(\xi)| = \Big|\Big(\frac{1+e^{-i\xi}}{2}\Big)^n \mathcal{L}(\xi)\Big| = \cos^n(\xi/2)|\mathcal{L}(\xi)|, \quad \xi\in[-\pi,\pi].$$

THEOREM 3.7 ([**61**]). *Let h_0 be the refinement mask of the refinable function ϕ of the form*

$$|\widehat{h}_0(\xi)| = \cos^n(\xi/2)|\mathcal{L}(\xi)|, \quad \xi\in[-\pi,\pi].$$

Suppose that

$$\begin{aligned} |\mathcal{L}(\xi)| &\le |\mathcal{L}(\tfrac{2\pi}{3})| && \text{for } |\xi|\le \tfrac{2\pi}{3}, \\ |\mathcal{L}(\xi)\mathcal{L}(2\xi)| &\le |\mathcal{L}(\tfrac{2\pi}{3})|^2 && \text{for } \tfrac{2\pi}{3}\le|\xi|\le\pi. \end{aligned} \tag{3.38}$$

Then $|\widehat{\phi}(\xi)| \le C(1+|\xi|)^{-n+\kappa}$, with $\kappa = \log(|\mathcal{L}(\frac{2\pi}{3})|)/\log 2$, and this decay is optimal.

This theorem allows us to estimate the decay of the Fourier transform of a refinable function via its refinement mask. Since $|{}_1\widehat{\phi}|^2 = |{}_2\widehat{\phi}|$, the decay rate of $|{}_1\widehat{\phi}|$ is half of that of $|{}_2\widehat{\phi}|$. Thus we can focus on the analysis of the decay of the Fourier transforms of pseudo-splines of Type II. Based on (1) of Lemma 3.3, we show that $P_{m,l}(y)$, defined in (3.4), satisfies (3.38). This directly leads to the estimate of the regularity of pseudo-splines. Note that the corresponding result for $l=m-1$ was proven in [**54**] which led to the optimal estimates for the decay of the Fourier transforms of the orthogonal and interpolatory refinable functions. Here, the more general result for pseudo-splines is obtained by a simpler proof than the original one of [**54**] and [**61**].

PROPOSITION 3.9. [**75**] *Let $P_{m,l}(y)$ be defined as in (3.4), where l,m are nonnegative integers with $l\le m-1$. Then*

$$P_{m,l}(y) \le P_{m,l}(\tfrac{3}{4}), \qquad \text{for } y\in[0,\tfrac{3}{4}], \tag{3.39}$$

$$P_{m,l}(y)P_{m,l}(4y(1-y)) \le \Big(P_{m,l}(\tfrac{3}{4})\Big)^2, \qquad \text{for } y\in[\tfrac{3}{4},1]. \tag{3.40}$$

PROOF. Since $P_{m,l}(y)$ is monotonically increasing (using item 1 of Lemma 3.3), (3.39) is obviously true. Hence, we focus on the proof of (3.40).

Throughout this proof, we let m be fixed. Let

$$W_{m,l}(y) := P_{m,l}(y)P_{m,l}(4y(1-y)) - \left(P_{m,l}(\frac{3}{4})\right)^2.$$

Then, the inequality (3.40) is equivalent to

$$W_{m,l}(y) \leq 0 \quad \text{for all } y \in [\frac{3}{4}, 1]. \tag{3.41}$$

Note that when $l = 0$, $P_{m,0}(y) = 1$ for all $y \in [0,1]$; then (3.41) is obviously true for $l = 0$. Therefore, if we can show that

$$W_{m,l+1}(y) - W_{m,l}(y) \leq 0, \quad \text{for all } y \in [\frac{3}{4}, 1], \quad l = 0, 1, \ldots, m-2, \tag{3.42}$$

then (3.41) follows from (3.42) which is equivalent to (3.40). We now focus on the inequality (3.42).

Using item 1 of Lemma 3.3, we have

$$\begin{aligned} W_{m,l+1}(y) - W_{m,l}(y) = &\left(\sum_{j=0}^{l+1}\binom{m-1+j}{j}y^j\right)\left(\sum_{j=0}^{l+1}\binom{m-1+j}{j}(4y(1-y))^j\right) \\ &- \left(\sum_{j=0}^{l}\binom{m-1+j}{j}y^j\right)\left(\sum_{j=0}^{l}\binom{m-1+j}{j}(4y(1-y))^j\right) \\ &+ P_{m,l}^2(\frac{3}{4}) - P_{m,l+1}^2(\frac{3}{4}). \end{aligned}$$

Splitting the sum $\sum_{j=0}^{l+1}\binom{m-1+j}{j}y^j$, one obtains

$$\begin{aligned} W_{m,l+1}(y) - W_{m,l}(y) = &\left(\sum_{j=0}^{l}\binom{m-1+j}{j}y^j\right)\left(\sum_{j=0}^{l+1}\binom{m-1+j}{j}(4y(1-y))^j\right) \\ &+ \binom{m+l}{l+1}y^{l+1}\sum_{j=0}^{l+1}\binom{m-1+j}{j}(4y(1-y))^j \\ &- \left(\sum_{j=0}^{l}\binom{m-1+j}{j}y^j\right)\left(\sum_{j=0}^{l}\binom{m-1+j}{j}(4y(1-y))^j\right) \\ &+ P_{m,l}^2(\frac{3}{4}) - P_{m,l+1}^2(\frac{3}{4}). \end{aligned}$$

Combining the first and the third term, one obtains

$$\begin{aligned}
W_{m,l+1}(y) - W_{m,l}(y) &= \binom{m+l}{l+1}(4y(1-y))^{l+1}\sum_{j=0}^{l}\binom{m-1+j}{j}y^j \\
&\quad +\binom{m+l}{l+1}y^{l+1}\sum_{j=0}^{l+1}\binom{m-1+j}{j}(4y(1-y))^j \\
&\quad +P^2_{m,l}(\tfrac{3}{4}) - P^2_{m,l+1}(\tfrac{3}{4}) \\
&= \binom{m+l}{l+1}\Bigg((4y(1-y))^{l+1}\sum_{j=0}^{l}\binom{m-1+j}{j}y^j \\
&\quad +y^{l+1}\sum_{j=0}^{l+1}\binom{m-1+j}{j}(4y(1-y))^j\Bigg) \qquad (3.43) \\
&\quad +P^2_{m,l}(\tfrac{3}{4}) - P^2_{m,l+1}(\tfrac{3}{4}).
\end{aligned}$$

Since $W_{m,l+1}(\frac{3}{4}) - W_{m,l}(\frac{3}{4}) = 0 - 0 = 0$, it suffices to show that $W_{m,l+1}(y) - W_{m,l}(y)$ monotonically decreases on $[\frac{3}{4}, 1]$, which is equivalent to showing that

$$G(y) := (4y(1-y))^{l+1}\sum_{j=0}^{l}\binom{m-1+j}{j}y^j + y^{l+1}\sum_{j=0}^{l+1}\binom{m-1+j}{j}(4y(1-y))^j$$

monotonically decreases on $[\frac{3}{4}, 1]$. We now compute G' as follows:

$$\begin{aligned}
G'(y) &= (l+1)(4-8y)(4y(1-y))^l\sum_{j=0}^{l}\binom{m-1+j}{j}y^j \\
&\quad +(4y(1-y))^{l+1}\sum_{j=1}^{l}\binom{m-1+j}{j}jy^{j-1} \\
&\quad +(l+1)y^l\sum_{j=0}^{l+1}\binom{m-1+j}{j}(4y(1-y))^j \\
&\quad +y^{l+1}(4-8y)\sum_{j=1}^{l+1}\binom{m-1+j}{j}j(4y(1-y))^{j-1} \\
&= (l+1)(4-8y)(4y(1-y))^l\sum_{j=0}^{l}\binom{m-1+j}{j}y^j \\
&\quad +(4y(1-y))^{l+1}\sum_{j=0}^{l-1}\binom{m+j}{j+1}(j+1)y^j \\
&\quad +(l+1)y^l\sum_{j=0}^{l+1}\binom{m-1+j}{j}(4y(1-y))^j \\
&\quad +y^{l+1}(4-8y)\sum_{j=0}^{l}\binom{m+j}{j+1}(j+1)(4y(1-y))^j.
\end{aligned}$$

Applying item 1 of Lemma 3.2 to the second and the fourth term above, one obtains

$$\begin{aligned} G'(y) &= (l+1)(4-8y)(4y(1-y))^l \sum_{j=0}^{l} \binom{m-1+j}{j} y^j \\ &+ (4y(1-y))^{l+1} \sum_{j=0}^{l} \binom{m-1+j}{j} (m+j) y^j \\ &- (m+l)\binom{m-1+l}{l} y^l (4y(1-y))^{l+1} \\ &+ (l+1) y^l \sum_{j=0}^{l} \binom{m-1+j}{j} (4y(1-y))^j \\ &+ (l+1)\binom{m+l}{l+1} y^l (4y(1-y))^{l+1} \\ &+ y^{l+1}(4-8y) \sum_{j=0}^{l} \binom{m-1+j}{j} (m+j)(4y(1-y))^j. \end{aligned}$$

By $(l+1)\binom{m+l}{l+1} = (m+l)\binom{m-1+l}{l}$ (item 1 of Lemma 3.2), we have

$$(l+1)\binom{m+l}{l+1} y^l (4y(1-y))^{l+1} - (m+l)\binom{m-1+l}{l} y^l (4y(1-y))^{l+1} = 0.$$

Hence,

$$\begin{aligned} G'(y) = &\sum_{j=0}^{l} \binom{m-1+j}{j} \Big((l+1)(4-8y)(4y(1-y))^l y^j + (m+j)(4y(1-y))^{l+1} y^j \\ &+ (l+1) y^l (4y(1-y))^j + (m+j)(4-8y) y^{l+1} (4y(1-y))^j \Big). \end{aligned}$$

Noting the common factor $y^j(4y(1-y))^j$ in the right hand side of the above equation, one obtains

$$\begin{aligned} G'(y) = &\sum_{j=0}^{l} \binom{m-1+j}{j} y^j (4y(1-y))^j \Big((l+1)(4-8y)(4y(1-y))^{l-j} \\ &+ (m+j)(4y(1-y))^{l+1-j} + (l+1) y^{l-j} + (m+j)(4-8y) y^{l+1-j} \Big). \end{aligned}$$

For $0 \le j \le l \le m-2$, consider

$$\begin{aligned} g_{l,j}(y) &:= (l+1)(4-8y)\big(4y(1-y)\big)^{l-j} + (m+j)\big(4y(1-y)\big)^{l+1-j} \\ &\quad + (l+1) y^{l-j} + (m+j)(4-8y) y^{l+1-j} \\ &= (l+1)(4y(1-y))^{l-j} \Big(4y(1-y) - (8y-4) \Big) \\ &\quad + (l+1) y^{l-j} \Big(1 - (8y-4) y \Big) \\ &\quad + (m+j-l-1) \Big((4y(1-y))^{l+1-j} - (8y-4) y^{l+1-j} \Big). \end{aligned}$$

The inequality $4y(1-y) \le y$ and $8y-4 \ge 2$ for $y \in [3/4, 1]$ show that $g_{l,j}(y) \le 0$ and $G'(y) \le 0$ on this interval. ∎

REMARK 3.3. It is clear that $W_{m,0}(y) = 0$, $y \in [\frac{3}{4}, 1]$, because $P_{m,0} = 1$. It was also proven by [**54**] that $W_{m,m-1}(y) \leq 0$, $y \in [\frac{3}{4}, 1]$, which is equivalent to (3.40). The decreasing of $W_{m,l}(y)$, for $y \in [\frac{3}{4}, 1]$, as l increases shown above indicates some difficulties in directly proving (3.40) for an arbitrary l, $0 < l < m - 1$, because it has a smaller margin than the case when $l = m - 1$. In fact, to some extent, the proof of (3.40) for the case when $l = m - 1$ relies on a numerical check for $m \leq 12$ (see [**61**]). Inequality (3.40) for the case $l = m - 1$ as proven in [**54**] (also see [**61**]) is one of the cornerstones of the wavelet theory, because it immediately leads to the optimal estimate of the decay of the Fourier transforms (hence, an estimate of the regularity) of both interpolatory and orthogonal refinable functions. Here, we take a different approach by proving that $W_{m,l}(y)$, $y \in [\frac{3}{4}, 1]$, decreases as l increases. As a result, we obtain (3.40) for all $0 \leq l \leq m - 1$ by the fact that $W_{m,0}(y) = 0$, $y \in [\frac{3}{4}, 1]$. This shows that introducing the concepts of the pseudo-splines gives a better understanding and a more complete picture of the proof of (3.40) and also, we hope, enriches the theory of wavelets. Note that the proof of (3.40) for all $0 \leq l \leq m - 1$ given here does not rely on any numerical computation and is simpler than the original proof of [**54**] and [**61**]. More recently, the authors in [**136**] provided a simpler proof of Proposition 3.9.

With this proposition, one obtains the regularity of pseudo-splines by applying Theorem 3.7.

THEOREM 3.8 ([**75**]). *Let ${}_2\phi$ be the pseudo-spline of Type II with order (m, l). Then*

$$|{}_2\widehat{\phi}(\xi)| \leq C\big(1 + |\xi|\big)^{-2m+\kappa},$$

where $\kappa = \log(P_{m,l}(\frac{3}{4}))/\log 2$. Consequently, ${}_2\phi \in C^{\alpha_2 - \varepsilon}$ with $\alpha_2 = 2m - \kappa - 1$. Furthermore, let ${}_1\phi$ be the pseudo-spline of Type I with order (m, l). Then

$$|{}_1\widehat{\phi}(\xi)| \leq C\big(1 + |\xi|\big)^{-m+\frac{\kappa}{2}}.$$

Consequently, ${}_1\phi \in C^{\alpha_1 - \varepsilon}$ with $\alpha_1 = m - \frac{\kappa}{2} - 1$.

PROOF. Since

$$\begin{aligned} {}_2\widehat{a}(\xi) &= \cos^{2m}(\xi/2) \sum_{j=0}^{l} \binom{m+l}{j} \sin^{2j}(\xi/2) \cos^{2(l-j)}(\xi/2) \\ &= \big(\cos(\xi/2)\big)^{2m} P_{m,l}(\sin^2(\xi/2)), \end{aligned}$$

the function $|\mathcal{L}(\xi)|$ in Theorem 3.7 is exactly $P_{m,l}(\sin^2(\xi/2))$. Applying (3.39) of Proposition 3.9

$$P_{m,l}(y) \leq P_{m,l}(\frac{3}{4}), \quad y \in [0, \frac{3}{4}],$$

and using $y = \sin^2(\xi/2)$, we have

$$\begin{aligned} |\mathcal{L}(\xi)| &= P_{m,l}(\sin^2(\xi/2)) \\ &= P_{m,l}(y) \leq P_{m,l}(\frac{3}{4}) = P_{m,l}(\sin^2(\frac{\pi}{3})) \quad \text{for } |\xi| \leq \frac{2\pi}{3}. \end{aligned}$$

Note that

$$|\mathcal{L}(2\xi)| = P_{m,l}(\sin^2(\xi)) = P_{m,l}(4\sin^2(\xi/2)(1 - \sin^2(\xi/2))) = P_{m,l}(4y(1-y)).$$

Applying (3.40) of Proposition 3.9

$$P_{m,l}(y)P_{m,l}(4y(1-y)) \le \big(P_{m,l}(\frac{3}{4})\big)^2, \quad y \in [\frac{3}{4}, 1],$$

we have

$$\begin{aligned} |\mathcal{L}(\xi)\mathcal{L}(2\xi)| &= P_{m,l}(\sin^2(\xi/2))P_{m,l}(4\sin^2(\xi/2)(1-\sin^2(\xi/2))) \\ &= P_{m,l}(y)P_{m,l}(4y(1-y)) \\ &\le \big(P_{m,l}(\frac{3}{4})\big)^2 = \big(P_{m,l}(\sin^2(\frac{\pi}{3}))\big)^2, \quad \text{for } \frac{2\pi}{3} \le |\xi| \le \pi. \end{aligned}$$

Hence, by Theorem 3.7, ${}_2\widehat{\phi}$ satisfies

$$|{}_2\widehat{\phi}(\xi)| \le C\big(1+|\xi|\big)^{-2m+\kappa},$$

where $\kappa = \log(P_{m,l}(\frac{3}{4}))/\log 2$. This leads to ${}_2\phi \in C^{\alpha_2-\varepsilon}$, where $\alpha_2 = 2m-\kappa-1$.

Since the decay of $|{}_1\widehat{\phi}|$ is exactly half of $|{}_2\widehat{\phi}|$, we have

$$|{}_1\widehat{\phi}(\xi)| \le C\big(1+|\xi|\big)^{-m+\frac{\kappa}{2}},$$

consequently ${}_1\phi \in C^{\alpha_1-\varepsilon}$, where $\alpha_2 = m - \frac{\kappa}{2} - 1$. ■

Table 1 gives the decay rates $\beta_{m,l}$ of the Fourier transform of pseudo-splines of Type II with order (m,l), for $2 \le m \le 8$ and $1 \le l \le m-1$. The regularity exponent of the corresponding pseudo-spline is, at least, $\alpha_2 = \beta_{m,l} - 1 - \varepsilon$. The decay rate of the Fourier transform of the pseudo-spline of Type I with the same order is $\frac{\beta_{m,l}}{2}$ and its regularity exponent α_1 is $\frac{\alpha_2-1}{2}$. Therefore, the table shows that for either type of the pseudo-splines and fixed order m, the decay rate of their Fourier transform decreases as l increases, while for fixed l, it increases as m increases. This is true indeed as shown in the following proposition.

TABLE 1. Decay rates $\beta_{m,l} = 2m - \kappa$ of pseudo-splines of Type II with order (m,l), for $2 \le m \le 8$ and $1 \le l \le m-1$.

(m,l)	$l=1$	$l=2$	$l=3$	$l=4$	$l=5$	$l=6$	$l=7$
$m=2$	2.67807						
$m=3$	4.29956	3.27208					
$m=4$	6.00000	4.73321	3.82507				
$m=5$	7.75207	6.27890	5.19506	4.35316			
$m=6$	9.54057	7.88626	6.64465	5.66363	4.86449		
$m=7$	11.35614	9.54057	8.15608	7.04717	6.13261	5.36349	
$m=8$	13.19265	11.23182	9.71691	8.48992	7.46770	6.59988	5.85310

PROPOSITION 3.10 ([**75**]). *Let $\beta_{m,l} = 2m - \kappa$ with $\kappa = \log P_{m,l}(\frac{3}{4})/\log 2$ as given in Theorem 3.8 and $0 \le l \le m-1$. Then:*

(1) *For fixed m, $\beta_{m,l}$ decreases as l increases.*
(2) *For fixed l, $\beta_{m,l}$ increases as m increases.*
(3) *When $l = m-1$, $\beta_{m,l}$ increases as m increases.*

Consequently, the decay rate $\beta_{2,1} = 2.67807$ is the smallest among all $\beta_{m,l}$, with $m \ge 2$ and $0 \le l \le m-1$.

PROOF. Part (1) follows directly from item 1 of Lemma 3.3, which shows that $P_{m,l}(\frac{3}{4})$ increases with l for fixed m.

For part (2), note that

$$\beta_{m,l} = 2m - \frac{\log P_{m,l}(\frac{3}{4})}{\log 2}.$$

Consider

$$2^{\beta_{m,l}} = 2^{2m - \frac{\log P_{m,l}(\frac{3}{4})}{\log 2}} = \frac{4^m}{P_{m,l}(\frac{3}{4})} = \frac{1}{4^{-m}P_{m,l}(\frac{3}{4})}.$$

Hence, part (2) is equivalent to the fact that

$$I_m := 4^{-m}P_{m,l}(\frac{3}{4})$$

decreases as m increases for fixed l, which is equivalent to that for fixed $0 \leq l \leq m-1$,

$$I_{m+1} - I_m < 0. \tag{3.44}$$

Note that

$$\begin{aligned} I_{m+1} - I_m &= 4^{-m-1}P_{m+1,l}(\frac{3}{4}) - 4^{-m}P_{m,l}(\frac{3}{4}) \\ &= 4^{-m-1}\sum_{j=0}^{l}\left(\binom{m+j}{j} - 4\binom{m-1+j}{j}\right)(\frac{3}{4})^j. \end{aligned}$$

Inequality (3.44) follows from the fact that for $0 \leq j \leq m-1$,

$$\binom{m+j}{j} = \frac{m+j}{m}\binom{m-1+j}{j} = (1+\frac{j}{m})\binom{m-1+j}{j} < 4\binom{m-1+j}{j}. \tag{3.45}$$

This concludes the proof of part (2).

Following an argument similar to that of part (2), one can show that part (3) is equivalent to having that

$$J_m := 4^{-m}P_{m,m-1}(\frac{3}{4})$$

decreases as m increases, which is equivalent to

$$J_{m+1} - J_m < 0 \quad \text{for } m \geq 1. \tag{3.46}$$

Following a direct computation, we have

$$J_{m+1} - J_m = 4^{-m-1}\left(\sum_{j=0}^{m}\binom{m+j}{j}(\frac{3}{4})^j - 4\sum_{j=0}^{m-1}\binom{m-1+j}{j}(\frac{3}{4})^j\right).$$

Let

$$M := \sum_{j=0}^{m}\binom{m+j}{j}(\frac{3}{4})^j - 4\sum_{j=0}^{m-1}\binom{m-1+j}{j}(\frac{3}{4})^j.$$

Then, (3.46) is equivalent to $M < 0$ for $m \geq 1$. It is easy to see that $M < 0$, when $m = 1$. For the case $m \geq 2$, we first note that

$$\begin{aligned} M &= \sum_{j=0}^{m-1} \binom{m+j}{j} (\frac{3}{4})^j - 4 \sum_{j=0}^{m-1} \binom{m-1+j}{j} (\frac{3}{4})^j + \binom{2m}{m} (\frac{3}{4})^m \\ &= \sum_{j=1}^{m-1} \binom{m-1+j}{j-1} (\frac{3}{4})^j - 3 \sum_{j=0}^{m-1} \binom{m-1+j}{j} (\frac{3}{4})^j + \binom{2m}{m} (\frac{3}{4})^m, \end{aligned}$$

where the last identity follows from item 1 of Lemma 3.2. Substituting j for $j-1$ in the first term, one obtains that

$$M = \frac{3}{4} \sum_{j=0}^{m-2} \binom{m+j}{j} (\frac{3}{4})^j - 3 \sum_{j=0}^{m-1} \binom{m-1+j}{j} (\frac{3}{4})^j + \binom{2m}{m} (\frac{3}{4})^m, \tag{3.47}$$

Splitting the second term in (3.47), one obtains

$$\begin{aligned} M = &\frac{3}{4} \sum_{j=0}^{m-2} \binom{m+j}{j} (\frac{3}{4})^j - 3 \sum_{j=0}^{m-2} \binom{m-1+j}{j} (\frac{3}{4})^j \\ &+ \binom{2m}{m} (\frac{3}{4})^m - 3 \binom{2m-2}{m-1} (\frac{3}{4})^{m-1}. \end{aligned} \tag{3.48}$$

For the last two terms of (3.48), we have

$$\begin{aligned} \binom{2m}{m} (\frac{3}{4})^m - 3 \binom{2m-2}{m-1} (\frac{3}{4})^{m-1} &= (\frac{3}{4})^m \left(\binom{2m}{m} - 4 \binom{2m-2}{m-1} \right) \\ &= (\frac{3}{4})^m \left((4 - \frac{2}{m}) \binom{2m-2}{m-1} - 4 \binom{2m-2}{m-1} \right) \\ &< 0. \end{aligned}$$

Therefore,

$$\begin{aligned} M &< \frac{3}{4} \sum_{j=0}^{m-2} \binom{m+j}{j} (\frac{3}{4})^j - 3 \sum_{j=0}^{m-2} \binom{m-1+j}{j} (\frac{3}{4})^j \\ &< \sum_{j=0}^{m-2} \binom{m+j}{j} (\frac{3}{4})^j - 3 \sum_{j=0}^{m-2} \binom{m-1+j}{j} (\frac{3}{4})^j \\ &= \sum_{j=0}^{m-2} \left(\binom{m+j}{j} - 3 \binom{m-1+j}{j} \right) (\frac{3}{4})^j. \end{aligned}$$

Applying (3.45), one obtains, for $0 \leq j \leq m-2$,

$$\binom{m+j}{j} = (1 + \frac{j}{m}) \binom{m-1+j}{j} < 3 \binom{m-1+j}{j}.$$

Therefore, we conclude that $M < 0$ and part (3) follows.

Finally, note that the decay rate of the Fourier transform of the pseudo-spline of Type I with order $(2, 1)$ is $\frac{\beta_{m,l}}{2} \approx 1.33903$. Hence, it follows from parts (1)-(3) that the decay rate of an arbitrary pseudo-spline of either type with order (m, l), $m > 2$, $0 \leq l \leq m-1$ is higher than 1.33903. ∎

Proposition 3.10 reveals that the decay rates of the Fourier transforms of either type of pseudo-splines increase as m increases for fixed l and decrease as l increases for fixed m. Here, we give an asymptotical analysis of the decay rate which, in turn, gives an asymptotical analysis of the regularity of ${}_1\phi$ and ${}_2\phi$ as the order $(m,l)\to\infty$.

THEOREM 3.9 ([**75**]). *Let ${}_1\phi$ and ${}_2\phi$ be the pseudo-splines of Type I and II respectively with order (m,l). Fix $l=\lfloor\lambda m\rfloor$, $0\le\lambda\le 1$, where $\lfloor\lambda m\rfloor$ denotes the largest integer which is smaller than or equal to λm. Then, we have*

$$|{}_1\widehat{\phi}(\xi)|\le C(1+|\xi|)^{-\frac{\mu}{2}m} \quad \text{and} \quad |{}_2\widehat{\phi}(\xi)|\le C(1+|\xi|)^{-\mu m},$$

where $\mu=\frac{\log\left(\frac{4}{1+\lambda}\right)^{\lambda+1}\left(\frac{\lambda}{3}\right)^{\lambda}}{\log 2}$, asymptotically for large m. This means that the asymptotic rates of the pseudo-spline of Type I and Type II are $\frac{\mu}{2}$ and μ respectively.

PROOF. We will only analyze pseudo-splines of type II, and the analysis for pseudo-splines of type I is similar. We first show the following identity:

$$x^{-l}P_{m,l}(x)\ge y^{-l}P_{m,l}(y), \quad \text{for } 0<x\le y\le 1. \tag{3.49}$$

Indeed, item 1 of Lemma 3.3 implies that

$$x^{-l}P_{m,l}(x)=\sum_{j=0}^{l}\binom{m-1+j}{j}x^{j-l}\ge\sum_{j=0}^{l}\binom{m-1+j}{j}y^{j-l}=y^{-l}P_{m,l}(y),$$

for $0<x\le y\le 1$.

In order to compute the asymptotic rate, we need to estimate the upper and lower bound of $P_{m,l}(\frac{3}{4})$ in terms of m and l. For this, we let $x=\frac{3}{4}$ and $y=1$ in (3.49) and obtain

$$P_{m,l}(\frac{3}{4})\ge(\frac{3}{4})^{l}P_{m,l}(1)=(\frac{3}{4})^{l}\binom{m+l}{l}. \tag{3.50}$$

Next, let $x=\frac{1}{2}$ and $y=\frac{3}{4}$ in (3.49), we obtain

$$P_{m,l}(\frac{3}{4})\le(\frac{3}{2})^{l}P_{m,l}(\frac{1}{2}).$$

Since

$$P_{m,l}(\frac{1}{2})=\sum_{j=0}^{l}\binom{m+l}{j}2^{-j}2^{j-l}=2^{-l}\sum_{j=0}^{l}\binom{m+l}{j},$$

one obtains

$$P_{m,l}(\frac{3}{4})\le(\frac{3}{4})^{l}\sum_{j=0}^{l}\binom{m+l}{j}. \tag{3.51}$$

Putting (3.50) and (3.51) together, we obtain the following estimates of $P_{m,l}(\frac{3}{4})$,

$$(\frac{3}{4})^{l}\binom{m+l}{l}\le P_{m,l}(\frac{3}{4})\le(\frac{3}{4})^{l}\sum_{j=0}^{l}\binom{m+l}{j}.$$

For $l\le m-1$, we have

$$\sum_{j=0}^{l}\binom{m+l}{j}\le m\binom{m+l}{l}.$$

Hence,

$$\left(\frac{3}{4}\right)^l \binom{m+l}{l} \le P_{m,l}\left(\frac{3}{4}\right) \le m\left(\frac{3}{4}\right)^l \binom{m+l}{l}. \tag{3.52}$$

Next, we use this estimate to analyze the decay of $|_2\widehat{\phi}|$ as m goes to infinity. The upper bound of $P_{m,l}(\frac{3}{4})$ in (3.52) implies that

$$2m - \frac{\log P_{m,l}(\frac{3}{4})}{\log 2} \ge 2m - \frac{\log\left(m(\frac{3}{4})^l \binom{m+l}{l}\right)}{\log 2}.$$

We estimate the right hand side of the above inequality asymptotically for large (m,l) to obtain the asymptotical lower bound of $2m - \frac{\log P_{m,l}(\frac{3}{4})}{\log 2}$. For this, we first recall the Stirling approximation, i.e. $m! \sim \sqrt{2\pi} e^{(m+\frac{1}{2})\log m - m}$ (see e.g. [**88**]), where $a_m \sim b_m$ means that $\frac{a_m}{b_m} \to 1$, $m \to \infty$. By the Stirling approximation, we have

$$\begin{aligned} \log m! &\sim \log \sqrt{2\pi} e^{(m+\frac{1}{2})\log m - m} \\ &\sim m \log m - m. \end{aligned} \tag{3.53}$$

Applying (3.53), one obtains

$$\begin{aligned} \log \binom{m+l}{l} &= \log(m+l)! - \log m! - \log l! \\ &\sim (m+l)\log(m+l) - (m+l) - (m\log m - m) - (l\log l - l) \\ &\sim (m+l)\log(m+l) - m\log m - l\log l. \end{aligned}$$

Thus,

$$\begin{aligned} 2m - \frac{\log\left(m\frac{3}{4}^l \binom{m+l}{l}\right)}{\log 2} &= 2m - \frac{\log m + l\log\frac{3}{4} + \log\binom{m+l}{l}}{\log 2} \\ &\sim m\left(2 - \frac{\frac{l}{m}\log\frac{3}{4} + (1+\frac{l}{m})\log(m+l) - \log m - \frac{l}{m}\log l}{\log 2}\right). \end{aligned}$$

By the assumption, $l = \lfloor \lambda m \rfloor$, $0 \le \lambda \le 1$. Hence, when m is sufficiently large, $\frac{l}{m} \sim \lambda$ and therefore,

$$\begin{aligned} 2m - \frac{\log\left(m\frac{3}{4}^l \binom{m+l}{l}\right)}{\log 2} &\sim m\left(2 - \frac{\log(1+\lambda)\left(\frac{3+3\lambda}{4\lambda}\right)^\lambda}{\log 2}\right) \\ &= m\left(\frac{\log\left(\frac{4}{1+\lambda}\right)^{\lambda+1}\left(\frac{\lambda}{3}\right)^\lambda}{\log 2}\right). \end{aligned}$$

Now we obtain the asymptotical lower bound of $2m - \frac{\log P_{m,l}(\frac{3}{4})}{\log 2}$, i.e. asymptotically, for large m with $l = \lfloor \lambda m \rfloor$,

$$2m - \frac{\log|P_{m,l}(\frac{3}{4})|}{\log 2} \ge m\left(\frac{\log\left(\frac{4}{1+\lambda}\right)^{\lambda+1}\left(\frac{\lambda}{3}\right)^\lambda}{\log 2}\right). \tag{3.54}$$

Next, we use the left hand side of (3.52) to obtain the asymptotical upper bound of $2m - \frac{\log P_{m,l}(\frac{3}{4})}{\log 2}$. First note that (3.52) gives

$$2m - \frac{\log P_{m,l}(\frac{3}{4})}{\log 2} \le 2m - \frac{l\log\frac{3}{4} + \log\binom{m+l}{l}}{\log 2}.$$

Applying arguments similar to the estimate of the lower bound by using (3.53), we obtain the following

$$
\begin{aligned}
2m-&\frac{l\log\frac{3}{4}+\log\binom{m+l}{l}}{\log 2}\\
&\sim m\left(2-\frac{\frac{l}{m}\log\frac{3}{4}+(1+\frac{l}{m})\log(m+l)-\log m-\frac{l}{m}\log l}{\log 2}\right)\\
&\sim m\left(\frac{\log\left(\frac{4}{1+\lambda}\right)^{\lambda+1}\left(\frac{\lambda}{3}\right)^{\lambda}}{\log 2}\right).
\end{aligned}
$$

This leads to the asymptotical lower bound of $2m-\frac{\log P_{m,l}\left(\frac{3}{4}\right)}{\log 2}$, i.e. asymptotically, for large m with $l=\lfloor\lambda m\rfloor$,

$$
2m-\frac{\log P_{m,l}\left(\frac{3}{4}\right)}{\log 2}\leq m\left(\frac{\log\left(\frac{4}{1+\lambda}\right)^{\lambda+1}\left(\frac{\lambda}{3}\right)^{\lambda}}{\log 2}\right). \tag{3.55}
$$

Combining (3.54) and (3.55), we conclude that for large m, the asymptotical upper and lower bounds coincide and equal to

$$
2m-\frac{\log P_{m,l}\left(\frac{3}{4}\right)}{\log 2}\sim m\left(\frac{\log\left(\frac{4}{1+\lambda}\right)^{\lambda+1}\left(\frac{\lambda}{3}\right)^{\lambda}}{\log 2}\right)=\mu m. \tag{3.56}
$$

Therefore the equation (3.56) gives that, fixing $l=\lfloor\lambda m\rfloor$ and asymptotically, for large m, we have

$$
|{}_2\widehat{\phi}(\xi)|\leq C(1+|\xi|)^{-\mu m}\quad\text{and}\quad|{}_1\widehat{\phi}(\xi)|\leq C(1+|\xi|)^{-\frac{\mu}{2}m},
$$

where $\mu=\frac{\log\left(\frac{4}{1+\lambda}\right)^{\lambda+1}\left(\frac{\lambda}{3}\right)^{\lambda}}{\log 2}$. ∎

REMARK 3.4. The above theorem shows that, asymptotically for large m, the smoothness of the pseudo-splines of type I and II increases at a rate $\mu/2$ and μ respectively. The proof of Theorem 3.9 also leads to the following two observations:

(1) Consider pseudo-splines of Type II with order $(m,m-p)$, where p is a fixed positive integer independent of m. The asymptotic rate is $2-\frac{\log 3}{\log 2}\approx 0.4150$. Indeed, when $l=m-p$, $\lambda\sim\frac{l}{m}=\frac{m-p}{m}\sim 1$ for sufficiently large m. Similarly, for pseudo-splines of Type I with order $(m,m-p)$, the corresponding asymptotic rate is $1-\frac{\log 3}{2\log 2}\approx 0.2075$.

(2) Assume that l is fixed for all m. The asymptotic rates of pseudo-splines of Type I and II with order (m,l) are 1 and 2 respectively. This is simply because, for the fixed integer l, $\lambda\sim\frac{l}{m}\sim 0$ for sufficiently large m.

TABLE 2. Asymptotically for large m, the smoothness of ${}_2\phi$ increases at rate μ, which is given in the following table with some choices of l.

$m\to\infty$	$l=0$	$l=\frac{m}{10}$	$l=\frac{m}{8}$	$l=\frac{m}{6}$	$l=\frac{m}{4}$	$l=\frac{m}{2}$	$l=m-1$
$\mu\approx$	2.0000	1.5581	1.4857	1.3789	1.2013	0.8301	0.4150

EXAMPLE 3.2. In Table 2, we give μ, the asymptotical rate of pseudo-splines of Type II with order $(m, \lfloor \lambda m \rfloor)$, as m goes to infinity and the parameter $\lambda = \frac{1}{10}, \frac{1}{8}, \frac{1}{6}, \frac{1}{4}, \frac{1}{2}, 1$. The asymptotic rate μ_0 for pseudo-splines of Type I with the same order is just $\mu_0 = \frac{\mu}{2}$.

A similar discussion can lead the regularity analysis for the dual pseudo-splines as it was done in [**72**].

THEOREM 3.10 ([**72**]). *Let $\tilde{\phi}_{m,l}$ be the dual pseudo-spline of order (m,l) with $0 \le l \le m-1$. Then*

$$|\widehat{\tilde{\phi}}_{m,l}| \le C(1+|\xi|)^{-\gamma_{m,l}}$$

with

$$\gamma_{m,l} = 2m + 1 - \log\big(P_{m-\frac{1}{2},l}(\frac{3}{4})\big)/\log 2,$$

and the decay rate $\gamma_{m,l}$ is optimal. Consequently, $\tilde{\phi}_{m,l} \in C^{\alpha}$ with $\alpha = \gamma_{m,l} - 1 - \varepsilon$.

4. Two lemmata

This section gives the two key technical lemmata that were used in the previous sections.

LEMMA 3.2. *For given nonnegative integers m, j, l, we have:*

(1) $\binom{m+1}{j} = \binom{m}{j} + \binom{m}{j-1}$ *for $j \ge 1$ and* $(j+1)\binom{m+j}{j+1} = (m+j)\binom{m-1+j}{j}$.

(2) $2(m+1)\sum_{j=0}^{l-1}\binom{m+l}{j} - l\sum_{j=0}^{l}\binom{m+l}{j} \ge 0$, *for $m \ge 1$ and $1 \le l \le m-1$.*

(3) $\dfrac{2^l\binom{m+l}{l}^{\frac{1}{2}}}{\sum_{j=0}^{l}\binom{m+l}{j}} \le 1$, *for all $m \ge 1$ and $0 \le l \le m-1$.*

PROOF. The identities in (1) are well known and can be proven directly by the definition of the binomial coefficients.

For (2), since $m > l$, we have

$$(m+1)\sum_{j=0}^{l-1}\binom{m+l}{j} - l\sum_{j=0}^{l-1}\binom{m+l}{j} \ge 0.$$

Subtracting this inequality from (2), we conclude that it remains to check if

$$(m+1)\sum_{j=0}^{l-1}\binom{m+l}{j} - l\binom{m+l}{l} \ge 0$$

holds, in order to verify (2). Since $(m+1)\binom{m+l}{l-1} = l\binom{m+l}{l}$, we have

$$(m+1)\sum_{j=0}^{l-1}\binom{m+l}{j} > (m+1)\binom{m+l}{l-1} = l\binom{m+l}{l}.$$

This gives (2) immediately.

Finally, we prove (3) by induction with respect to m. Since (3) is obviously true for $l = 0$, we now focus on $1 \le l \le m-1$. When $m = 1$, the inequality trivially holds. Assume (3) holds when $m = m_0$, i.e.

$$2^{2l}\binom{m_0+l}{l} \le \Big(\sum_{j=0}^{l}\binom{m_0+l}{j}\Big)^2,$$

for all $1 \le l \le m_0 - 1$. Consider the case $m = m_0 + 1$. We first show that (3) holds for all l, where $1 \le l \le m_0 - 1$. For $1 \le l \le m_0 - 1$, we have

$$\begin{aligned}
2^{2l}\binom{m_0+l+1}{l} &= \frac{m_0+l+1}{m_0+1} 2^{2l}\binom{m_0+l}{l} \\
&\le \frac{m_0+l+1}{m_0+1}\left(\sum_{j=0}^{l}\binom{m_0+l}{j}\right)^2 \quad \text{(by induction hypothesis)} \\
&= \left(\sum_{j=0}^{l}\binom{m_0+l}{j} + \left(\sqrt{\frac{m_0+l+1}{m_0+1}} - 1\right)\sum_{j=0}^{l}\binom{m_0+l}{j}\right)^2 \\
&= \left(\sum_{j=0}^{l}\binom{m_0+l}{j}\right. \\
&\qquad \left. + \frac{l}{m_0+1+\sqrt{(m_0+l+1)(m_0+1)}}\sum_{j=0}^{l}\binom{m_0+l}{j}\right)^2 \\
&< \left(\sum_{j=0}^{l}\binom{m_0+l}{j} + \frac{l}{2m_0+2}\sum_{j=0}^{l}\binom{m_0+l}{j}\right)^2 \\
&\le \left(\sum_{j=0}^{l}\binom{m_0+l}{j} + \sum_{j=0}^{l-1}\binom{m_0+l}{j}\right)^2 \quad \text{(from (2))} \\
&= \left(1 + \sum_{j=1}^{l}\binom{m_0+l}{j} + \sum_{j=1}^{l}\binom{m_0+l}{j-1}\right)^2 \\
&= \left(\sum_{j=0}^{l}\binom{m_0+l+1}{j}\right)^2 \quad \text{(from (1))}.
\end{aligned}$$

This shows that (3) holds for all $1 \le l \le m_0 - 1$, it remains to show (3) holds for $l = m_0$, i.e. to show

$$2^{2m_0}\binom{2m_0+1}{m_0} \le \left(\sum_{j=0}^{m_0}\binom{2m_0+1}{j}\right)^2. \tag{3.57}$$

Applying $\binom{n}{j} = \binom{n}{n-j}$, we have

$$\begin{aligned}
\sum_{j=0}^{m_0}\binom{2m_0+1}{j} &= \frac{1}{2}\left(\sum_{j=0}^{m_0}\binom{2m_0+1}{j} + \sum_{j=0}^{m_0}\binom{2m_0+1}{j}\right) \\
&= \frac{1}{2}\left(\sum_{j=0}^{m_0}\binom{2m_0+1}{j} + \sum_{j=m_0+1}^{2m_0+1}\binom{2m_0+1}{j}\right) \\
&= \frac{1}{2}\sum_{j=0}^{2m_0+1}\binom{2m_0+1}{j} = 2^{2m_0}.
\end{aligned}$$

Then (3.57) is equivalent to $\binom{2m_0+1}{m_0} \le \sum_{j=0}^{m_0}\binom{2m_0+1}{j}$, which is obviously true. This concludes the proof of (3). ■

Next, we give several basic properties of the polynomials $P_{m,l}(y)$ and $R_{m,l}(y)$ defined by (3.4) and (3.5). Part (2)-(4) of the following lemma were mainly used in the proof of Theorem 3.2.

LEMMA 3.3. *For nonnegative integers m and l with $l \le m-1$, let $P_{m,l}(y)$ and $R_{m,l}(y)$ be the polynomials defined in (3.4) and (3.5). Then:*

(1) $P_{m,l}(y) = \sum_{j=0}^{l} \binom{m-1+j}{j} y^j$.

(2) $R'_{m,l}(y) = -(m+l)\binom{m+l-1}{l} y^l (1-y)^{m-1}$.

(3) *Define* $Q(y) := R_{m,l}(y) + R_{m,l}(1-y)$. *Then,*

$$\min_{y\in[0,1]} Q(y) = Q(\frac{1}{2}) = 2^{1-m-l} \sum_{j=0}^{l} \binom{m+l}{j}.$$

(4) *Define* $S(y) := R^2_{m,l}(y) + R^2_{m,l}(1-y)$. *Then,*

$$\min_{y\in[0,1]} S(y) = S(\frac{1}{2}) = 2^{1-2m-2l} (\sum_{j=0}^{l} \binom{m+l}{j})^2.$$

PROOF. For fixed m, we prove (1) by induction with respect to l. It is obviously true for $l = 0$. Now suppose (1) holds for l_0. Consider $l = l_0 + 1$,

$$\begin{aligned} P_{m,l}(y) &= \sum_{j=0}^{l_0+1} \binom{m+l_0+1}{j} y^j (1-y)^{l_0-j+1} \\ &= (1-y)^{l_0+1} + \sum_{j=1}^{l_0+1} \binom{m+l_0+1}{j} y^j (1-y)^{l_0-j+1}. \end{aligned}$$

Applying the first identity in (1) of Lemma 3.2, we have,

$$\begin{aligned} P_{m,l}(y) &= (1-y)^{l_0+1} + \sum_{j=1}^{l_0+1} \binom{m+l_0}{j} y^j (1-y)^{l_0-j+1} \\ &\qquad + \sum_{j=1}^{l_0+1} \binom{m+l_0}{j-1} y^j (1-y)^{l_0-j+1} \\ &= \sum_{j=0}^{l_0+1} \binom{m+l_0}{j} y^j (1-y)^{l_0-j+1} + \sum_{j=1}^{l_0+1} \binom{m+l_0}{j-1} y^j (1-y)^{l_0-j+1} \\ &= \sum_{j=0}^{l_0} \binom{m+l_0}{j} y^j (1-y)^{l_0-j+1} + \binom{m+l_0}{l_0+1} y^{l_0+1} \\ &\quad + \sum_{j=0}^{l_0} \binom{m+l_0}{j} y^{j+1} (1-y)^{l_0-j} \\ &= (1-y) P_{m,l_0}(y) + \binom{m+l_0}{l_0+1} y^{l_0+1} + y P_{m,l_0}(y) \\ &= P_{m,l_0}(y) + \binom{m+l_0}{l_0+1} y^{l_0+1}. \end{aligned}$$

Then, by the inductive hypothesis,

$$\begin{aligned} P_{m,l}(y) &= \sum_{j=0}^{l_0} \binom{m-1+j}{j} y^j + \binom{m+l_0}{l_0+1} y^{l_0+1} \\ &= \sum_{j=0}^{l_0+1} \binom{m-1+j}{j} y^j. \end{aligned}$$

We prove (2) by induction with respect to l for given m. It is obviously true when $l=0$. Suppose (2) holds for l_0, i.e. $R'_{m,l_0}(y) = -(m+l_0)\binom{m+l_0-1}{l_0} y^{l_0}(1-y)^{m-1}$, and consider the case $l = l_0+1 \leq m-1$. Using (1) and definition of $R_{m,l}(y)$ in (3.5), we have

$$\begin{aligned} R_{m,l_0+1}(y) &= (1-y)^m P_{m,l_0+1}(y) \\ &= (1-y)^m \Big(P_{m,l_0}(y) + \binom{m+l_0}{l_0+1} y^{l_0+1} \Big). \end{aligned}$$

Since $R_{m,l_0}(y) = (1-y)^m P_{m,l_0}(y)$, we have

$$R_{m,l_0+1}(y) = \binom{m+l_0}{l_0+1} y^{l_0+1}(1-y)^m + R_{m,l_0}(y).$$

Then,

$$\begin{aligned} R'_{m,l_0+1}(y) &= (l_0+1)\binom{m+l_0}{l_0+1} y^{l_0}(1-y)^m - m\binom{m+l_0}{l_0+1} y^{l_0+1}(1-y)^{m-1} \\ &\quad + R'_{m,l_0}(y) \\ &= (l_0+1)\binom{m+l_0}{l_0+1} y^{l_0}(1-y)^m - m\binom{m+l_0}{l_0+1} y^{l_0+1}(1-y)^{m-1} \\ &\quad -(m+l_0)\binom{m+l_0-1}{l_0} y^{l_0}(1-y)^{m-1}. \end{aligned}$$

Pulling the common factor $y^{l_0}(1-y)^{m-1}$ out, one obtains

$$\begin{aligned} R'_{m,l_0+1}(y) &= y^{l_0}(1-y)^{m-1}\Big((l_0+1)\binom{m+l_0}{l_0+1}(1-y) - m\binom{m+l_0}{l_0+1} y \\ &\quad -(m+l_0)\binom{m+l_0-1}{l_0} \Big) \\ &= y^{l_0}(1-y)^{m-1}\Big((l_0+1)\binom{m+l_0}{l_0+1} - (l_0+1)\binom{m+l_0}{l_0+1} y \\ &\quad -m\binom{m+l_0}{l_0+1} y - (m+l_0)\binom{m+l_0-1}{l_0} \Big). \end{aligned}$$

Combining the second and the third term, one obtains

$$\begin{aligned} R'_{m,l_0+1}(y) &= y^{l_0}(1-y)^{m-1}\Big((l_0+1)\binom{m+l_0}{l_0+1} - (m+l_0+1)\binom{m+l_0}{l_0+1} y \\ &\quad -(m+l_0)\binom{m+l_0-1}{l_0} \Big). \end{aligned}$$

By the second identity in (1) of Lemma 3.2, one obtains $(l_0+1)\binom{m+l_0}{l_0+1} = (m+l_0)\binom{m+l_0-1}{l_0}$. Hence, $R'_{m,l_0+1}(y) = -(m+l_0+1)\binom{m+l_0}{l_0+1}y^{l_0+1}(1-y)^{m-1}$. This concludes the proof of (2).

For (3), we compute $Q'(y)$, i.e.

$$Q'(y) = R'_{m,l}(y) + (R_{m,l}(1-y))' = R'_{m,l}(y) - R'_{m,l}(1-y).$$

Applying (2), one obtains

$$Q'(y) = (m+l)\binom{m+l-1}{l}\Big(y^{m-1}(1-y)^l - (1-y)^{m-1}y^l\Big).$$

Now, we show that $Q'(y) \le 0$ on $[0, \frac{1}{2}]$, $Q'(y) \ge 0$ on $[\frac{1}{2}, 1]$. Note that

$$y^{m-l-1} \le (1-y)^{m-l-1}, \qquad \text{for all } y \in [0, \frac{1}{2}].$$

Multiplying both sides by $y^l(1-y)^l$,

$$y^{m-1}(1-y)^l \le (1-y)^{m-1}y^l, \qquad \text{for all } y \in [0, \frac{1}{2}].$$

Similarly we have

$$y^{m-1}(1-y)^l \ge (1-y)^{m-1}y^l, \qquad \text{for all } y \in [\frac{1}{2}, 1].$$

We conclude that

$$Q'(y)\begin{cases} \le 0, & y \in [0, \frac{1}{2}] \\ \ge 0, & y \in [\frac{1}{2}, 1]. \end{cases}$$

This means that $Q(y)$ reaches its minimum value at point $y = \frac{1}{2}$. Now, we compute $Q(\frac{1}{2})$. Note that $Q(\frac{1}{2}) = 2R_{m,l}(\frac{1}{2}) = 2^{1-m}P_{m,l}(\frac{1}{2})$. Recall that $P_{m,l}(y)$ is defined in (3.4), i.e. $P_{m,l}(y) = \sum_{j=0}^{l}\binom{m+l}{j}y^j(1-y)^{l-j}$. Then $Q(\frac{1}{2}) = 2^{1-m}2^{-l}\sum_{j=0}^{l}\binom{m+l}{j} = 2^{1-m-l}\sum_{j=0}^{l}\binom{m+l}{j}$.

With (3), the proof of (4) is simpler. Since

$$S'(y) = 2R_{m,l}(y)R'_{m,l}(y) + 2R_{m,l}(1-y)\big(R_{m,l}(1-y)\big)',$$

using the identities $R_{m,l}(y) = (1-y)^m P_{m,l}(y)$, $R'_{m,l}(y) = -(m+l)\binom{m+l-1}{l}y^l(1-y)^{m-1}$ and $\big(R_{m,l}(1-y)\big)' = (m+l)\binom{m+l-1}{l}y^{m-1}(1-y)^l$, we obtain

$$\begin{aligned}\frac{S'(y)}{2(m+l)\binom{m+l-1}{l}} &= y^m P_{m,l}(1-y)y^{m-1}(1-y)^l - (1-y)^m P_{m,l}(y)y^l(1-y)^{m-1} \\ &= y^{2m-1}\sum_{j=0}^{l}\binom{m-1+j}{j}(1-y)^{l+j} \\ &\qquad - (1-y)^{2m-1}\sum_{j=0}^{l}\binom{m-1+j}{j}y^{l+j} \\ &= \sum_{j=0}^{l}\binom{m-1+j}{j}\Big((1-y)^{l+j}y^{2m-1} - y^{l+j}(1-y)^{2m-1}\Big).\end{aligned}$$

Since, for each $0 \leq j \leq l$, when $y \in [0, \frac{1}{2}]$, $y^{2m-l-j-1} \leq (1-y)^{2m-l-j-1}$, and when $y \in [\frac{1}{2}, 1]$, $y^{2m-l-j-1} \geq (1-y)^{2m-l-j-1}$, then by similar arguments in (2) we conclude,

$$S'(y)\begin{cases} \leq 0, & y \in [0, \frac{1}{2}], \\ \geq 0, & y \in [\frac{1}{2}, 1]. \end{cases}$$

Thus $\min_{y\in[0,1]} S(y) = S(\frac{1}{2})$. Since $R_{m,l}(\frac{1}{2}) = 2^{-m-l}\sum_{j=0}^{l}\binom{m+l}{j}$, we have

$$S(\frac{1}{2}) = 2R_{m,l}^2(\frac{1}{2}) = 2^{1-2m-2l}\Big(\sum_{j=0}^{l}\binom{m+l}{j}\Big)^2. \qquad \blacksquare$$

REMARK 3.5. From (1) of Lemma 3.3 we know that the refinement mask of the pseudo-spline of Type I in (3.2) can be written as

$$|{}_1\widehat{a}(\xi)|^2 = \cos^{2m}(\xi/2)\sum_{j=0}^{l}\binom{m-1+j}{j}\sin^{2j}(\xi/2).$$

Hence, the pseudo-spline of Type I with order $(m, m-1)$ is indeed the refinable function whose shifts form an orthonormal system constructed in [**60**] and the pseudo-spline of Type II with order $(m, m-1)$ is indeed the autocorrelation of the orthogonal refinable function, which is interpolatory.

LECTURE 4

Frame based image restorations

Image restoration is often formulated as an inverse problem. For simplicity of notation, we denote images as vectors in $\mathbb{R}^n$ with n equal to the total number of pixels. The objective is to find the unknown true image $u \in \mathbb{R}^n$ from an observed image (or measurements) $f \in \mathbb{R}^l$ defined by

$$f = Au + \eta, \tag{4.1}$$

where η is a white Gaussian noise with variance σ^2, and $A \in \mathbb{R}^{l\times n}$ is a linear operator, typically a convolution operator for image deconvolution problems, a projection operator for image inpainting and the identity for image denoising. This lecture is devoted to frame based image restorations.

As we discussed in Lecture 2, fast framelet decomposition and reconstruction algorithms are linear operators. When we consider signals living in $\mathbb{R}^n$, these linear operators have matrix representations. Here we simply denote the framelet decomposition as a matrix $W \in \mathbb{R}^{m\times n}$ with $m \geq n$, and reconstruction as $W^\top$. Then based on our analysis in Lecture 2, we have $W^T W = I$, where I is the identity matrix. Thus, for every vector $u \in \mathbb{R}^n$,

$$u = W^\top(Wu). \tag{4.2}$$

The components of the vector Wu are called the canonical coefficients representing u. The matrix W is generated from the masks $\{h_\ell : \ell = 0, 1, \dots r\}$ constructed from the univariate UEP (Theorem 2.6). We emphasize that we view the image as a column vector and the corresponding tight wavelet frame transform as the matrix W action just for simplicity of notation. In computations of image restorations, we use 2D wavelet decomposition and reconstruction algorithms given by Algorithm 2.3 instead of matrix multiplications.

Since tight wavelet frame systems are redundant systems (i.e. $m \geq n$), the mapping from the image u to its coefficients is not one-to-one, i.e., the representation of u in the frame domain is not unique. Therefore, there are three formulations for the sparse approximation of the underlying images; namely, the analysis based approach, the synthesis based approach and the balanced approach. The analysis based approach was first proposed in [**84, 170**]. In this approach, we assume that the coefficient vector Wu can be sparsely approximated, and is usually formulated as a minimization problem involving a penalty on the term $\|Wu\|_1$. The synthesis based approach was first introduced in [**66, 86, 87, 90, 91**]. In that approach, the underlying image u is assumed to be synthesized from a sparse coefficient vector α with $u = W^\top\alpha$, and it is usually formulated as a minimization problem involving a penalty on the term $\|\alpha\|_1$. The balanced approach was first used in [**34, 36**] for high resolution image reconstruction. It was further developed for various image restoration problems in [**14, 15, 16, 17, 25, 32, 38**]. In this approach, the underlying image u is assumed to be synthesized from some sparse coefficient vector α via $u = W^\top \alpha$ which is penalized by $\|\alpha\|_1$, and the distance of α to the range of W is minimized at the same time. Although the synthesis based, analysis based and balanced approaches are developed independently in the literature, the balanced approach can be viewed as a way to balance the analysis and synthesis based approach.

First, Section 1 gives a general introduction. Then Section 2 illustrates the ideas of the development of the balanced approach, as well as details of a fast algorithm, under the scenario of image inpainting [**16, 168**]. Then an accelerated algorithm of [**168**] for both balanced and synthesis based approach for general image restorations (4.1) is discussed. Within this section we discuss the linearized Bregman algorithm and its connection to the Uzawa's algorithm as an alternative to either the balanced or the synthesis based approaches. In Section 3, we present the analysis based approach [**24**], and its applications in image deblurring.

1. Modeling

This section establishes the models for frame based image restorations, especially, image inpainting, denoising, and deblurring.

1.1. Preliminaries

We start with some basic notations. For any $x \in \mathbb{R}^n$,

$$\|x\|_p := \left(\sum_{j=1}^n |x_j|^p\right)^{1/p},$$

$1 \le p < \infty$. We denote the inner product $\langle \cdot, \cdot \rangle : \mathbb{R}^n \times \mathbb{R}^n \mapsto \mathbb{R}$ as

$$\langle x, y \rangle = x^\top y = \sum_{j=1}^n x_j y_j.$$

Then $\|x\|_2^2 = \langle x, x \rangle$. Let $\|x\|_D$ denote the D-norm, where D is a symmetric positive definite matrix, defined by $\|x\|_D = \sqrt{x^\top D x}$. For any real symmetric matrix H, $\rho_{\max}(H)$ denotes the maximum eigenvalue of H in magnitude. For any $m \times n$ real matrices A, $\|A\|_2 = \sqrt{\rho_{\max}(A^\top A)}$.

Now we recall some concepts and facts from general convex analysis, which are used throughout this lecture. A function $F : \mathbb{R}^n \mapsto \bar{\mathbb{R}}$ is convex if the domain of F is convex and for all u in the domain of F,

$$F(tu + (1-t)v) \le tF(u) + (1-t)F(v), \quad \text{for all } t \in [0, 1],$$

and it is said to be strictly convex if

$$F(tu + (1-t)v) < tF(u) + (1-t)F(v), \quad \text{for all } t \in (0, 1) \text{ and } u \ne v.$$

Here $\bar{\mathbb{R}} := \mathbb{R} \cup \{\pm\infty\}$.

A function F is said to be subdifferentiable at u if there exists $u^* \in \mathbb{R}^n$ such that

$$F(v) \ge F(u) + \langle v - u, u^* \rangle, \quad \text{for all } v \in \mathbb{R}^n, \tag{4.3}$$

and u^* is called a subgradient of F at u. The set of subgradients of F at u is denoted as $\partial F(u)$, which, in general, is not a singleton. By definition of subgradients, it is easy to see that for $u^* \in \partial F(u)$ and $v^* \in \partial F(v)$ with $F(u)$ and $F(v)$ being finite, we have

$$\langle u^* - v^*, u - v\rangle \geq 0. \tag{4.4}$$

The same as derivatives, we have the following chain rule for subgradients, where the equality is the set equality,

$$\partial(F(Tu)) = T^\top(\partial F)(Tu),$$

where T is some linear operator on $\mathbb{R}^n$, whenever both sides of equality are well defined. If F is convex and differentiable at u, then we have $\partial F(u) = \{\nabla F(u)\}$, i.e. $\partial F(u)$ contains $\nabla F(u)$ as its unique element. If F is 2nd order differentiable, then F is convex if and only if the Hessian of F, $\nabla^2 F$ is semi-positive definite; and F is strictly convex if and only if $\nabla^2 F$ is positive definite.

It is well known (see e.g. [**82**]) that if a convex function F is bounded in a neighborhood of some point $v \in \{u \in \mathbb{R}^n : F(u) < \infty\}$, then it is continuous and subdifferentiable everywhere in the interior of $\{u \in \mathbb{R}^n : F(u) < \infty\}$. In particular if the convex function F nowhere assumes the values $\pm\infty$, then it is continuous and subdifferentiable everywhere in $\mathbb{R}^n$ (this statement is not true if F is defined on a Banach space, instead of a finite dimensional space).

The general optimality condition for a convex subdifferentiable functional $F(u)$ can be characterized by subgradients which is known as the Fermat's rule [**154**, Theorem 10.1]. The statement is as follows:

$$u^\star = \arg\min_u F(u) \quad \text{if and only if } 0 \in \partial F(u^\star). \tag{4.5}$$

In other words, $u^\star$ minimizes $F(u)$ if and only if 0 is a subgradient of F at $u^\star$.

A convex function $F : \mathbb{R}^n \mapsto \bar{\mathbb{R}}$ is said to be proper, if it nowhere takes the value $-\infty$ and is not identically equal to $+\infty$. We call F lower semi-continuous at v if

$$\liminf_{u \to v} F(u) \geq F(v).$$

Note that if a convex function F nowhere assumes $\pm\infty$, then it is proper and lower semi-continuous (in fact it is continuous based on the above discussions). The reason to introduce the concept of proper and lower semi-continuity is to include functions F that actually assumes value ∞. For example, given a set $\Omega \in \mathbb{R}^n$, we can define the corresponding indicator function $\mathcal{C}_\Omega$ as

$$\mathcal{C}_\Omega(u) = \begin{cases} 0, & u \in \Omega, \\ \infty, & u \in \Omega^c. \end{cases}$$

The function $\mathcal{C}_\Omega$ is proper, but it is lower semi-continuous if and only if Ω is closed. If the set Ω is convex, then $\mathcal{C}_\Omega$ is a convex function. Note that we can use indicator functions to move convex constraints for an optimization problem into the corresponding objective function. For example, minimizing $F(u)$ subject to $u \in \Omega$ is equivalent to minimizing $F(u)+\mathcal{C}_\Omega(u)$, which is still a convex optimization problem whenever Ω is convex.

The function F is said to be coercive, if

$$F(u) \to \infty \quad \text{when } \|u\|_2 \to \infty.$$

Then the general existence theorem of convex optimization problems states as follows (see e.g. [**82**]): the problem

$$\min_{u\in\mathbb{R}^n} F(u)$$

has at least one solution if F is proper, convex, lower semi-continuous and coercive; if, in addition, F is strictly convex, then the solution is unique.

A function F is said to have an L-Lipschitz continuous gradient if

$$\|\nabla F(u) - \nabla F(v)\| \leq L\|u - v\|, \quad \forall\ u,\ v \tag{4.6}$$

for some $L > 0$. We call L the Lipschitz constant of F.

Finally, we recall the following result on convergence of minimizing sequences, which is a special case of one of the results of [**130**]. The proof of this result is taken from [**24**].

PROPOSITION 4.1. *Let $F(u)$ be a convex function defined on $\mathbb{R}^n$ and nowhere assumes the values $\pm\infty$. Suppose F has a unique minimizer $u^\star \in \mathbb{R}^n$. Then any minimizing sequence $\{u_k\}_{k=0}^\infty$, i.e. $F(u_k) \to F(u^\star)$ as $k \to \infty$, converges to $u^\star$ in any Euclidean norm of $\mathbb{R}^n$.*

PROOF. Since all norms of $\mathbb{R}^n$ are equivalent, we shall focus on the ℓ_2-norm and prove that

$$\lim_{k\to\infty} \|u_k - u^\star\|_2 = 0. \tag{4.7}$$

Since $u^\star \in \mathbb{R}^n$ is the unique minimizer of $F(u)$, we have $F(u) > F(u^\star)$ for all $u \neq u^\star$. Now, we suppose that (4.7) does not hold, so there exists a subsequence u_{k_i} such that $\|u_{k_i} - u^\star\|_2 > \epsilon$ for some $\epsilon > 0$ and for all i. Then, $F(u_{k_i}) > \min\{F(u) : \|u - u^\star\|_2 = \epsilon\}$. Indeed, let v be the intersection of the sphere $\{u : \|u - u^\star\|_2 = \epsilon\}$ and the line segment from $u^\star$ to u_{k_i}; then there exists a positive number $t \in (0, 1)$ such that $v = tu^\star + (1 - t)u_{k_i}$. By the convexity of F and the definition of $u^\star$, we have

$$\begin{aligned} F(u_{k_i}) &> tF(u^\star) + (1-t)F(u_{k_i}) \geq F(tu^\star + (1-t)u_{k_i}) = F(v) \\ &\geq \min\{F(u) : \|u - u^\star\|_2 = \epsilon\}. \end{aligned}$$

Denote $\tilde{u} = \arg\min\{F(u) : \|u - u^\star\|_2 = \epsilon\}$, which exists because spheres in $\mathbb{R}^n$ are compact and F is continuous. By noting that $\tilde{u} \neq u^\star$, we have

$$F(u^\star) = \lim_{i\to+\infty} F(u_{k_i}) \geq F(\tilde{u}) > F(u^\star),$$

which is a contradiction. ∎

REMARK 4.1. The conclusion of Proposition 4.1 is no longer true if F is a functional defined on a general normed space. The reason is that $\inf_u\{F(u) : \|u - u^\star\| = \epsilon\}$ may not be attainable for any $u \in \{\|u - u^\star\| = \epsilon\}$, and $\inf_u\{F(u) : \|u - u^\star\| = \epsilon\}$ could be equal to $\inf_u F(u)$ (see [**130**] for a counterexample, as well as general theories on the convergence of minimizing sequences).

1.2. Image restoration models

The balanced approach, synthesis based approach and analysis based approach can be integrated into one minimization problem as follows:

$$\min_{\alpha\in\mathbb{R}^m} \frac{1}{2}\|AW^\top\alpha - f\|_D^2 + \frac{\kappa}{2}\|(I - WW^\top)\alpha\|_2^2 + \|\mathrm{diag}(\lambda)\alpha\|_1, \tag{4.8}$$

where $0 \leq \kappa \leq \infty$ and λ is a given vector defined as

$$\lambda = (\lambda_1, \lambda_2, \cdots, \lambda_m)^\top, \tag{4.9}$$

with $\lambda_i > 0$ for $i = 1, \cdots, m$. We denote $\lambda_{\min} := \min_i \lambda_i > 0$. The matrix D is some symmetric positive definite matrix. It is usually chosen as a preconditioner to the equation $AW^\top\alpha = f$. Throughout this lecture, we take $D = (AA^\top + \theta I)^{-1}$ for some properly chosen $\theta > 0$. We refer to the model (4.8) as the *(single-system) balanced approach.*

When $0 < \kappa < \infty$, the problem (4.8) is called the balanced approach. When $\kappa = 0$, the problem (4.8) is reduced to the *synthesis based approach*:

$$\min_{\alpha \in \mathbb{R}^m} \ \frac{1}{2}\|AW^\top\alpha - f\|_D^2 + \|\mathrm{diag}(\lambda)\alpha\|_1. \tag{4.10}$$

When $\kappa = \infty$, the problem (4.8) is reduced to the analysis based approach. To see this, we note that the distance $\|(I - WW^\top)\alpha\|$ must be 0 when $\kappa = \infty$. This implies that α is in the range of W, i.e., $\alpha = Wu$ for some $u \in \mathbb{R}^n$, so we can rewrite (4.8) as

$$\min_{\alpha \in \mathrm{Range}(W)} \ \frac{1}{2}\|AW^\top\alpha - f\|_D^2 + \|\mathrm{diag}(\lambda)\alpha\|_1 = \min_{u \in \mathbb{R}^n} \ \frac{1}{2}\|Au - f\|_D^2 + \|\mathrm{diag}(\lambda)Wu\|_1.$$

Now, we refer to the problem

$$\min_{u \in \mathbb{R}^n} \ \frac{1}{2}\|Au - f\|_D^2 + \|\mathrm{diag}(\lambda)Wu\|_1, \tag{4.11}$$

as the *analysis based approach.* We remark that the analysis based approach does not require using the inverse of W. Hence, it can be generalized to any linear transform W.

The objective functions in (4.8) with $0 \leq \kappa \leq \infty$ and in (4.11) are convex (but not strictly convex in general), proper and continuous on $\mathbb{R}^m$. Furthermore, as Theorem 4.2 shows that they are also coercive. Therefore, there exists at least one solution for both problems.

It is clear that when $0 < \kappa < \infty$, (4.8) balances between (4.10) and (4.11); hence, is called the balanced approach. Since the algorithms and convergence analysis on the balanced approach are still valid when $\kappa = 0$, we include discussion of the synthesis based approach into our presentation of the balanced approach. We even intend to call the model (4.10) the general balanced approach for $0 \leq \kappa < \infty$ and include the synthesis based approach as a special case.

We note that when the rows of W form an orthonormal basis, instead of being a redundant tight frame, the above three approaches are exactly the same, since in this case, $WW^\top = I$. However, for redundant tight frame system W, the analysis based, synthesis based and balanced approach cannot be derived from one another. In fact, it was observed in [**47, 83**] for example, that there is a gap between the analysis based and synthesis based approach. Both of them have their own favorable data sets and applications. In general, it is hard to draw definitive conclusions on which approach is better without specifying the applications and data sets. We further note that the ℓ_1-minimization problems in compressed sensing is akin to the synthesis based approach in nature. On the other hand, the TV-norm minimization problem in imaging restoration is, for many cases, an analysis based approach. For frame based image restoration, numerical simulations in [**24**] show that the analysis based approach tends to generate smoother images than the synthesis based approach. This is because the coefficient Wu is quite often linked to the smoothness of the underlying image [**10, 9, 97, 106, 110**]. However, the synthesis based approach tends to exploit more of the sparse representation of the underlying solution in terms of the given frame system by utilizing the redundancy. Therefore, the synthesis based approach usually enhances and sharpens edges, although it may introduce some artifacts as shown in [**16**]. The balanced approach bridges the analysis based and synthesis based approach and it balances the smoothness and the sparsity provided by frames as shown in [**14, 15, 16, 25, 32, 34, 36, 38**].

To solve the synthesis based approach (4.10), the proximal forward and backward splitting algorithm was used in [**62, 66, 86, 87, 90, 91**]. Also, the accelerated proximal gradient (APG) algorithms of [**168**], designed for the general balanced approach, can be applied to get a fast algorithm for the synthesis based approach.

For the analysis based approach (4.11), the coordinate descent method was used in [**84, 170**]. More recently, the split Bregman algorithm was used to develop a fast algorithm for the analysis based approach in frame based image restoration in [**24**], where numerical simulations showed that the split Bregman algorithm is efficient for image deblurring, decomposition, denoise, and inpainting. The split Bregman algorithm was first proposed in [**96**] and was shown to be powerful in [**96, 185**] when it was applied to various PDE based image restoration approaches, e.g., ROF and nonlocal variational models. Convergence analysis of the split Bregman was given in [**24**].

For the balanced approach in frame based image restoration, the model and algorithm were first developed in [**32, 34, 36, 38**]. The balanced approach was reformulated as the proximal forward-backward splitting algorithm in [**14, 15, 16, 25**]. The balanced approach gives satisfactory simulation results, as shown in [**14, 15, 16, 25, 32, 34, 36, 38**]. Recently, fast algorithms for the balanced approach in frame based image restoration whose convergence speeds are much faster than those of the proximal forward-backward splitting algorithm were developed in [**168**]. The accelerated proximal gradient algorithms proposed in [**168**] are based on and extended from several variants of accelerated proximal gradient algorithms that were studied in [**4, 140, 141, 142, 143, 175**]. These accelerated proximal gradient algorithms have an attractive iteration complexity of $O(1/\sqrt{\epsilon})$ for achieving ϵ-optimality. Also these accelerated proximal gradient algorithms are simple and only use the soft-thresholding operator, just like algorithms such as the linearized Bregman algorithm, the split Bregman algorithm and the proximal forward-backward splitting algorithm.

Finally, we note that the formulation of (4.8) can be extended to image restoration of two-layered images [**24, 84, 170**]. Real images usually have two layers, referring to cartoons (the piecewise smooth part of the image) and textures (the oscillating pattern part of the image). Usually, different layers have sparse approximations under different tight frame systems. Therefore, these two different layers should be considered separately. One natural idea is to use two tight frame systems that can sparsely represent cartoons and textures respectively. The corresponding image restoration problem can be formulated as the following ℓ_1-minimization problem:

(4.12)

$$\min_{\alpha^1,\alpha^2\in\mathbb{R}^m} \frac{1}{2}\|A(\sum_{i=1}^{2} W_i^\top\alpha^i) - f\|_D^2 + \sum_{i=1}^{2}\frac{\kappa_i}{2}\|(I - W_iW_i^\top)\alpha^i\|_2^2 + \sum_{i=1}^{2}\|\mathrm{diag}(\lambda^i)\alpha^i\|_1,$$

where, for $i = 1, 2$, $W_i^T W_i = I$, $\kappa_i > 0$, λ_i is a given positive weight vector, and D is a given symmetric positive definite matrix. We refer to model (4.12) as the *two-system balanced approach.*

2. Balanced approach

The balanced approach for frame based image restorations was first developed in [**34, 36, 38**] for high resolution image reconstruction from a few low resolution images. The problem of high resolution image constructions is converted to the problem of filling in the missing wavelet frame coefficients, i.e. inpainting in wavelet frame transformation domain, by designing a proper wavelet tight frame [**34, 36, 38**]. The ideas of [**34, 36**] are used in [**37**] to develop the balanced approach for frame based image inpainting (in pixel domain) whose complete analysis of convergence and optimal properties of the solution are given in [**16**]. Analysis of the convergence and optimal properties of the solutions of algorithms in [**34, 36, 38**] is given in [**14, 25, 32**].

In this section, we use image inpainting to illustrate how the ideas of the balanced approach are formed and developed. Also, we prove the convergence of the image inpainting algorithm. Then we present an accelerated algorithm of [**168**] for the balanced approach (which includes the synthesis based approach) and the proof of convergence given by [**168**] is included. After that, we recall the linearized Bregman algorithm as an alternative to the balanced and synthesis approaches. We prove the convergence of the linearized Bregman algorithm by connecting it to the Uzawa's algorithm. Numerical experiments on both image inpainting and deblurring are presented at the end of this section.

2.1. Balanced approach for image inpainting

The mathematical model for image inpainting can be stated as follows. Let the original image $u \in \mathbb{R}^n$ be defined on the domain $\Omega = \{1, 2, \cdots, n\}$ and the nonempty set $\Lambda \subsetneq \Omega$ be the given observed region. Then the observed (incomplete) image f is

$$f(i) = \begin{cases} u(i) + \eta(i), & i \in \Lambda, \\ \text{arbitrary}, & i \in \Omega \setminus \Lambda, \end{cases} \tag{4.13}$$

where $\eta(i)$ is the noise. The goal is to find u from f. When $\eta(i) = 0$ for all $i \in \Lambda$, we require that $u(i) = f(i)$ and u is just the solution of an interpolation problem. Otherwise, we seek a smooth solution u that satisfies $|u(i) - f(i)| \leq \eta(i)$ for all $i \in \Lambda$. In both cases, variational approaches will penalize some cost functionals (which normally are weighted function norms of the underlying solution) to control the roughness of the solution, see for instance [**7, 42**].

Image inpainting is to recover data by interpolation. There are many interpolation schemes available, e.g., spline interpolation, but the majority of them are only good for smooth functions. Images are either piecewise smooth function or formed by textures which do not have the global smoothness required by most interpolation schemes. The major challenge in image inpainting is to keep the features, e.g. edges, of images which cannot be well preserved by many of the available interpolation algorithms. Furthermore, since images are usually contaminated by noise, a good inpainting algorithm should have a built-in denoising component.

The simple idea of the balanced approach for frame based image inpainting comes as follows: one may use any simple interpolation scheme to interpolate the given data that leads to an inpainted image. Edges might be blurred and noises are still present in this inpainted image. One of the simplest ways to sharpen the image and remove noises at the same time is to set small coefficients under a tight wavelet frame transform to zero. When we reconstruct the image using the modified frame coefficients, it will no longer interpolate the data, and the simplest way to make it interpolate the given data is to put the given data back. One may iterate this procedure till convergence.

To be precise, let $\mathcal{P}_\Lambda$ be the diagonal matrix with diagonal entries 1 for the indices in Λ and 0 otherwise. Starting with the initial guess u_0, the iterative procedure mentioned above can be expressed as

$$u_{k+1} = \mathcal{P}_\Lambda f + (I - \mathcal{P}_\Lambda) W^\top \mathcal{T}_\lambda (W u_k). \tag{4.14}$$

Here

$$\mathcal{T}_\lambda([\beta_1, \beta_2, \ldots, \beta_m]^\top) := [t_{\lambda_1}(\beta_1), t_{\lambda_2}(\beta_2), \ldots, t_{\lambda_m}(\beta_m)]^\top \tag{4.15}$$

is the soft-thresholding function [**77**]:

$$t_{\lambda_i}(\beta_i) := \begin{cases} \operatorname{sgn}(\beta_i)(|\beta_i| - \lambda_i), & \text{if } |\beta_i| > \lambda_i, \\ 0, & \text{if } |\beta_i| \leq \lambda_i, \end{cases} \tag{4.16}$$

Note that by using soft-thresholding instead of hard-thresholding which is traditionally used to sharpen edges, we reduce artifacts and obtain the desired minimization property in each iteration. The thresholding operator $\mathcal{T}_\lambda$ also plays two other important roles, namely, removing noises in the image and perturbing the frame coefficients Wu_k so that information contained in the given region Λ can permeate into the missing region $\Omega \setminus \Lambda$.

Now, define $\alpha_k = \mathcal{T}_\lambda(Wu_k)$. Then (4.14) can be rewritten as an iteration in terms of α_k:

$$\alpha_{k+1} = \mathcal{T}_\lambda\Big(W\mathcal{P}_\Lambda f + W(I - \mathcal{P}_\Lambda)W^\top \alpha_k\Big).$$

Then we have the following frame based image inpainting algorithm:

ALGORITHM 4.1.

(i) *Set an initial guess* α_0.

(ii) *Iterate on* k *until convergence:*

$$\alpha_{k+1} = \mathcal{T}_\lambda\Big(W\mathcal{P}_\Lambda f + W(I - \mathcal{P}_\Lambda)W^\top \alpha_k\Big). \tag{4.17}$$

(iii) *Let* $\alpha^\star$ *to the output of Step (ii). If* $\eta(i) = 0$ *for all* $i \in \Lambda$ *in (4.13), we set* $u^\star := \mathcal{P}_\Lambda f + (I - \mathcal{P}_\Lambda)W^\top\alpha^\star$ *to be the solution (to the inpainting problem); otherwise, since* $\mathcal{T}_\lambda$ *can remove noise, we set* $u^\diamond = W^\top\alpha^\star$ *to be the solution (to the inpainting-plus-denoising problem).*

Algorithm 4.1 was first proposed in [**37**], and a complete analysis of its convergence was given in [**16**]. It was proved in [**16**] that α_k generated by Algorithm 4.1 is a converging sequence that minimizes

$$\min_{\alpha \in \mathbb{R}^m} \frac{1}{2}\|\mathcal{P}_\Lambda W^\top \alpha - \mathcal{P}_\Lambda f\|_2^2 + \frac{1}{2}\|(I - WW^\top)\alpha\|_2^2 + \|\operatorname{diag}(\lambda)\alpha\|_1. \tag{4.18}$$

Note that the problem (4.18) is a special case of (4.8) with $D = I$, $A = \mathcal{P}_\Lambda$ and $\kappa = 1$. In fact, (4.18) was discovered first and was later generalized to (4.8) and (4.12). Here, we consider a slightly more general problem than (4.18), with $\kappa \in [0, \infty)$,

$$\min_{\alpha \in \mathbb{R}^m} \frac{1}{2}\|\mathcal{P}_\Lambda W^\top \alpha - \mathcal{P}_\Lambda f\|_2^2 + \frac{\kappa}{2}\|(I - WW^\top)\alpha\|_2^2 + \|\text{diag}(\lambda)\alpha\|_1. \tag{4.19}$$

Now, we follow arguments that is similar to those in [**16**] to show that Algorithm 4.1 converges to a minimizer of (4.18) (i.e. (4.19) with $\kappa = 1$). Define

$$F_1(\alpha) := \|\text{diag}(\lambda)\alpha\|_1 \quad \text{and} \quad F_2^\kappa(\alpha) := \frac{1}{2}\|\mathcal{P}_\Lambda(W^\top\alpha) - \mathcal{P}_\Lambda \mathbf{f}\|_2^2 + \frac{\kappa}{2}\|(I - WW^\top)\alpha\|_2^2. \tag{4.20}$$

Here for the case $\kappa = 1$ we denote $F_2^1(\alpha)$ simply as $F_2(\alpha)$. Then the minimization problem (4.18) can be written as

$$\min_{\alpha \in \mathbb{R}^m} F_1(\alpha) + F_2(\alpha).$$

For any proper, convex, lower semi-continuous function φ which takes values in $(-\infty, +\infty]$, its *proximal operator* (see e.g. [**109, 137, 138**]) is defined by

$$\text{prox}_\varphi(\mathbf{x}) := \arg\min_{\mathbf{y}}\{\frac{1}{2}\|\mathbf{x} - \mathbf{y}\|_2^2 + \varphi(\mathbf{y})\}. \tag{4.21}$$

It is easy to show that (see e.g. [**58**])

$$\text{prox}_{F_1}(\beta) = \arg\min_{\alpha \in \mathbb{R}^m}\{\frac{1}{2}\|\beta - \alpha\|_2^2 + \|\text{diag}(\lambda)\alpha\|_1\} = \mathcal{T}_\lambda(\beta). \tag{4.22}$$

Now, we can rewrite the iterations of α_k in (4.17) in terms of proximal operators defined in (4.21).

LEMMA 4.1. *Iteration of α_k given by* (4.17) *is equivalent to*

$$\alpha_{k+1} = \text{prox}_{F_1}(\alpha_k - \nabla F_2(\alpha_k)). \tag{4.23}$$

PROOF. Indeed, by (4.22), we have

$$\begin{aligned} \alpha_{k+1} &= \text{prox}_{F_1}[W\mathcal{P}_\Lambda f + W(I - \mathcal{P}_\Lambda)W^\top \alpha_k] \\ &= \text{prox}_{F_1}[\alpha_k - \alpha_k + W\mathcal{P}_\Lambda f + WW^\top\alpha_k - W\mathcal{P}_\Lambda W^\top \alpha_k] \\ &= \text{prox}_{F_1}[\alpha_k - ((I - WW^\top)\alpha_k + W\mathcal{P}_\Lambda(\mathcal{P}_\Lambda W^\top \alpha_k - \mathcal{P}_\Lambda f))]. \end{aligned}$$

Since $(I - WW^\top)^2 = I - WW^\top$, we have

$$\begin{aligned} \alpha_{k+1} &= \text{prox}_{F_1}[\alpha_k - ((I - WW^\top)^2\alpha_k + W\mathcal{P}_\Lambda(\mathcal{P}_\Lambda W^\top \alpha_k - \mathcal{P}_\Lambda f))] \\ &= \text{prox}_{F_1}[\alpha_k - \nabla(\frac{1}{2}\|\mathcal{P}_\Lambda W^\top \alpha_k - \mathcal{P}_\Lambda f\|_2^2 + \frac{1}{2}\|(I - WW^\top)\alpha_k\|_2^2)] \\ &= \text{prox}_{F_1}(\alpha_k - \nabla F_2(\alpha_k)). \end{aligned}$$

This concludes the proof of the lemma. ∎

The new form of iteration (4.23) of α_k is known as the proximal forward-backward splitting. A rather comprehensive convergence analysis of this type of iterations was given in [**58**]. Let us now recall the main convergence theorem in [**58**] for the finite dimensional case.

THEOREM 4.1. *Consider the minimization problem*

$$\min_{x \in \mathbb{R}^m} F_1(x) + F_2(x), \tag{4.24}$$

where $F_1 : \mathbb{R}^m \mapsto \bar{\mathbb{R}}$ is a proper, convex, lower semi-continuous function, and $F_2 : \mathbb{R}^m \mapsto \mathbb{R}$ is a convex, differentiable function with an L-Lipschitz continuous *gradient. Assume a minimizer of* (4.24) *exists. Then for any initial guess* x_0, *the iteration (called the proximal forward-backward splitting):*

$$x_{k+1} = \mathrm{prox}_{F_1/L}(x_k - \nabla F_2(x_k)/L)$$

converges to a minimizer of $F_1(x) + F_2(x)$.

We do not provide a proof for Theorem 4.1 because it does not reveal any information on the speed of convergence of the sequence α_k. In the next section, we prove another version of Theorem 4.1 (Theorem 4.4) for a general balanced approach (4.8).

By Lemma 4.1, proving convergence of Algorithm 4.1 is now reduced to proving that F_1 and F_2 defined in (4.20) satisfy all the assumptions in Theorem 4.1.

THEOREM 4.2. *The sequence* α_k *defined by* (4.17) *converges to a minimizer of the minimization problem* (4.18) *for any initial guess* α_0.

PROOF. Obviously F_1 is a proper, convex, lower semi-continuous (in fact continuous) function, and F_2 is a convex and differentiable function. Therefore, all we need to prove are:

(1) F_2 is 1-Lipschitz.
(2) $F_1(\alpha) + F_2(\alpha)$ is coercive, i.e. $F_1(\alpha) + F_2(\alpha) \to \infty$ as $\|\alpha\|_2 \to \infty$, which implies that $\min_\alpha\{F_1(\alpha) + F_2(\alpha)\}$ has at least one solution.

For (1), noting that $W(I - \mathcal{P}_\Lambda)W^\top \succeq 0$, we have

$$\begin{aligned}\nabla^2 F_2 &= (I - WW^\top) + W\mathcal{P}_\Lambda W^\top \\ &\preceq (I - WW^\top) + WW^\top \\ &= I.\end{aligned}$$

Therefore, F_2 is 1-Lipschitz.

For (2), we have

$$F_1(\alpha) + F_2(\alpha) \geq F_1(\alpha) = \sum_{j=1}^{m} \lambda_j|\alpha_j| \geq \lambda_{\min}\|\alpha\|_1 \geq \lambda_{\min}\|\alpha\|_2,$$

which proves coercivity of $F_1(\alpha) + F_2(\alpha)$. ∎

Theorem 4.2 proves convergence of Algorithm 4.1 to (4.19) for the case $\kappa = 1$. In fact, Theorem 4.1 suggests the following Algorithm 4.2 for the general inpainting model (4.19) with $0 \leq \kappa < \infty$.

ALGORITHM 4.2.

(i) *Set an initial guess* α_0.
(ii) *Iterate on* k *until convergence:*

$$\alpha_{k+1} = \mathcal{T}_{\lambda/L}\Big(\alpha_k - \nabla F_2^\kappa(\alpha_k)/L\Big), \tag{4.25}$$

where L *is the Lipschitz constant of* F_2^κ *defined by* (4.6).
(iii) *Let* $\alpha^\star$ *to the output of Step (ii). If* $\eta(i) = 0$ *for all* $i \in \Lambda$ *in (4.13), we set* $u^\star := \mathcal{P}_\Lambda f + (I - \mathcal{P}_\Lambda)W^\top\alpha^\star$ *to be the solution (to the inpainting problem); otherwise, since* $\mathcal{T}_{\lambda/L}$ *can remove noise, we set* $u^\diamond = W^\top\alpha^\star$ *to be the solution (to the inpainting-plus-denoising problem).*

Then we have the following convergence theorem.

THEOREM 4.3. *The sequence α_k defined by (4.25) converges to a minimizer of the minimization problem (4.19) with $0 \le \kappa < \infty$ for any initial guess α_0.*

PROOF. All we need to show is that L, the Lipschitz constant of F_2^κ, is bounded. In fact, we will show that $L \le \max\{1, \kappa\}$. Indeed, we have

$$\begin{aligned}\nabla^2 F_2^\kappa &= \kappa(I - WW^\top) + W\mathcal{P}_\Lambda W^\top \\ &\preceq \kappa(I - WW^\top) + WW^\top \\ &\preceq \kappa I + (1-\kappa)WW^\top \\ &\preceq \begin{cases} I, & \text{for } 0 \le \kappa \le 1, \\ \kappa I, & \text{for } \kappa \ge 1, \end{cases}\end{aligned}$$

which implies $L \le \max\{1, \kappa\}$. ∎

2.2. Proximal forward-backward splitting for general balanced approach

This section presents a proximal forward-backward splitting algorithm for the general balanced approach (4.8) with $0 \le \kappa < \infty$, as well as some convergence analysis of the algorithm. The corresponding algorithm for the balanced approach involving multiple frame systems (4.12) can be obtained similarly. Let

$$F_1(\alpha) = \|\text{diag}(\lambda)\alpha\|_1 \text{ and } F_2(\alpha) = \frac{1}{2}\|AW^\top\alpha - f\|_D^2 + \frac{\kappa}{2}\|(I - WW^\top)\alpha\|_2^2.$$

Then, the proximal forward-backward splitting algorithm can be written in the following short form (see [**14, 15, 16, 17, 25, 32, 34, 36, 38, 58**])

$$\alpha_{k+1} = \text{prox}_{F_1/L}(\alpha_k - \nabla F_2(\alpha_k)/L).$$

The detailed form of the above algorithm is given as follows:

ALGORITHM 4.3. *For a given vector $\lambda \in \mathbb{R}_+^m$, choose any $\alpha_0 \in \mathbb{R}^m$. For $k = 0, 1, 2, \ldots$, generate α_{k+1} from α_k according to the following iterations:*

(i) *Set $g_k = \alpha_k - \nabla F_2(\alpha_k)/L$;*
(ii) *Set $\alpha_{k+1} = \mathcal{T}_{\lambda/L}(g_k)$.*

Here, the gradient of F_2 is given by

$$\nabla F_2(\alpha) = WA^T D(AW^\top\alpha - f) + \kappa(I - WW^\top)\alpha,$$

and the Hessian matrix of F_2 is

$$\nabla^2 F_2 = WA^T DAW^\top + \kappa(I - WW^\top).$$

It is obvious that F_2 is L-Lipschitz, and $L \le \rho_{\max}(A^\top DA) + \kappa$. This generalizes the inpainting algorithms given in the previous section to algorithms for various image restoration problems. Although the original development of algorithms took a different path, this idea is used in the proof of the convergence of the balanced approach frame based algorithms given in [**14, 15, 16, 17, 25, 32, 34, 36, 38**].

First, we prove the following lemma ([**168**, Lemma 2.1]) which characterizes the optimal solutions of (4.8).

LEMMA 4.2. *For each $\lambda > 0$, the optimal solution $\alpha^\star$ of (4.8) satisfies*

$$\|\alpha^\star\|_1 \le C,$$

where

$$C = \begin{cases} \min\{\|f\|_D^2/2, \|\text{diag}(\lambda)\alpha_{\text{LS}}\|_1\}/\lambda_{\min} & \textit{if } A \textit{ is surjective}, \\ \|f\|_D^2/(2\lambda_{\min}) & \textit{otherwise}, \end{cases}$$

with $\alpha_{\rm LS} := WA^\top(AA^\top)^{-1}f$.

PROOF. Let $\alpha^\star$ be an optimal solution of (4.8). Then

$$\lambda_{\min}\|\alpha^\star\|_1 \le F_1(\alpha^\star) + F_2(\alpha^\star) \le F_1(0) + F_2(0) = \frac{1}{2}\|f\|_D^2.$$

Hence $\|\alpha^\star\|_1 \le \|f\|_D^2/(2\lambda_{\min})$. In addition, if A is surjective then we have

$$\lambda_{\min}\|\alpha^\star\|_1 \le F_1(\alpha^\star) + F_2(\alpha^\star) \le F_1(\alpha_{\rm LS}) + F_2(\alpha_{\rm LS}) = \|\mathrm{diag}(\lambda)\alpha_{\rm LS}\|_1.$$

This concludes the proof of the lemma. ∎

Now, we prove that the sequence $\{\alpha_k\}_{k\ge 0}$ generated by Algorithm 4.3 converges to a minimizer of:

$$\min_{\alpha\in\mathbb{R}^m} F_1(\alpha) + F_2(\alpha),$$

under suitable conditions. A similar proof can be found in e.g. [**4**]. For notational convenience we denote $F(\alpha) := F_1(\alpha) + F_2(\alpha)$ and

$$\ell_F(\alpha;\beta) := F_2(\beta) + \langle \nabla F_2(\beta), \alpha - \beta\rangle + \|\mathrm{diag}(\lambda)\alpha\|_1,$$

where the sum of the first two terms is the linear approximation of F_2 at β. Since F_2 has an L-Lipschitz continuous gradient and is convex, we have the following inequality

$$F(\alpha) - \frac{L}{2}\|\alpha - \beta\|_2^2 \le \ell_F(\alpha;\beta), \quad \text{for all } \alpha,\beta\in\mathbb{R}^m. \tag{4.26}$$

THEOREM 4.4. *Consider the minimization problem*

$$\min_{\alpha\in\mathbb{R}^m} F_1(\alpha) + F_2(\alpha), \tag{4.27}$$

where $F_1 = \|\mathrm{diag}(\lambda)\alpha\|_1$ *and* $F_2 : \mathbb{R}^m \mapsto \mathbb{R}$ *is a convex, differentiable function with an L-Lipschitz continuous gradient. Let* $F := F_1 + F_2$ *and denote* $\alpha^\star$ *be a solution of* (4.27). *Then the sequence* $\{\alpha_k\}_{k\ge 0}$ *generated by Algorithm* 4.3 *satisfies*

$$F(\alpha_k) - F(\alpha^\star) \le \frac{L\|\alpha^\star - \alpha_0\|_2^2}{2k}. \tag{4.28}$$

As a consequence, for given $\epsilon > 0$, *we have*

$$F(\alpha_k) - F(\alpha^\star) \le \epsilon, \quad \textit{whenever } k \ge \frac{L(C + \|\alpha_0\|_2)^2}{2\epsilon}. \tag{4.29}$$

When (4.27) *has a unique solution* $\alpha^\star$, *we have*

$$\lim_{k\to\infty} \|\alpha_k - \alpha^\star\|_2 = 0.$$

PROOF. For a given $k \ge 1$ and any optimal solution $\alpha^\star$, let

$$\alpha_{k+1} := \mathcal{T}_{\lambda/L}(\alpha_k - \nabla F_2(\alpha_k)/L).$$

First, we show that

$$\alpha_{k+1} \in \arg\min_{\alpha}\{\ell_F(\alpha;\alpha_k) + L\langle \alpha_{k+1} - \alpha_k, \alpha\rangle\}. \tag{4.30}$$

By letting $g_k = \alpha_k - \nabla F_2(\alpha_k)/L$ and removing constant terms in (4.30), it is easy to see that (4.30) is equivalent to

$$\alpha_{k+1} \in \arg\min_{\alpha}\{\langle \alpha_{k+1} - g_k, \alpha\rangle + \|\mathrm{diag}(\lambda)\alpha\|_1/L\}. \tag{4.31}$$

Now by the definition of α_{k+1}, we have

$$g_k[j] - \alpha_{k+1}[j] = \begin{cases} \operatorname{sgn}(g_k[j])\lambda_j/L, & \text{when } |g_k[j]| > \lambda_j/L, \\ g_k[j], & \text{when } |g_k[j]| \le \lambda_j/L. \end{cases}$$

This implies that

$$g_k - \alpha_{k+1} \in \partial\|\operatorname{diag}(\lambda)\alpha_{k+1}\|_1/L,$$

or equivalently,

$$0 \in \alpha_{k+1} - g_k + \partial\|\operatorname{diag}(\lambda)\alpha_{k+1}\|_1/L,$$

which implies (4.31) and hence (4.30). By (4.30), we now have

$$\ell_F(\alpha_{k+1};\alpha_k) + L\langle\alpha_{k+1} - \alpha_k, \alpha_{k+1}\rangle \le \ell_F(\alpha^\star;\alpha_k) + L\langle\alpha_{k+1} - \alpha_k, \alpha^\star\rangle. \tag{4.32}$$

Letting $\alpha = \alpha_{k+1}$ and $\beta = \alpha_k$ in (4.26), we have

$$F(\alpha_{k+1}) \le \ell_F(\alpha_{k+1};\alpha_k) + \frac{L}{2}\|\alpha_{k+1} - \alpha_k\|_2^2. \tag{4.33}$$

Applying (4.32) to (4.33), we have

$$\begin{aligned} F(\alpha_{k+1}) &\le \ell_F(\alpha^\star;\alpha_k) + L\langle\alpha_{k+1} - \alpha_k, \alpha^\star - \alpha_{k+1}\rangle + \frac{L}{2}\|\alpha_{k+1} - \alpha_k\|_2^2 \\ &= \ell_F(\alpha^\star;\alpha_k) + \frac{L}{2}\|\alpha^\star - \alpha_k\|_2^2 - \frac{L}{2}\|\alpha^\star - \alpha_{k+1}\|_2^2, \end{aligned}$$

where the last equality follows from straightforward calculations. Now, we have

$$\begin{aligned} F(\alpha_{k+1}) &\le \ell_F(\alpha^\star;\alpha_k) + \frac{L}{2}\|\alpha^\star - \alpha_k\|_2^2 - \frac{L}{2}\|\alpha^\star - \alpha_{k+1}\|_2^2, \\ &\le F(\alpha^\star) + \frac{L}{2}\|\alpha^\star - \alpha_k\|_2^2 - \frac{L}{2}\|\alpha^\star - \alpha_{k+1}\|_2^2. \end{aligned} \tag{4.34}$$

Here, the last inequality follows from the definition of ℓ_F and the convexity of F_2.

Subtracting $F(\alpha^\star)$ from both sides of the last inequality of (4.34) gives

$$F(\alpha_{k+1}) - F(\alpha^\star) \le \frac{L}{2}\|\alpha^\star - \alpha_k\|_2^2 - \frac{L}{2}\|\alpha^\star - \alpha_{k+1}\|_2^2.$$

Telescoping on the above inequality, we have

$$\sum_{j=1}^{k+1} F(\alpha_j) - (k+1)F(\alpha^\star) \le \frac{L}{2}\|\alpha^\star - \alpha_0\|_2^2. \tag{4.35}$$

Now, by using (4.30) with $\alpha = \alpha_k$, we have

$$\begin{aligned} \ell_F(\alpha_{k+1};\alpha_k) + L\langle\alpha_{k+1} - \alpha_k, \alpha_{k+1}\rangle &\le \ell_F(\alpha_k;\alpha_k) + L\langle\alpha_{k+1} - \alpha_k, \alpha_k\rangle \\ &= F(\alpha_k) + L\langle\alpha_{k+1} - \alpha_k, \alpha_k\rangle. \end{aligned}$$

By applying the above inequality to (4.33) we obtain

$$F(\alpha_{k+1}) - F(\alpha_k) \le -\frac{L}{2}\|\alpha_{k+1} - \alpha_k\|_2^2.$$

After multiplying k at both sides of the above inequality and then telescoping, we have

$$(k+1)F(\alpha_{k+1}) - \sum_{j=1}^{k+1} F(\alpha_j) = kF(\alpha_{k+1}) - \sum_{j=1}^{k} F(\alpha_j) \le -\frac{L}{2}\sum_{j=1}^{k} j\|\alpha_{j+1} - \alpha_j\|_2^2.$$

Combining the above inequality with (4.35), we have

$$(k+1)(F(\alpha_{k+1}) - F(\alpha^\star)) \leq \frac{L}{2}\|\alpha^\star - \alpha_0\|_2^2,$$

and thus (4.28) follows. In addition, by applying the following inequalities

$$\|\alpha^\star - \alpha_0\|_2 \leq \|\alpha^\star\|_2 + \|\alpha_0\|_2 \leq \|\alpha^\star\|_1 + \|\alpha_0\|_2,$$

and Lemma 4.2, we obtain (4.29).

The conclusion that $\alpha_k \to \alpha^\star$ whenever $\alpha^\star$ is the unique minimizer of F follows directly from Proposition 4.1. ∎

REMARK 4.2. Theorem 4.1 whose proof is given in [**58**] says that the sequence α_k itself converges to an optimal solution $\alpha^\star$ whenever there exists one as given in Theorem 4.1, which is stronger than what we have shown above. The interested reader should consult [**58**] for details. However, the speed of convergence, which is more important in practice, is not clear from the proof of [**58**]. Here, we focused on showing that α_k is a minimizing sequence that reaches to an ϵ-optimal solution after $O(L/\epsilon)$ iterations. Furthermore, in image restorations we are more interested in finding a solution at which the cost functional has a value that is close to the minimal value, rather than knowing whether the solution is close to a minimizer or not. In this regard, the proof here is more informative.

2.3. Accelerated algorithm

As shown in Theorem 4.4, the proximal forward-backward splitting algorithm generates an ϵ-optimal solution of (4.8) in $O(L/\epsilon)$ iterations, which is reasonably efficient. However, in practice, faster algorithms are always desired. For example when the matrix A is huge (e.g. Radon transform for tomography in 3D), each iteration of Algorithm 4.3 can be expensive. Therefore, one always wishes to reduce the total number of iterations. As we will see from Theorem 4.5 the accelerated proximal gradient (APG) algorithm of [**168**] (see also the FISTA algorithm of [**4**]) is much more efficient than the proximal forward-backward splitting algorithm because it only requires $O(\sqrt{L/\epsilon})$ iterations to obtain an ϵ-optimal solution.

The APG algorithm of [**168**] is obtained by adjusting the step $g_k = \alpha_k - \nabla F_2(\alpha_k)/L$ in the proximal forward-backward splitting algorithm. This idea has already appeared in [**4, 173**]. Next, we describe the APG algorithm for solving (4.8).

ALGORITHM 4.4 (APG for (4.8)). *For a given vector $\lambda \in \mathbb{R}^m_+$, choose $\alpha_0 = \alpha_{-1} \in \mathbb{R}^m$, $t_0 = 1$, and $t_{-1} = 0$. For $k = 0, 1, 2, \ldots$, generate α_{k+1} from α_k according to the following iteration:*

(i) *Set* $\beta_k = \alpha_k + \frac{t_{k-1}-1}{t_k}(\alpha_k - \alpha_{k-1})$.
(ii) *Set* $g_k = \beta_k - \nabla F_2(\beta_k)/L$.
(iii) *Set* $\alpha_{k+1} = \mathcal{T}_{\lambda/L}(g_k)$.
(iv) *Compute* $t_{k+1} = \frac{1+\sqrt{1+4t_k^2}}{2}$.

Note that, when $\kappa = 0$, i.e. $F_2(\alpha) = \frac{1}{2}\|AW^\top\alpha - f\|_D^2$, Algorithm 4.4 leads to an efficient algorithm for the synthesis based approach. When we take $t_k = 1$ for all k in the APG algorithm, it is reduced to the proximal forward-backward splitting algorithm for the balanced approach in frame based image restorations, and it is the popular iterative shrinkage/thresholding algorithms proposed in [**62, 66, 90, 91**].

The iterative shrinkage/thresholding algorithms and the proximal forward-backward splitting algorithms have been developed and analyzed independently by many researchers. These algorithms only require gradient evaluations and soft-thresholding operations, so the computation cost for each iteration is very cheap. But, for any $\epsilon > 0$, these algorithms terminate in $O(L/\epsilon)$ iterations with an ϵ-optimal solution as proven in Theorem 4.4. The APG algorithm proposed here, however, reaches an ϵ-optimal solution in $O(\sqrt{L/\epsilon})$ iterations. Thus APG algorithm of [**168**] greatly accelerates the proximal forward-backward splitting algorithms used in [**14, 15, 16, 17, 25, 32, 34, 36, 38**] for the balanced approach in frame based image restorations.

Here, we recall the proof of convergence of Algorithm 4.4 given by [**168**, Theorem 2.1], which gives an upper bound on the number of iterations for the APG algorithm for solving (4.8) to achieve ϵ-optimality. The proof follows a similar idea to that of [**4, 173**] and Theorem 4.4.

We note that for the sequence t_k defined in Algorithm 4.4, we have the following inequality which can be easily verified by induction

$$t_k \geq \frac{k+1}{2}. \tag{4.36}$$

THEOREM 4.5. *Let $\{\alpha_k\}$, $\{\beta_k\}$, and $\{t_k\}$ be the sequences given by Algorithm 4.4. Then for any $k \geq 1$ and any optimal solution $\alpha^\star$ to the minimization problem (4.8) with $0 \leq \kappa < \infty$, we have*

$$F(\alpha_k) - F(\alpha^\star) \leq \frac{2L\|\alpha^\star - \alpha_0\|_2^2}{(k+1)^2}. \tag{4.37}$$

Hence

$$F(\alpha_k) - F(\alpha^\star) \leq \epsilon, \quad \textit{whenever } k \geq \sqrt{\frac{2L}{\epsilon}}(\|\alpha_0\|_2 + C) - 1, \tag{4.38}$$

where C is given by Lemma 4.2. Furthermore, if $\alpha^\star$ is the unique minimizer of $F(\alpha)$, then $\alpha_k \to \alpha^\star$ as $k \to \infty$.

PROOF. For a given $k \geq 1$ and any optimal solution $\alpha^\star$, let $s_k := \mathcal{T}_{\lambda/L}(g_k)$ and $\tilde{\alpha} := ((t_k - 1)\alpha_k + \alpha^\star)/t_k$. We first show that

$$s_k \in \arg\min_{\alpha}\{\ell_F(\alpha;\beta_k) + L\langle s_k - \beta_k, \alpha\rangle\}. \tag{4.39}$$

By definition of g_k in Algorithm 4.4 and removing constant terms in (4.39), it is easy to see that (4.39) is equivalent to

$$s_k \in \arg\min_{\alpha}\{\langle s_k - g_k, \alpha\rangle + \|\mathrm{diag}(\lambda)\alpha\|_1/L\}. \tag{4.40}$$

Now by the definition of s_k, we have

$$g_k[j] - s_k[j] = \begin{cases} \mathrm{sgn}(g_k[j])\lambda_j/L, & \text{when } |g_k[j]| > \lambda_j/L, \\ g_k[j], & \text{when } |g_k[j]| \leq \lambda_j/L. \end{cases}$$

This implies that

$$g_k - s_k \in \partial\|\mathrm{diag}(\lambda)s_k\|_1/L,$$

or equivalently

$$0 \in s_k - g_k + \partial\|\mathrm{diag}(\lambda)s_k\|_1/L,$$

which implies (4.40) and hence (4.39). By (4.39), we now have

$$\ell_F(s_k;\beta_k) + L\langle s_k - \beta_k, s_k\rangle \leq \ell_F(\tilde{\alpha};\beta_k) + L\langle s_k - \beta_k, \tilde{\alpha}\rangle. \tag{4.41}$$

Letting $\alpha = s_k$ and $\beta = \beta_k$ in (4.26) and noting that $\alpha_{k+1} = s_k$, we have

$$F(\alpha_{k+1}) = F(s_k) \le \ell_F(s_k; \beta_k) + \frac{L}{2}\|s_k - \beta_k\|_2^2.$$

Applying (4.41) to the above inequality, we have

$$\begin{aligned} F(\alpha_{k+1}) &\le \ell_F(\tilde{\alpha}; \beta_k) + L\langle s_k - \beta_k, \tilde{\alpha} - s_k\rangle + \frac{L}{2}\|s_k - \beta_k\|_2^2 \\ &= \ell_F(\tilde{\alpha}; \beta_k) + \frac{L}{2}\|\tilde{\alpha} - \beta_k\|_2^2 - \frac{L}{2}\|\tilde{\alpha} - s_k\|_2^2 \\ &= \ell_F(\tilde{\alpha}; \beta_k) + \frac{L}{2}\|\tilde{\alpha} - \beta_k\|_2^2 - \frac{L}{2}\|\tilde{\alpha} - \alpha_{k+1}\|_2^2, \end{aligned}$$

where the second to the last equality follows from straightforward calculations. Plug the definition of $\tilde{\alpha}$ into the above inequality and denoting $\gamma_k := (t_{k-1} - 1)\alpha_{k-1} - t_{k-1}\alpha_k$, one obtains

(4.42)

$$\begin{aligned} F(\alpha_{k+1}) &\le \frac{t_k - 1}{t_k}\ell_F(\alpha_k; \beta_k) + \frac{1}{t_k}\ell_F(\alpha^\star; \beta_k) \\ &+ \frac{L}{2t_k^2}\|(t_k - 1)\alpha_k + \alpha^\star - t_k\beta_k\|_2^2 - \frac{L}{2t_k^2}\|(t_k - 1)\alpha_k + \alpha^\star - t_k\alpha_{k+1}\|_2^2, \\ &= \frac{t_k - 1}{t_k}\ell_F(\alpha_k; \beta_k) + \frac{1}{t_k}\ell_F(\alpha^\star; \beta_k) + \frac{L}{2t_k^2}\|\alpha^\star - \gamma_k\|_2^2 - \frac{L}{2t_k^2}\|\alpha^\star - \gamma_{k+1}\|_2^2 \\ &\le \frac{t_k - 1}{t_k}F(\alpha_k) + \frac{1}{t_k}F(\alpha^\star) + \frac{L}{2t_k^2}\|\alpha^\star - \gamma_k\|_2^2 - \frac{L}{2t_k^2}\|\alpha^\star - \gamma_{k+1}\|_2^2. \end{aligned}$$

Here, the first inequality follows from the fact that ℓ_F is convex and $t_k \ge 1$ by (4.36). The last inequality follows from the convexity of F_2.

Subtracting $F(\alpha^\star)$ from both sides of the last inequality of (4.42), multiplying t_k^2 at both sides and noticing that $t_{k-1}^2 = t_k(t_k - 1)$, we have

$$t_k^2(F(\alpha_{k+1}) - F(\alpha^\star)) \le t_{k-1}^2(F(\alpha_k) - F(\alpha^\star)) + \frac{L}{2}\|\alpha^\star - \gamma_k\|_2^2 - \frac{L}{2}\|\alpha^\star - \gamma_{k+1}\|_2^2.$$

Telescoping on the above inequality, and using $t_{-1} = 0$ and $\gamma_0 = \alpha_0$, we have

$$t_k^2(F(\alpha_{k+1}) - F(\alpha^\star)) \le \frac{L}{2}\|\alpha^\star - \alpha_0\|_2^2.$$

By (4.36), the inequality (4.37) follows; and using Lemma 4.2, we obtain (4.38).

The conclusion that $\alpha_k \to \alpha^\star$ whenever $\alpha^\star$ is the unique minimizer of F follows directly from Proposition 4.1. ∎

REMARK 4.3.

(1) Theorem 4.5 shows that the number of iterations to get an ϵ-optimal solution is of order $O(\sqrt{L/\epsilon})$. Therefore, less iterations are required if we have a smaller Lipschitz constant L. For the general balanced approach, we have $L \le \rho_{\max}(A^\top DA) + \kappa$. Then for given A and D, the smaller κ is, the faster α_k from the APG algorithm converges.

(2) For the inpainting problem considered in the previous section, we have $A = \mathcal{P}_\Lambda$ and $D = I$. We have shown in the proof of Theorem 4.3 that $L = \max\{1, \kappa\}$. Thus APG converges fastest whenever $\kappa \le 1$ and $\kappa = 1$ is the case where we have fast convergence while the reconstructed image has good regularity.

(3) It is clear that the Algorithm (4.4) can be used for both the balanced approach and the synthesis based approach (the combination of the two is called the general balanced approach here). In general, the balanced approach and the synthesis based approach can have the same algorithm by using different cost functionals, while algorithms for the analysis based approach need to be developed separately.

We finally present the APG algorithm for the two-system balanced approach (4.12) whose convergence can be proved in a manner similarly to Theorem 4.5. We denote

$$F_2(\alpha^1,\alpha^2) := \frac{1}{2}\|A(\sum_{i=1}^{2} W_i^\top \alpha^i) - f\|_D^2 + \sum_{i=1}^{2}\frac{\kappa_i}{2}\|(I - W_iW_i^\top)\alpha^i\|_2^2$$

and denote its Lipschitz constant as L (see [**168**] for an estimate of L). We note that the two-system balanced approach generally works better than (4.8) when the given image has two different layers, e.g. cartoon and textures that can be sparsely approximated by two different tight frame systems.

ALGORITHM 4.5 (APG for (4.12)). *Let* $i = 1, 2$. *For given vectors* $\lambda^i \in \mathbb{R}_+^m$, *choose* $\alpha_0^i = \alpha_{-1}^i \in \mathbb{R}^m$, $t_0 = 1$, *and* $t_{-1} = 0$. *For* $k = 0, 1, 2, \ldots$, *generate* α_{k+1}^i *from* α_k^i *according to the following iteration:*

(i) *Set*

$$\beta_k^1 = \alpha_k^1 + \frac{t_{k-1}-1}{t_k}(\alpha_k^1 - \alpha_{k-1}^1) \quad \textit{and} \quad \beta_k^2 = \alpha_k^2 + \frac{t_{k-1}-1}{t_k}(\alpha_k^2 - \alpha_{k-1}^2).$$

(ii) *Set*

$$g_k^1 = \beta_k^1 - \nabla_{\alpha^1} F_2(\beta_k^1, \beta_k^2)/L \quad \textit{and} \quad g_k^2 = \beta_k^2 - \nabla_{\alpha^2} F_2(\beta_k^1, \beta_k^2)/L.$$

(iii) *Set*

$$\alpha_{k+1}^1 = \mathcal{T}_{\lambda^1/L}(g_k^1) \quad \textit{and} \quad \alpha_{k+1}^2 = \mathcal{T}_{\lambda^2/L}(g_k^2).$$

(iv) *Compute* $t_{k+1} = \frac{1+\sqrt{1+4t_k^2}}{2}$.

2.4. Linearized Bregman algorithm

Recently, the linearized Bregman algorithm has been applied to develop a fast algorithm for frame based image deblurring in [**23**], which converges to the minimizer of the following minimization problem:

$$\min_{\alpha\in\mathbb{R}^m}\{\frac{\kappa}{2}\|\alpha\|_2^2 + \|\mathrm{diag}(\lambda)\alpha\|_1 : AW^\top\alpha = f\}, \tag{4.43}$$

when A is surjective. Furthermore, the optimal solution $u_\lambda^\star$ converges to the minimizer of $\min_{\alpha\in\mathbb{R}^m}\{\|\alpha\|_1 : AW^\top\alpha = f\}$ as $\lambda \to \infty$, where, for simplicity, each entry of the vector λ is set to be the same (see [**23**]). Hence, the linearized Bregman algorithm is used to solve a variation of the synthesis based approach when λ is sufficiently large. The linearized Bregman algorithm was first proposed to solve ℓ_1-minimization problems in compressed sensing by [**184**] and was made efficient in [**147**]. The convergence analysis of the linearized Bregman algorithm was given in [**21, 22**]. Then, it was used in the nuclear norm minimization for matrix completion by [**13**], where the authors realized that the linearized Bregman can be reformulated as the Uzawa's algorithm (see e.g. [**82**] for more details about the Uzawa algorithm).

A simple computation of [**23**] shows that when A is invertible, (4.43) is equivalent to

$$\min_{\alpha\in\mathbb{R}^m}\{\frac{\kappa}{2}\|(I-WW^\top)\alpha\|_2^2+\|\mathrm{diag}(\lambda)\alpha\|_1 : AW^\top\alpha=f\}.$$

This looks like a variation of the balanced approach. However, when the large parameter vector λ is chosen, which happens when one applies linearized Bregman, it is closer to a variation of the synthesis based approach. A detailed discussion of the case when A is not invertible is provided in [**23**].

The advantage of Bregman iterations (either linearized Bregman or split Bregman iterations) in frame based image restorations is that big coefficients come back at first after few iterations and stay. This, in particular, is important to image deblurring, since big wavelet frame coefficients contain information of edges and features of images. The main goal of deblurring is to restore blurred edges and features.

Iterative algorithms involving Bregman distance were introduced to image and signal processing by many authors including [**30, 31**]. See [**145**] for an overview. In [**145**], a Bregman iteration was proposed for the nondifferentiable TV energy for image restoration. Then, in [**184**], it was shown to be remarkably successful for ℓ_1-norm minimization problems in compressive sensing. To further improve the performance of the Bregman iteration, a linearized Bregman algorithm was invented in [**59**]; see also [**184**]. More details and an improvement called "kicking" of the linearized Bregman algorithm were described in [**147**], and a rigorous theory was given in [**21, 22**]. Wavelet-based denoising using the Bregman iteration was introduced in [**182**], and was further extended by using translation invariant wavelets in [**129**]. Here, we focus on the linearized Bregman algorithm for wavelet frame based image restorations of [**23**].

The linearized Bregman algorithm is given as follows.

ALGORITHM 4.6 (Linearized Bregman).

(i) *Set initial guess* $\beta_0\in\mathbb{R}^m$.

(ii) *For* $k=0,1,\ldots,$ *perform the following two-line-iteration until convergence:*

$$\begin{aligned}\alpha_{k+1}&=\frac{1}{\kappa}\mathcal{T}_\lambda(\beta_k),\\ \beta_{k+1}&=\beta_k+WA^\top(f-AW^\top\alpha_{k+1}).\end{aligned}\tag{4.44}$$

The linearized Bregman algorithm looks similar to proximal forward-backward splitting algorithm. For simplicity, we suppose $\kappa=1$. If we let $F_1(\alpha)=\|\mathrm{diag}(\lambda)\alpha\|_1$ and $F_2(\alpha)=\frac{1}{2}\|AW^\top\alpha-f\|_2^2$, then from (4.44) we have

$$\begin{aligned}\alpha_{k+1}=\mathcal{T}_\lambda(\beta_k)&=\mathrm{proxy}_{F_1}(\beta_{k-1}-\nabla F_2(\alpha_k))\\ &=\mathrm{proxy}_{F_1}(\beta_0-\sum_{j=1}^{k}\nabla F_2(\alpha_j)).\end{aligned}$$

Therefore, one essential difference between the linearized Bregman algorithm (4.44) and the proximal forward-backward splitting algorithm is that the linearized Bregman uses all $\{\alpha_j : j=1,2,\ldots,k\}$ to update α_{k+1}, while the proximal forward-backward splitting algorithm only uses the approximation of the previous iteration α_k. On the other hand, the linearized Bregman algorithm is in fact a primal-dual algorithm, where the variable β_k links to the dual variable of α_k. We further explain this in the following discussions.

It was proved in [**22**] that if the sequence $\{\alpha_k\}_{k\in\mathbb{N}}$ from (4.44) converges, its limit is the unique solution of the following optimization problem

$$\min_{u\in\mathbb{R}^n}\left\{\|\mathrm{diag}(\lambda)\alpha\|_1+\frac{\kappa}{2}\|\alpha\|_2^2:\ AW^\top\alpha=f\right\}, \tag{4.45}$$

where $0<\kappa<\infty$. Later, in [**21**], it was proved that the sequence $\{u_k\}_{k\in\mathbb{N}}$ converges. A rather crucial discovery that makes it possible to link the linearized Bregman algorithm to the Uzawa's algorithm is the discovery of the optimization problem (4.45) corresponding to the linearized Bregman algorithm [**22**].

The linearized Bregman algorithm coincides with the Uzawa's algorithm (see e.g. [**13, 82, 151**] for details on Uzawa's algorithm) applied to the problem (4.45). In this section, we first show the equivalence of the linearized Bregman algorithm with Uzawa's algorithm when applied to (4.45). Then we prove convergence of Uzawa's method in a more general setting, which in turn gives a convergence proof of linearized Bregman algorithm.

For simplicity, we denote $B=AW^\top$ and denote the objective function in (4.45) as $F_\lambda(\alpha)$, i.e.

$$F_\lambda(\alpha):=\|\mathrm{diag}(\lambda)\alpha\|_1+\frac{\kappa}{2}\|\alpha\|_2^2.$$

The Lagrangian for the problem (4.45) is given by

$$\mathcal{L}(\alpha,p)=F_\lambda(\alpha)+\langle p,f-B\alpha\rangle.$$

If strong duality holds (which is indeed the case for (4.45)), then $\alpha^\star$ and $p^\star$ are called primal-dual optimal if the pair $(\alpha^\star,p^\star)$ is a saddle point of the Lagrangian $\mathcal{L}(\alpha,p)$, i.e. the pair satisfies

$$\sup_p\inf_\alpha\mathcal{L}(\alpha,p)=\mathcal{L}(\alpha^\star,p^\star)=\inf_\alpha\sup_p\mathcal{L}(\alpha,p). \tag{4.46}$$

The Uzawa's algorithm is an iterative procedure that solves the above saddle point problem as follows

$$\begin{aligned}\alpha_{k+1}&=\arg\min_\alpha\mathcal{L}(\alpha,p_k),\\ p_{k+1}&=p_k+\delta_k(f-B\alpha_{k+1}),\end{aligned} \tag{4.47}$$

where $\{\delta_k\}_{k\in\mathbb{N}}$ are positive step sizes. Note that the second step in (4.47) is one step of gradient ascend solving the maximization problem $\sup_p\mathcal{L}(\alpha_{k+1},p)$, because $\nabla\mathcal{L}(\alpha_{k+1},p)=f-Bu_{k+1}$. Therefore, Uzawa's algorithm is an alternative minimization and maximization method solving the saddle point problem (4.46).

By adding some terms independent of variable α, we have

$$\begin{aligned}\arg\min_\alpha\mathcal{L}(\alpha,p_k)&=\arg\min_\alpha F_\lambda(\alpha)+\langle p_k,f-B\alpha\rangle\\ &=\arg\min_\alpha\|\mathrm{diag}(\lambda)\alpha\|_1+\frac{\kappa}{2}\|\alpha-\frac{1}{\kappa}B^\top p_k\|_2^2\\ &=\frac{1}{\kappa}\mathcal{T}_\lambda(B^\top p_k).\end{aligned}$$

Applying $B^\top$ on both sides of the second equation in (4.47), letting $\beta_k=B^\top p_k$ and $\delta_k=1$, we obtain the linearized Bregman algorithm (4.44). In fact, the sequence α_k for the linearized Bregman algorithm is exactly the same as Uzawa's method. Therefore, we focus on analyzing the convergence of Uzawa's algorithm (4.47).

First, we establish the strong convexity of $F_\lambda(\alpha)$ in the following lemma.

LEMMA 4.3. *Let* $v_1 \in \partial F_\lambda(\alpha_1)$ *and* $v_2 \in \partial F_\lambda(\alpha_2)$. *Then*

$$\langle v_1 - v_2, \alpha_1 - \alpha_2 \rangle \geq \kappa \|\alpha_1 - \alpha_2\|_2^2. \tag{4.48}$$

PROOF. An element in $v_i \in \partial F_\lambda(\alpha_i)$, $i = 1, 2$, takes the form $v_i = w_i + \kappa\alpha_i$, with $w_i \in \partial\|\mathrm{diag}(\lambda)\alpha_i\|_1$. Therefore, we have

$$\langle v_1 - v_2, \alpha_1 - \alpha_2 \rangle = \langle w_1 - w_2, \alpha_1 - \alpha_2 \rangle + \kappa\|\alpha_1 - \alpha_2\|_2^2.$$

Thus we only need to show that the first term of the right-hand side is nonnegative. By the definition of subgradient, it is easy to see that $\langle w_i, \alpha_i \rangle = \|\mathrm{diag}(\lambda)\alpha_i\|_1$. Furthermore, we have

$$\begin{aligned}\langle w_1, \alpha_2 \rangle &= \langle \mathrm{diag}(\lambda)^{-1}w_1, \mathrm{diag}(\lambda)\alpha_2 \rangle \\ &\leq \|\mathrm{diag}(\lambda)^{-1}w_1\|_\infty \|\mathrm{diag}(\lambda)\alpha_2\|_1 \leq \|\mathrm{diag}(\lambda)\alpha_2\|_1.\end{aligned}$$

Similarly we have $\langle w_2, \alpha_1 \rangle \leq \|\mathrm{diag}(\lambda)\alpha_1\|_1$. Therefore, we have

$$\begin{aligned}\langle w_1 - w_2, \alpha_1 - \alpha_2 \rangle &= \langle w_1, \alpha_1 \rangle + \langle w_2, \alpha_2 \rangle - \langle w_1, \alpha_2 \rangle - \langle w_2, \alpha_1 \rangle \\ &= \|\mathrm{diag}(\lambda)\alpha_1\|_1 + \|\mathrm{diag}(\lambda)\alpha_2\|_1 - \langle w_1, \alpha_2 \rangle - \langle w_2, \alpha_1 \rangle \\ &\geq 0.\end{aligned}$$

This concludes the proof of the lemma. ∎

THEOREM 4.6. *Suppose the step sizes obey* $0 < \inf_k \delta_k \leq \sup_k \delta_k < 2\kappa/\|B\|_2^2$. *Then the sequence* α_k *obtained from Uzawa's algorithm* (4.47) *converges to the unique solution of* (4.45).

PROOF. Let $(\alpha^\star, p^\star)$ be the optimal primal-dual pair for the problem (4.45). Then the optimality conditions in (4.46) and (4.47) give

$$\begin{aligned}v_{k+1} - B^\top p_k &= 0, \\ v^\star - B^\top p^\star &= 0,\end{aligned}$$

where $v_{k+1} \in \partial F_\lambda(\alpha_{k+1})$, and $v^\star \in \partial F_\lambda(\alpha^\star)$. Then we have,

$$(v_{k+1} - v^\star) - B^\top(p_k - p^\star) = 0,$$

and it follows from Lemma 4.3 that

$$\langle \alpha_{k+1} - \alpha^\star, B^\top(p_k - p^\star) \rangle = \langle \alpha_{k+1} - \alpha^\star, v_{k+1} - v^\star \rangle \geq \kappa\|\alpha_{k+1} - \alpha^\star\|_2^2.$$

Thus,

$$\langle B(\alpha_{k+1} - \alpha^\star), p_k - p^\star \rangle \geq \kappa\|\alpha_{k+1} - \alpha^\star\|_2^2.$$

Noting that $B\alpha^\star = f$ and using the second equation in (4.47), we have

$$\begin{aligned}\|p_{k+1} - p^\star\|_2^2 &= \|p_k - p^\star + \delta_k(f - B\alpha_{k+1})\|_2^2 \\ &= \|p_k - p^\star + \delta_k B(\alpha^\star - \alpha_{k+1})\|_2^2 \\ &= \|p_k - p^\star - \delta_k B(\alpha_{k+1} - \alpha^\star)\|_2^2.\end{aligned}$$

If we let $r_k := \|p_k - p^\star\|_2$, we then have

$$\begin{aligned}r_{k+1}^2 &= r_k^2 - 2\delta_k\langle p_k - p^\star, B(\alpha_{k+1} - \alpha^\star)\rangle + \delta_k^2\|B(\alpha_{k+1} - \alpha^\star)\|_2^2 \\ &\leq r_k^2 - 2\kappa\delta_k\|\alpha_{k+1} - \alpha^\star\|_2^2 + \delta_k^2\|B\|_2^2\|\alpha_{k+1} - \alpha^\star\|_2^2.\end{aligned}$$

From the assumptions of δ_k, we have $2\kappa\delta_k - \delta_k^2\|B\|_2^2 \geq C$, for some $C > 0$. Therefore,

$$r_{k+1}^2 \leq r_k^2 - C\|\alpha_{k+1} - \alpha^\star\|_2^2.$$

The above inequality shows that r_k is a monotonically decreasing sequence and thus converges. Applying limits to both sides of the above inequality gives $\lim_{k\to\infty} \|\alpha_{k+1} - \alpha^\star\|_2 = 0$ and this concludes the proof of the theorem. ■

Now, we have the following convergence theorem for the linearized Bregman algorithm (4.44), which is a direct consequence of Theorem 4.6.

THEOREM 4.7. *The sequence α_k obtained from Algorithm* 4.6 *converges to the unique solution of* (4.45), *provided that $\kappa > \|B\|_2^2/2$.*

PROOF. Convergence follows from Theorem 4.6, and the uniqueness of the solution of (4.45) follows from the fact that

$$\|\mathrm{diag}(\lambda)\alpha\|_1 + \frac{\kappa}{2}\|\alpha\|_2^2$$

is strictly convex for any $\kappa > 0$. ■

Finally, we note that in order to speed-up Algorithm 4.6, i.e. to reduce the total number of iterations required for a given stopping criterion, a preconditioned version of Algorithm 4.6 was proposed in [**23**]. The idea is that instead of considering equation $AW^\top\alpha = f$ in (4.45), we consider the equivalent equation

$$D^{\frac{1}{2}}AW^\top\alpha = D^{\frac{1}{2}}f,$$

where D is a symmetric positive definite matrix that approximates the pseudoinverse of A. Then by replacing A with $D^{\frac{1}{2}}A$ and f with $D^{\frac{1}{2}}f$ in Algorithm 4.6, one obtains the preconditioned linearized Bregman algorithm. Here, we use the same notation D as in (4.8) for example, because they serve the same purpose. Therefore, just as in (4.8), we choose $D = (AA^\top + \theta I)^{-1}$.

2.5. Role of the redundancy

Tight frames are different from orthonormal systems because tight frames are redundant. What does the redundancy bring us here? We start with a philosophical point about the algorithm and then give some quantitative analysis on the error reduction with each iteration.

Assume that some blocks of pixels are missing in a given image and we want to solve the inpainting problem in the wavelet frame domain as mentioned before. Since the framelets used are compactly supported, the coefficients of those framelets whose supports fall in the missing blocks are missing and the coefficients of those framelets whose supports overlap with the missing blocks are inaccurate. The main step of Algorithm 4.1 perturbs the frame coefficients Wu_n via thresholding so that information contained in the available coefficients will permeate into the missing frame coefficients. Here, the redundancy is very important, since the available coefficients and its associated atoms in the system contain information of the missing coefficients and missing atoms only if the system is redundant, as the atoms in an orthonormal basis are orthogonal to each other and do not contain information of other atoms in L_2-sense.

While applying the thresholding operator on the frame coefficients is a very important step in Algorithm 4.1, to remove noise, perturb the coefficients and sharpen the edges, it also introduces new errors and artifacts. We use image inpainting as an example to show how the numerical errors and artifacts introduced by thresholding can be reduced by the redundancy of the system W. We take the computed solution $u^\star$ for the case $\eta = 0$ as an example. Similar analysis holds for the computation of each iteration. Our computed solution $u^\star$ that interpolates the given data satisfies $\alpha^\star = \mathcal{T}_\lambda(Wu^\star)$ and

$$u^\star = \mathcal{P}_\Lambda f + (I - \mathcal{P}_\Lambda)W^\top \mathcal{T}_\lambda W u^\star.$$

That is, on Λ, $W^\top \mathcal{T}_\lambda W u^\star$ is replaced by f. But since $\mathcal{P}_\Lambda f = \mathcal{P}_\Lambda u^\star = \mathcal{P}_\Lambda W^\top W u^\star$, we are actually replacing $\mathcal{P}_\Lambda W^\top \mathcal{T}_\lambda W u^\star$ by $\mathcal{P}_\Lambda W^\top W u^\star$, which generates artifacts. Hence to reduce the artifacts, we require that the norm of

$$\mathcal{P}_\Lambda W^\top W u^\star - \mathcal{P}_\Lambda W \mathcal{T}_\lambda W u^\star = \mathcal{P}_\Lambda W^\top (W u^\star - \mathcal{T}_\lambda W u^\star)$$

is small.

Clearly, the smaller the norm of $W^\top \mathbf{e} := W^\top (W u^\star - \mathcal{T}_\lambda W u^\star)$ is, the smaller the artifact is. Note that the reconstruction operator $W^\top$ can eliminate the error components sitting in the kernel of $W^\top$. In fact, since $W^\top$ projects all sequences down to the orthogonal complement of the kernel of $W^\top$, which is the range of W, the component of $\mathbf{e}$ in the kernel of $W^\top$ does not contribute. The redundant system reduces the errors as long as the component of $\mathbf{e}$ in the kernel of $W^\top$ is not zero. Therefore, in general, the larger the kernel of $W^\top$ is, i.e. the more redundant the frame system is, the greater the error reduction. To increase the redundancy, we use quasi-affine tight frame system (which is also called an undecimated tight frame system since no downsampling is required during decompositions). In contrast, if W is an orthonormal system (not redundant at all), then the kernel of $W^\top$ is just $\{0\}$ and we have $\|W^\top \mathbf{e}\| = \|\mathbf{e}\|$. The point here is that redundancy reduces the norm of $(\alpha^\star - Wf)$.

2.6. Some simulations

First, we provide a comparison of computation efficiency between the PFBS and the APG algorithm under the scenario of image inpainting. Qualities of inpainting and deblurring using both single- and two-system balanced approach (4.8) and (4.12) are presented as well. In particular, we present results using the synthesis based approach (i.e. (4.8) and (4.12) with $\kappa = 0$). Finally, we present deblurring results using the linearized Bregman algorithm given by Algorithm 4.6 with preconditioning.

For the balanced approach (4.8), we take W corresponding to the piecewise linear B-spline quasi-affine tight frame system. For the two-system balanced approach (4.12), we take W_1 the same as W for (4.8), while W_2 is generated by a local discrete cosine transform.

Throughout this section, we adopt the following stopping criterion for Algorithm 4.4 (we use a similar stopping criterion for Algorithm 4.5):

$$\min\left\{\frac{\|\alpha_k - \alpha_{k-1}\|_2}{\max\{1, \|\alpha_k\|_2\}}, \gamma\frac{\big|\|AW^\top\alpha_k - f\|_D - \|AW^\top\alpha_{k-1} - f\|_D\big|}{\|AW^\top\alpha_k - f\|_D}\right\} < 5\times 10^{-4},$$

where $\gamma = 1$ for image inpainting and $\gamma = 4$ for deblurring (in order to prevent the algorithm from stopping too early since for deblurring the residual $\|AW^\top\alpha_k - f\|_D$ usually changes slowly with k). To measure the quality of the restored image, we use the PSNR value defined by

$$\text{PSNR} := -20\log_{10}\frac{\|u - \tilde{u}\|}{255n},$$

where u and $\tilde{u}$ are the original and restored images respectively, and n is total number of pixels in u. Finally, we note that all algorithms are programmed in MATLAB and run on a 64-bits Windows laptop with a Quad Core 1.73GHz CPU and 8.0G RAM.

2.6.1. *Balanced Approach for Image Inpainting.* In Table 1, we compare the efficiency of the APG and the PFBS algorithms for image inpainting without noise, i.e. $A = \mathcal{P}_\Lambda$, $D = I$ and $\eta = 0$. Level of framelet decomposition is chosen to be 1 for optimal quality of restored images. One can see that the APG algorithm is more efficient than the PFBS algorithm while the quality of restored images is comparable. Some image results are shown in Figure 1.

Inpainting results using the two-system balanced approach (4.12) are also presented in Table 1. It is worth noticing that for images containing both cartoons and textures, e.g. "barbara512", the two-system balanced approach (4.12) works better than the single-system balanced approach (4.8). Such improvement is shown in Figure 2, where one can see that textures are better restored using the two-system balanced approach (4.12), because in general, textures can be sparsely approximated by local cosine transforms.

Table 1: Comparisons of PFBS and APG algorithms: image inpainting without noise (i.e., $\eta = 0$ in (4.1)). Level of framelet decomposition is taken to be 1.

Inpainting	APG for (4.8)			PFBS for (4.8)			APG for (4.12)		
$\sigma = 0$	$\kappa = 1, \lambda = 0.03$			$\kappa = 1, \lambda = 0.03$			$\kappa^i = 1, \lambda^i = 0.01$		
	iter	psnr	time (s)	iter	psnr	time (s)	iter	psnr	time (s)
peppers256	19	39.38	3.07	59	38.09	9.17	29	39.09	6.11
goldhill256	21	35.17	3.34	60	34.79	10.51	30	35.34	6.47
boat256	22	33.79	3.42	64	33.64	11.30	29	33.78	6.34
camera256	20	32.95	3.23	62	32.62	9.67	29	33.23	7.04
bridge256	24	34.85	3.80	74	33.93	11.84	33	34.39	7.64
bowl256	19	36.93	2.97	59	36.45	9.11	35	38.54	7.90
barbara512	23	34.48	24.44	88	33.49	91.27	30	39.08	39.66
baboon512	24	31.68	25.28	73	30.98	76.28	32	31.65	41.61
fingerprint512	23	32.89	24.25	73	30.41	77.25	34	33.46	44.79
zebra512	23	34.38	24.42	67	33.42	69.18	33	34.91	42.44

Results of the synthesis based approach, i.e. (4.8) and (4.12) with $\kappa = 0$, are presented in Table 2. The level of framelet decomposition is chosen to be 2 for optimal quality of restored images. When comparing Table 2 with Table 1 one can see that the balanced approach usually produces better results than the synthesis based approach. This is due to the lack of smoothness for the synthesis based approach ($\kappa = 0$).

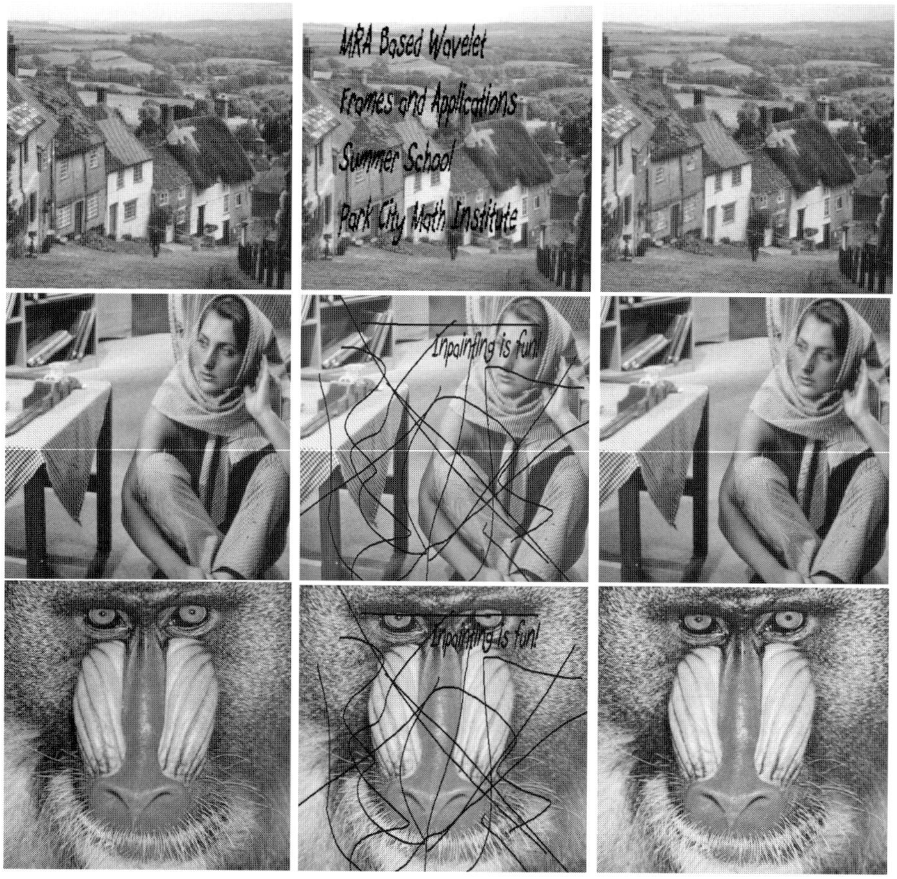

FIGURE 1. Inpainting results for "goldhill256" (first row), "barbara512" (second row) and "baboon512" (third row) using the single-system balanced approach (4.8) solved by APG algorithm (images taken from Table 1). Columns 1-3 are the original image, observed image and restored image. Gold Hill is licensed under Creative Commons License Attribution-ShareAlike 2.5 Wikimedia Commons by Sean Davis, 2004.

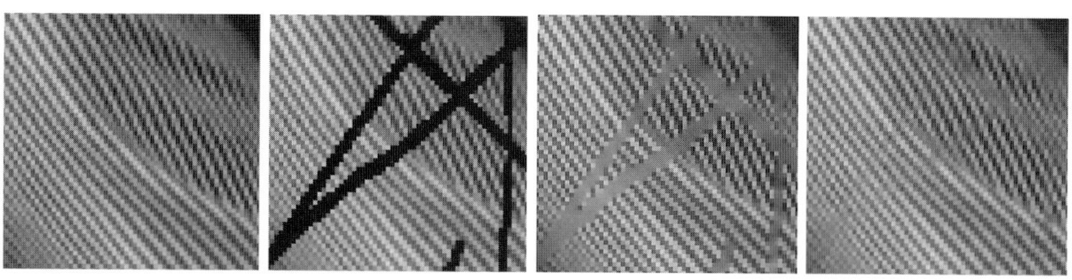

FIGURE 2. Comparison of image inpainting using the single-system balanced approach (4.8) and the two-system balanced approach (4.12). Images from left to right are: original image, observed image, and restored image using single- and two-system. The image presented is a zoom-in view of "barbara51" (taken from Table 1).

Table 2: Numerical results of APG algorithm in solving synthesis based method, i.e. (4.8) with $\kappa = 0$, for image inpainting without noise. Level of framelet decomposition is taken to be 2.

Inpainting	Single-System			Two-Systems		
$\sigma = 0$	$\lambda = 0.03$			$\lambda_1 = \lambda_2 = 0.01$		
	iter	psnr	time	iter	psnr	time
peppers256	51	39.00	13.51	83	39.36	26.02
goldhill256	62	33.45	16.71	78	34.40	24.42
boat256	61	32.93	16.65	29	33.52	9.22
camera256	67	31.27	17.82	29	32.53	9.39
bridge256	79	33.41	21.15	81	33.99	25.80
bowl256	59	35.78	15.98	29	37.35	9.22
barbara512	59	32.48	102.97	186	36.39	367.66
baboon512	81	29.86	143.44	29	30.36	57.35
fingerprint512	75	34.66	132.93	188	34.93	367.11
zebra512	80	33.90	141.98	84	34.95	165.97

2.6.2. *Image Deblurring.* Here, we present numerical results using the single- and two-system balanced approach (4.8) and (4.12) solved by the APG algorithm for image deblurring with noise. The matrix A in (4.1) is taken to be a convolution matrix with the corresponding kernel being a Gaussian function (generated in MATLAB by "fspecial('gaussian',15,1.5);" to be precise) and η is generated from a zero mean Gaussian distribution with variance $\sigma = 3$. Also, numerical results of image deblurring using the (preconditioned) linearized Bregman algorithm are presented.

For all the methods we used (i.e. the APG algorithm for the single- and two-system balanced approach, and the linearized Bregman algorithm), we choose $\theta = 0.01$ for the matrix $D = (AA^\top + \theta I)^{-1}$ which can be computed efficiently via fast Fourier transform (FFT). We choose the level of framelet decomposition to be 1 for the AGP algorithms and 4 for linearized Bregman algorithm for optimal quality of the restored images. Finally, we note that for the linearized Bregman algorithm, we use the following stopping criterion:

$$\mathrm{std}(AW^\top \alpha_k - f) < \sigma.$$

In Table 3, we summarize the deblurring results for all of three algorithms, and the corresponding images are presented in Figure 3. Note that all three models produce comparable results in terms of PSNR values. However, the APG algorithm for both the single- and two-system balanced approaches are faster than the linearized Bregman algorithm. Also, we show a zoom-in view for "barbara" in Figure 4 where we can see that texture components of the image are best recovered by using the two-system balanced approach (4.12).

Table 3: Comparisons of APG algorithm for both single-system balanced approach (4.8) and two-system balanced approach (4.12), and linearized Bregman algorithm for (4.45).

deblurring	Single-System		Two-Systems		Linearized Bregman	
$\sigma = 3$	$\kappa = 1, \lambda = 0.005$		$\kappa^i = 1, \lambda^i = 0.005$		$\kappa = 1, \lambda = 0.01$	
	psnr	time (s)	psnr	time (s)	psnr	time (s)
barbara	24.64	20.7	24.75	63.2	24.67	126.2
duck	31.15	17.5	31.04	41.8	30.87	100.1
wolf	30.83	25.9	30.6	72.6	30.66	186.5

FIGURE 3. Deblurring results for "barbara", "duck" and "wolf". Columns 1-5 are: the original image, observed image, results of single-system, results of two-system and results of the linearized Bregman algorithm.

FIGURE 4. Zoom-in to the texture part of "barbara". Images from left to right are: original image, observed image, results of single-system, results of two-system and results of linearized Bregman algorithm.

3. Analysis based approach

In this section, we present the split Bregman algorithm [**96**] for a more general analysis based approach:

$$\min_{u\in\mathbb{R}^n} \quad H(u) + \|\text{diag}(\lambda)Wu\|_1, \tag{4.49}$$

where $H(\cdot)$ is some smooth convex function. In particular, when

$$H(u) = \frac{1}{2}\|Au - f\|_D^2,$$

(4.49) coincides with (4.11) given in the introduction. The optimization problem (4.49) can be solved by the split Bregman algorithm which was proposed by [**96**]. We provide a convergence analysis of the split Bregman algorithm following the proof given by [**24**].

3.1. Split Bregman algorithm

The ℓ_1 term involved in (4.49) is nonsmooth and nonseparable. This prevents us from using optimization methods for smooth functions. A natural idea to solve (4.49) is to use a smoothed ℓ_1 norm to approximate the actual ℓ_1 norm and then apply optimization methods for smooth functions. This approach is commonly used in some early algorithms for PDE based, particularly for total variation (TV) based, variational models for image processing

$$\min_{u\in\mathbb{R}^n} \quad H(u) + \lambda|\nabla u|,$$

(see e.g., [**180, 40**]). However, in these smoothing methods, the better the approximations (to the ℓ_1 norm) are, the slower the algorithms converge. Therefore, in order to make these algorithms converge faster, the smooth approximation to the ℓ_1 norm cannot be too accurate. Consequently, one cannot obtain a sparse solution by solving the minimization problem with a smoothed norm. However, sparsity is important in many cases of ℓ_1 regularization problems.

Another difficulty for the analysis based approach of (4.49) is that the term $\|\text{diag}(\lambda)Wu\|_1$ is not separable. Therefore, one cannot simply use soft thresholding as one normally does in the synthesis based approach, since it is impossible to keep any sequence in the range of W after applying a thresholding operator. An iterative method to find a solution in the range of W for a special case of (4.49), where $H(u) = \frac{1}{2}\|u - f\|^2$, was proposed by [**33**] by exploring duality.

As mentioned above, compared with the synthesis based approach, the ℓ_1 norm of $|Wu|$ involved in (4.49) is neither smooth nor separable. To overcome this, one transfers (4.49) to a problem involving only separable nonsmooth terms. This is the main idea of split Bregman. In particular, one replaces the term Wu in (4.49) by a new variable d first and then adds a new constraint $d = Wu$ into (4.49). Hence, (4.49) becomes

$$\min_{u,d} \; H(u) + \|\text{diag}(\lambda)d\|_1 \quad \text{subject to} \quad d = Wu, \tag{4.50}$$

In order to solve (4.50), an iterative algorithm based on the Bregman distance with an inexact solver was proposed in [**96**]. This leads to the alternating split Bregman algorithm for (4.50). The split Bregman algorithm for (4.50) was first demonstrated in [**96**] to be an efficient tool for solving problems arising from TV norm minimization problems of PDE based models for image restoration, such as denoising.

The derivation of the split Bregman algorithm in [**96, 24**] is based on Bregman distance. The split Bregman method can also be understood as the augmented Lagrangian method (see e.g. [**94**]) applying to (4.50) (see e.g.[**85, 171**]). Here, we use the augmented Lagrangian method to derive the split Bregman algorithm, since this connects to the Uzawa's algorithm as we discussed for the linearized Bregman algorithm. Note that (4.50) is equivalent to the following optimization problem

$$\min_{u,d} H(u) + \|\mathrm{diag}(\lambda)d\|_1 + \frac{\mu}{2}\|Wu - d\|_2^2 \quad \text{subject to} \quad d = Wu. \tag{4.51}$$

The Lagrangian for problem (4.51), which is called the augmented Lagrangian, is given as

$$\mathcal{L}_\mu(u,d,p) := H(u) + \|\mathrm{diag}(\lambda)d\|_1 + \langle p, d - Wu\rangle + \frac{\mu}{2}\|Wu - d\|_2^2.$$

The saddle points of $\mathcal{L}_\mu(u,d,p)$ can be obtained by the following iterative procedure

$$\begin{cases} (u_{k+1}, d_{k+1}) = \arg\min\limits_{u,d} H(u) + \|\mathrm{diag}(\lambda)d\|_1 + \langle p_k, d - Wu\rangle + \frac{\mu}{2}\|Wu - d\|_2^2, \\ p_{k+1} = p_k + \delta\mu(d_{k+1} - Wu_{k+1}), \end{cases}$$

which consists of one step of joint optimization of variables (u,d) followed by one step of gradient ascend of variable p. Now, letting $b_k = -p_k/\mu$, we have

$$\begin{cases} (u_{k+1}, d_{k+1}) = \arg\min\limits_{u,d} H(u) + \|\mathrm{diag}(\lambda)d\|_1 + \mu\langle b_k, Wu - d\rangle + \frac{\mu}{2}\|Wu - d\|_2^2, \\ b_{k+1} = b_k + \delta(Wu_{k+1} - d_{k+1}), \end{cases}$$

which is equivalent to

$$\begin{cases} (u_{k+1}, d_{k+1}) = \arg\min\limits_{u,d} H(u) + \|\mathrm{diag}(\lambda)d\|_1 + \frac{\mu}{2}\|Wu - d + b_k\|_2^2, \\ b_{k+1} = b_k + \delta(Wu_{k+1} - d_{k+1}). \end{cases}$$

Now, if one alternatively optimizes the variables u and d in the first equation above, we have the split Bregman algorithm as follows

$$\begin{cases} u_{k+1} = \arg\min\limits_{u} H(u) + \frac{\mu}{2}\|Wu - d_k + b_k\|_2^2, \\ d_{k+1} = \arg\min\limits_{d} \|\mathrm{diag}(\lambda)d\|_1 + \frac{\mu}{2}\|d - Wu_{k+1} - b_k\|_2^2, \\ b_{k+1} = b_k + \delta(Wu_{k+1} - d_{k+1}). \end{cases} \tag{4.52}$$

Since $H(u)$ is convex and differentiable, the subproblem in the first line is easy to solve. Further, noting that the first term of the subproblem in the second line is the ℓ_1 norm, the subproblem in the second line can be solved by a simple soft shrinkage. These features make the iteration efficient and fast for many problems that are difficult to solve by other means. Besides its speed, the split Bregman method has several advantages. It has a relatively small memory footprint when compared to second order methods that require explicit representations of the Hessian matrix. Also, the method is easy to code. Both of these characteristics make this split Bregman method a practical algorithm for large scale problems.

For image restoration problems (4.1), we choose $H(u) = \frac{1}{2}\|Au - f\|_D^2$ and then (4.50) becomes

$$\min_{u} \frac{1}{2}\|Au - f\|_D^2 + \|\mathrm{diag}(\lambda)d\|_1 \quad \text{subject to} \quad d = Wu, \tag{4.53}$$

which is equivalent to the analysis based approach (4.11). We summarize the split Bregman algorithm (4.52) for the analysis based approach in the following algorithm.

ALGORITHM 4.7 (Split Bregman).

(i) *Set initial guess* d_0 *and* b_0.

(ii) *For* $k = 0, 1, \dots$, *perform the following iterations until convergence*

$$\begin{aligned} u_{k+1} &= (A^\top DA + \mu I)^{-1}(A^\top Df + \mu W^\top(d_k - b_k)), \\ d_{k+1} &= \mathcal{T}_{\lambda/\mu}(Wu_{k+1} + b_k), \\ b_{k+1} &= b_k + \delta(Wu_{k+1} - d_{k+1}). \end{aligned} \tag{4.54}$$

Note that when A is diagonalizable by discrete Fourier transform, e.g. a convolution matrix, the first equation of (4.54) can be solved efficiently by FFT.

3.2. Convergence analysis of split Bregman algorithm

Now, we provide a convergence analysis of the split Bregman algorithm (4.52) for a general convex differentiable function $H(u)$. In particular, the convergence analysis applies to Algorithm 4.7 for image restoration problems. We note that, since all the subproblems involved in (4.52) are convex, the first order optimality condition gives

$$\begin{cases} 0 = \nabla H(u_{k+1}) + \mu W^\top(Wu_{k+1} - d_k + b_k), \\ 0 = p_{k+1} + \mu(d_{k+1} - Wu_{k+1} - b_k), \quad \text{with} \quad p_{k+1} \in \partial\|\mathrm{diag}(\lambda)d_{k+1}\|_1, \\ b_{k+1} = b_k + \delta(Wu_{k+1} - d_{k+1}). \end{cases} \tag{4.55}$$

This simple observation will be used in the proof of the convergence of the unconstrained split Bregman method (4.52). Note that we will not use the fact that $W^\top W = I$, because it is not required by the proof. Therefore, the following convergence analysis also applies when W is replaced by any other linear operator like ∇ (see [**24**] for details).

THEOREM 4.8. *Assume that there exists at least one solution* $u^\star$ *of* (4.49). *Assume that* $0 < \delta \leqslant 1$ *and* $\mu > 0$. *Then, we have the following properties for the unconstrained split Bregman algorithm* (4.52):

$$\lim_{k\to+\infty} \|\mathrm{diag}(\lambda)Wu_k\|_1 + H(u_k) = \|\mathrm{diag}(\lambda)Wu^\star\|_1 + H(u^\star). \tag{4.56}$$

Furthermore,

$$\lim_{k\to+\infty} \|u_k - u^\star\|_2 = 0 \tag{4.57}$$

whenever (4.49) *has a unique solution.*

PROOF. Let $u^\star$ be an arbitrary solution of (4.49). By the first order optimality condition, $u^\star$ must satisfy

$$0 = W^\top p^\star + \nabla H(u^\star), \tag{4.58}$$

where $p^\star \in \partial\|\mathrm{diag}(\lambda)d^\star\|_1$ with $d^\star = Wu^\star$. Let

$$b^\star = \frac{1}{\mu}p^\star.$$

We obtain

$$\begin{cases} 0 = \nabla H(u^\star) + \mu W^\top(Wu^\star - d^\star + b^\star), \\ 0 = p^\star + \mu(d^\star - Wu^\star - b^\star), \quad \text{with} \quad p^\star \in \partial\|\mathrm{diag}(\lambda)d^\star\|_1, \\ b^\star = b^\star + \delta(Wu^\star - d^\star). \end{cases} \tag{4.59}$$

Therefore, $(u^\star, d^\star, b^\star)$ is a fixed point of (4.55). Consequently, if the iteration (4.52) converges, it converges to a solution of (4.49) when (4.56) is proven.

Denote the errors by

$$u_k^e = u_k - u^\star, \quad d_k^e = d_k - d^\star, \quad b_k^e = b_k - b^\star.$$

Subtracting the first equation of (4.59) from the first equation of (4.55), we have

$$0 = \nabla H(u_{k+1}) - \nabla H(u^\star) + \mu W^\top(Wu_{k+1}^e - d_k^e + b_k^e).$$

Taking the inner product of the left- and right-hand sides with respect to u_{k+1}^e, we have

$$0 = \langle \nabla H(u_{k+1}) - \nabla H(u^\star), u_{k+1}^e\rangle + \mu\|Wu_{k+1}^e\|_2^2 - \mu\langle W^\top d_k^e, u_{k+1}^e\rangle + \mu\langle W^\top b_k^e, u_{k+1}^e\rangle. \tag{4.60}$$

When similar manipulations are applied to the second equation of (4.55) and the second equation of (4.59), we obtain

$$0 = \langle p_{k+1} - p^\star, d_{k+1} - d^\star\rangle + \mu\|d_{k+1}^e\|_2^2 - \mu\langle Wu_{k+1}^e, d_{k+1}^e\rangle - \mu\langle b_k^e, d_{k+1}^e\rangle, \tag{4.61}$$

where $p_{k+1} \in \partial\|\mathrm{diag}(\lambda)d_{k+1}\|_1$ and $p^\star = \mu b^\star \in \partial\|\mathrm{diag}(\lambda)d^\star\|_1$. By summing (4.60) and (4.61), we get

$$\begin{aligned} 0 = {} & \langle \nabla H(u_{k+1}) - \nabla H(u^\star), u_{k+1}^e\rangle + \langle p_{k+1} - p^\star, d_{k+1} - d^\star\rangle \\ & + \mu\big(\|Wu_{k+1}^e\|_2^2 + \|d_{k+1}^e\|_2^2 - \langle Wu_{k+1}^e, d_k^e + d_{k+1}^e\rangle + \langle b_k^e, Wu_{k+1}^e - d_{k+1}^e\rangle\big). \end{aligned} \tag{4.62}$$

Now, subtracting the third equation of (4.59) from the third equation of (4.55) we obtain

$$b_{k+1}^e = b_k^e + \delta(Wu_{k+1}^e - d_{k+1}^e).$$

Taking the inner product of the above identity with itself and reorganizing the terms, we have

$$\langle b_k^e, Wu_{k+1}^e - d_{k+1}^e\rangle = \frac{1}{2\delta}\big(\|b_{k+1}^e\|_2^2 - \|b_k^e\|_2^2\big) - \frac{\delta}{2}\|Wu_{k+1}^e - d_{k+1}^e\|_2^2. \tag{4.63}$$

Substituting (4.63) into (4.62), we have

(4.64)
$$
\begin{aligned}
&\frac{\mu}{2\delta}\left(\|b_k^e\|_2^2-\|b_{k+1}^e\|_2^2\right)\\
=&\langle \nabla H(u_{k+1})-\nabla H(u^\star), u_{k+1}^e\rangle+\langle p_{k+1}-p^\star, d_{k+1}-d^\star\rangle\\
&+\mu\left(\|Wu_{k+1}^e\|_2^2+\|d_{k+1}^e\|_2^2-\langle Wu_{k+1}^e, d_k^e+d_{k+1}^e\rangle-\frac{\delta}{2}\|Wu_{k+1}^e-d_{k+1}^e\|_2^2\right)\\
=&\langle \nabla H(u_{k+1})-\nabla H(u^\star), u_{k+1}^e\rangle+\langle p_{k+1}-p^\star, d_{k+1}-d^\star\rangle\\
&+\mu\left(\frac{1-\delta}{2}\|Wu_{k+1}^e-d_{k+1}^e\|_2^2+\frac{1}{2}\|Wu_{k+1}^e\|_2^2+\frac{1}{2}\|d_{k+1}^e\|_2^2-\langle Wu_{k+1}^e, d_k^e\rangle\right)\\
=&\langle \nabla H(u_{k+1})-\nabla H(u^\star), u_{k+1}^e\rangle+\langle p_{k+1}-p^\star, d_{k+1}-d^\star\rangle\\
&+\mu\left(\frac{1-\delta}{2}\|Wu_{k+1}^e-d_{k+1}^e\|_2^2+\frac{1}{2}\|Wu_{k+1}^e-d_k^e\|_2^2+\frac{1}{2}\|d_{k+1}^e\|_2^2-\frac{1}{2}\|d_k^e\|_2^2\right).
\end{aligned}
$$

By summing the above equation from $k=0$ to $k=K$, we get

(4.65)
$$
\begin{aligned}
&\frac{\mu}{2\delta}\left(\|b_0^e\|_2^2-\|b_{K+1}^e\|_2^2\right)+\frac{\mu}{2}\|d_0^e\|_2^2\\
=&\sum_{k=0}^{K}\langle \nabla H(u_{k+1})-\nabla H(u^\star), u_{k+1}-u^\star\rangle+\sum_{k=0}^{K}\langle p_{k+1}-p^\star, d_{k+1}-d^\star\rangle\\
&+\mu\left(\frac{1-\delta}{2}\sum_{k=0}^{K}\|Wu_{k+1}^e-d_{k+1}^e\|_2^2+\frac{1}{2}\sum_{k=0}^{K}\|Wu_{k+1}^e-d_k^e\|_2^2+\frac{1}{2}\|d_{K+1}^e\|_2^2\right).
\end{aligned}
$$

By (4.4), and that $0<\delta\le 1$, all terms involved in the right-hand side of the above equation are nonnegative. This observation leads to the following inequality:

$$
\frac{\mu}{2\delta}\|b_0^e\|_2^2+\frac{\mu}{2}\|d_0^e\|_2^2\ge\sum_{k=0}^{K}\langle \nabla H(u_{k+1})-\nabla H(u^\star), u_{k+1}-u^\star\rangle. \tag{4.66}
$$

By the assumption $\mu>0$, we have $\sum_{k=0}^{+\infty}\langle \nabla H(u_{k+1})-\nabla H(u^\star), u_{k+1}-u^\star\rangle<+\infty$, which leads to

$$
\lim_{k\to+\infty}\langle \nabla H(u_k)-\nabla H(u^\star), u_k-u^\star\rangle=0. \tag{4.67}
$$

By (4.3), we have

$$
H(u_k)-H(u^\star)-\langle u_k-u^\star, \nabla H(u^\star)\rangle\ge 0
$$

and

$$
H(u^\star)-H(u_k)-\langle u^\star-u_k, \nabla H(u_k)\rangle\ge 0,
$$

which leads to

$$
0\le H(u_k)-H(u^\star)-\langle u_k-u^\star, \nabla H(u^\star)\rangle\le\langle \nabla H(u_k)-\nabla H(u^\star), u_k-u^\star\rangle.
$$

This, together with (4.67), leads to

$$
\lim_{k\to+\infty}H(u_k)-H(u^\star)-\langle u_k-u^\star, \nabla H(u^\star)\rangle=0. \tag{4.68}
$$

Similarly, (4.65) also provides us the following inequalities

$$\frac{\mu}{2\delta}\|b_0^e\|_2^2 + \frac{\mu}{2}\|d_0^e\|_2^2 \geq \sum_{k=0}^{K}\langle p_{k+1} - p^\star, d_{k+1} - d^\star\rangle$$

and

$$\frac{\mu}{2\delta}\|b_0^e\|_2^2 + \frac{\mu}{2}\|d_0^e\|_2^2 \geq \frac{\mu}{2}\sum_{k=0}^{K}\|Wu_{k+1}^e - d_k^e\|_2^2.$$

Then by noticing that $Wu^\star = d^\star$ and following the same argument as above, we have

$$\lim_{k\to+\infty}\|\text{diag}(\lambda)d_k\|_1 - \|\text{diag}(\lambda)d^\star\|_1 - \langle d_k - d^\star, p^\star\rangle = 0 \tag{4.69}$$

and

$$\lim_{k\to+\infty}\|Wu_{k+1} - d_k\|_2 = 0. \tag{4.70}$$

Since $\|\cdot\|_1$ is continuous, by (4.69) and (4.70), we obtain

$$\lim_{k\to+\infty}\|\text{diag}(\lambda)Wu_k\|_1 - \|\text{diag}(\lambda)Wu^\star\|_1 - \langle Wu_k - Wu^\star, p^\star\rangle = 0. \tag{4.71}$$

Summing this and (4.68) yields

$$\begin{aligned}\lim_{k\to+\infty}\Big(&\big(\|\text{diag}(\lambda)Wu_k\|_1 + H(u_k)\big) - \big(\|\text{diag}(\lambda)Wu^\star\|_1 + H(u^\star)\big)\\ &- \langle u_k - u^\star, \nabla H(u^\star) + W^Tp^\star\rangle\Big) = 0.\end{aligned} \tag{4.72}$$

This, together with (4.58), proves (4.56), and (4.57) follows from Proposition 4.1. ∎

REMARK 4.4. The proof of convergence still works if $H(u)$ is merely convex and continuous (then $H(u)$ is subdifferentiable). In that case, one can simply replace the first line of (4.55) by

$$0 = q_{k+1} + \mu W^\top(Wu_{k+1} - d_k + b_k), \quad q_{k+1} \in \partial H(u_{k+1}), \tag{4.73}$$

and revise accordingly in the proof of Theorem 4.8. However, the equation (4.73) may be much harder to solve than the original one given in (4.55). For many applications, $H(u)$ corresponds to fidelity term and takes the form $H(u) = \frac{1}{2}\|Au - f\|_D^2$, which is differentiable. Then the first equation of (4.55) corresponds to a linear system of equations which can be solved efficiently whenever A is diagonalizable by discrete Fourier transform, e.g. when A is a convolution matrix.

3.3. Simulations

Now, we apply Algorithm 4.7 to image deblurring problems. The matrix A in (4.1) is taken to be a convolution matrix with the corresponding kernel being a Gaussian function (generated in MATLAB by "fspecial('gaussian',15,1.5);") and η is generated from a zero mean Gaussian distribution with variance $\sigma = 3$. For the matrix $D = (AA^\top + \theta I)^{-1}$ we choose $\theta = 0.01$, and we pick the level of framelet decomposition to be 4. For stopping criterion, we use

$$\frac{\|Wu_{k+1} - d_{k+1}\|_2}{\|f\|_2} < 5\times 10^{-4}.$$

Table 4 summarizes the deblurring results of Algorithm 4.7 and the corresponding images are shown in Figure 5. For convenience of comparison, we also present the results from Table 3 and Figure 3. As one can see, the single-system balanced approach outperform the analysis based approach in terms of both quality and speed. The two-system balanced approach is comparable to both Linearized Bregman and split Bregman algorithms, but it works better when images contain both cartoons and textures (see Figure 6).

FIGURE 5. Deblurring results for "barbara", "duck" and "wolf". Columns 1-5 are: the original image, results of single-system, results of two-system, results of the linearized Bregman algorithm and results of the analysis based approach.

FIGURE 6. Zoom-in to texture component of "barbara". Images from left to right are: the original image, results of single-system, results of two-system, results of linearized Bregman algorithm and results of the analysis based approach.

Table 4: Comparisons of APG algorithm for both single-system balanced approach (4.8) and two-system balanced approach (4.12), and linearized Bregman algorithm for (4.45).

Deblurring	Single-System		Two-Systems		Linearized Bregman		Analysis Approach	
$\sigma = 3$	$\kappa = 1, \lambda = 0.005$		$\kappa^i = 1, \lambda^i = 0.005$		$\kappa = 1, \lambda = 0.01$		$\mu = 0.1, \lambda = 0.05$	
	psnr	time (s)	psnr	time (s)	psnr	time (s)	psnr	time (s)
barbara	24.64	20.7	24.75	63.2	24.67	126.2	24.58	32.18
duck	31.15	17.5	31.04	41.8	30.87	100.1	31.00	26.57
wolf	30.83	25.9	30.6	72.6	30.66	186.5	30.75	81.40

LECTURE 5

Other applications of frames

1. Background and models

The previous lecture focused on using sparse approximations of wavelet frames to solve image restoration problems; however, these techniques have much broader applications. In this lecture, we consider some other interesting and important applications of tight frames. In Section 2, we discuss the model proposed in [**18**] on blind deblurring (motion deblurring to be specific) problems. Blind deblurring problems can still be modeled as (4.1), but the convolution kernel that determines A is unknown and needs to be solved simultaneously with u. In Section 3, we present a frame based image segmentation model of [**71**]. This model is no longer of the form (4.1) for image restoration problems. However, the concept of sparse approximation of tight frame systems still applies. More importantly, this model sheds some light on geometric interpretations of tight frames and grants links between tight frames and differential operators. In Section 4, we recall the model proposed by [**112**] on reconstruction of scenes (visible surfaces) from scattered, noisy and possibly sparse range data (point clouds).

Tight frames and the concept of sparse approximation of them have even more applications than described in this lecture. For example, researchers have recently applied frames to tomography (e.g. CT, PET and MRI). Medical images are similar to usual photographs (e.g. images of people, building or nature scenes) in the sense that they can be sparsely approximated by tight frames. Therefore, the basic idea of frame based tomography is to use the ℓ_1-norm to penalize frame coefficients of the image that needs to be recovered with an appropriate fidelity term that modes the specific signal acquisition technique (i.e. types of tomography) and type of noise. Also, applications of tight frames to geometric modeling have been developed.

1.1. Blind deconvolution

Motion blur caused by camera shake has been one of the prime causes of poor image quality in digital imaging, especially when using telephoto lenses or using long shutter speeds. In practice, using long shutter speeds is unavoidable when there is little light in the scene. Many researchers have worked on recovering clear images from motion-blurred images. The motion blur caused by camera shake is usually a spatially-invariant blurring process, which can be modeled as

$$f = u * p + \eta, \tag{5.1}$$

where $*$ is the convolution operator, u is the clear image to recover, f is the observed blurred image and p is the blur kernel (or *point spread function*), and η is the noise. If the blur kernel is given a priori, the model (5.1) is the same as (4.1), and the recovery of the clear image u is called a *non-blind deconvolution* problem; otherwise it is called a *blind deconvolution* problem.

It is known that the non-blind deconvolution problem is an ill-conditioned problem because of its high sensitivity to noise. Blind deconvolution is not only ill-conditioned but also ill-posed since both the blur kernel and the clear image are unknown. Thus the problem is highly under-determined as there are much more unknowns than available information. Since the relative motion between the camera and the scene can be arbitrary and usually hard to estimate a priori, motion deblurring is a typical blind deconvolution problem.

Early works on blind deblurring commonly use a single image and assume a parametric model of the blur kernel p, such that the blur kernel can be obtained by only estimating a few parameters (e.g. [**149**]). In practice, this is usually an oversimplification. In order to handle more complicated motion blurring, multi-image based approaches have been proposed to obtain more information about the blur kernel by either actively or passively capturing multiple images on the scene (e.g., [**3, 5, 46, 153, 172**]).

In recent years, there has been steady progress on removing motion blur from a single image for more general blur kernels. There are two typical approaches. One approach is to use some probabilistic priors on image's edge distribution to derive the blur kernel (e.g., [**89, 119, 127**]) or to manually select blurred edges to obtain the local blur kernel ([**113**]). The main issue with this type of method is that the assumed probabilistic priors do not always hold true for general images, especially for nature images of complex structures. The second approach is to formulate the blind deconvolution as a joint minimization problem with a certain regularization on both the blur kernel p and the clear image u:

$$E(u,p) = \min_{p,g} \Phi(u * p - f) + \lambda_1 \Psi_1(p) + \lambda_2 \Psi_2(u), \tag{5.2}$$

where $\Phi(p * u - f)$ is the fidelity term, $\Psi_1(p)$ and $\Psi_2(u)$ is the regularization term on the kernel and the clear image respectively.

In [**2, 44, 49**] TV-based approaches were proposed to solve general blind deconvolution problems, where the fidelity term in (5.2) is the usual ℓ_2 norm on image intensities; and both regularization terms in (5.2) are total variation measurement (ℓ_1 norm on function derivatives). In [**166**] a more sophisticated minimization model was presented where the fidelity term is a weighted ℓ_2 norm on both similarities in image intensity and similarities in the image's derivatives; and the regularization terms are the ℓ_1 norm of the blur kernel's derivatives and the ℓ_1 norm of the image intensities. This method demonstrated impressive performance on recovering motion-blurred images containing moderately complex structures. However, the minimization model proposed in [**166**] is quite complicated and requires a fairly accurate prior on the size of the blur kernel. Moreover, the recovered images from this method, as well as other TV-based methods, were lacking sufficient details and textures on the regions of complex structure, which limited their applications in practice.

In Section 2, we follow the tight frame based blind deconvolution model of [**18**]. This model takes into account both the sparsity of the image u and that of the kernel p. To be precise, both the image and the kernel are assumed to have sparse representation under two different quasi-affine tight frame systems. A fast algorithm is provided in [**18**] based on the linearized Bregman algorithm. Here, we recall the general model of [**18**] and leave the details to Section 2. The tight frame based blind deconvolution model can be found in [**19, 20**] as well.

Given a blurred image f, our goal is to recover the clear image u and the blur kernel p from (5.1):

$$f = u * p + \eta.$$

Let α denote the framelet coefficients of the clear image u with $u = W_1^\top \alpha$, and let β denote the framelet coefficients of the blur kernel p with $p = W_2^\top \beta$. Then,

$$f = (W_1^\top \alpha) * (W_2^\top \beta) + \eta.$$

Thus, we formulate the *frame based blind motion deblurring model* as:

$$\min_{\alpha,\beta} \|\alpha\|_1 + \lambda \|\beta\|_1$$

subject to

$$\|(W_1^\top \alpha) * (W_2^\top \beta) - f\|_2 \leqslant \sigma,$$

where W_i, $i = 1, 2$, are the decomposition operators of framelet systems, λ is the parameter which balances the sparseness between the kernel and the image, and σ is the noise parameter of the observed image f. The solution of this minimization is approximated by applying the linearized Bregman algorithm alternatively as shown in Section 2 with full details.

1.2. Segmentation

As we mentioned before, tight frames give redundant representations to signals and images, and the redundancy of tight frames usually leads to sparse approximation of images, which is known to be a desirable property for image restoration problems, like denoising, inpainting, deblurring, etc. Also, there is some research on texture classification and segmentation using wavelets or wavelet frames [**178, 1**]. However, utilizing the property of sparse approximation of tight frames for general image segmentation problems has not been considered in the literature. In Section 3 we recall a frame based segmentation model by [**71**].

The segmentation model of [**71**] is applicable to general images. However, we focus on medical images. Segmenting biological structures, e.g. cortical or subcortical structures, blood vessels, tumors etc., from various types of medical images (e.g. CT, MRI, ultrasound, etc.) is very important for detecting abnormalities, studying and tracking progress of diseases, and surgery planning. Medical image segmentation is a difficult problem due to medical images commonly having poor contrasts, different types of noise, and missing or diffusive boundaries. Numerous algorithms have been developed in the literature targeting either general segmentation problems or the segmentation of specific biological structures (see [**48, 133, 126, 176, 183, 128, 179, 150, 174, 92, 139, 122**] and the references therein).

Now, we give a brief preview of the frame based segmentation model of [**71**] and leave the details to Section 3. Since it is confusing to denote images as vectors in $\mathbb{R}^n$ for segmentation problems, we shall understand image f as s-dimensional arrays, i.e. $f \in \mathbb{R}^{N_1 \times N_2 \times \cdots \times N_s} =: \mathcal{S}_s$, with s typically 2 or 3. We still denote the fast framelet decomposition and reconstruction as W and $W^\top$, while they should be understood as linear operators, instead of matrices, defined precisely as in Algorithm 2.3. We note that we can use the same notation for image restoration problems because, mathematically speaking, it makes no difference.

For a given observed image $f \in \mathcal{S}_s$ (Figure 1(a)), a typical objective of segmentation is to divide the image domain into two sub-domains: one is the domain of the object of interest Ω (white region in Figure 1(b)) and the other is the background

Ω^c (black region in Figure 1(b)). The quality of segmentation is usually measured by: (1), how well the boundary of Ω (Figure 1(c)), denoted as $\partial\Omega$, fits the boundary of the object of interest in f; (2), the smoothness of $\partial\Omega$. Here (2) is an important requirement when noise or other types of error are present in the image f.

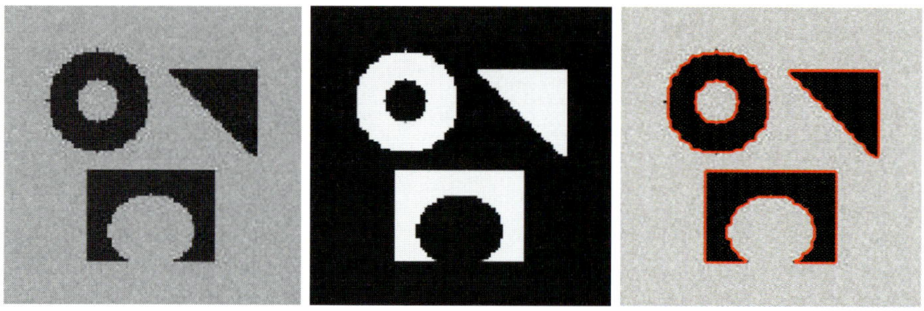

(a) (b) (c)

FIGURE 1. Figure (a) shows an observed image. Figure (b) shows the domain decomposition based on some segmentation where the white regions indicate the object of interest while the black regions indicate the background. Figure (c) shows the boundary of the segmented object (red curves).

One of the most popular ways of modeling segmentation problems is to use level set formulations. The key idea of the level set based segmentation models (see e.g. [**41, 146, 148, 164**]) is to find an appropriate level set function $u \in \mathcal{S}_s$ such that $\Omega := \{u \geq \alpha\}$ provides a desired segmentation of the given image f for some α ($\alpha = 0$ is commonly used). The beauty of the level set formulation is that it models general domains (not necessarily connected) with complicated boundaries, and the quality of segmentation is easily controlled by adding appropriate penalties to the level set function u.

The frame based segmentation model of [**71**] combines the property of sparse approximation of tight frame systems to piecewise smooth functions with the idea of the level set method. We now illustrate how the frame based segmentation model of [**71**] is motivated. The idea is motivated by that of [**43, 45, 12**] where TV-based segmentation models are considered.

Let χ_Ω denotes the characteristic function supporting on domain Ω. A desired segmentation of a given image $f \in \mathcal{S}_s$ should be a solution to the following non-convex optimization problem

$$\min_{\Omega; c^1, c^2 \in \mathbb{R}} \|\lambda \cdot W\chi_\Omega\|_1 + \mu\Big(\|\chi_\Omega(c^1 - f)\|_2^2 + \|\chi_{\Omega^c}(c^2 - f)\|_2^2\Big). \tag{5.3}$$

REMARK 5.1.

(1) The first term of (5.3) controls the regularity of $\partial\Omega$. The frame transform operator W can be understood as a certain discretization of some differential operator. For example in 2D, when W corresponds to the piecewise linear wavelet, W contains difference operators that approximate $\partial_x u$ and $\partial_y u$ with the corresponding 2D filters $[1,2,1]^\top \cdot [1,0,-1]$ and $[1,0,-1]^\top \cdot [1,2,1]$; also it contains difference operators that approximate $\partial_{xx} u$ and $\partial_{yy} u$ with the corresponding 2D filters $[1,2,1]^\top \cdot [-1,2,-1]$

and $[-1,2,-1]^\top \cdot [1,2,1]$. This observation indicates that with different choices of parameter λ and tight frames, the frame transform operator W approximates different differential operators. Therefore, the first term of (5.3) provides rather rich geometric information of $\partial\Omega$.

(2) The second term of (5.3) is the fidelity term that controls the accuracy of the segmentation. The two constants c^1 and c^2 approximate the average values of f inside and outside of Ω respectively. In general, one can consider c^1 and c^2 as arrays in $\mathcal{S}_s$ that approximate f itself inside and outside of Ω. However, for simplicity, we will only consider c^1 and c^2 as constants here.

(3) Note that when $W = \nabla$, the model (5.3) is known as the piecewise constant Mumford-Shah model studied by [**43**], where the authors employed the level set method and used gradient descent to solve the energy (5.3).

The major difficulty of solving (5.3) is that it is a nonconvex optimization problem which in general is hard to solve. Here, we describe a way that one can partially convexify the problem (5.3) following a similar idea as [**45, 12**].

If one replaces χ_Ω by u and χ_{Ω^c} by $1-u$ in (5.3), it is easy to rewrite the problem (5.3) equivalently as

$$\min_{u\in\{0,1\};c^1,c^2\in\mathbb{R}} \|\lambda\cdot Wu\|_1 + \mu\langle\zeta(c^1,c^2),u\rangle, \tag{5.4}$$

where $u \in \{0,1\}$ forces u to be a binary array, and $\zeta(c^1,c^2)$ is defined as

$$\zeta(c^1,c^2) := (c^1-f)^2 - (c^2-f)^2.$$

The problem (5.4) is equivalent to (5.3) in the sense that χ_Ω is a solution to (5.4) if and only if Ω is a solution to (5.3), for any given c^1 and c^2. Note that the objective function of (5.4) is now convex with respect to u, while the non-convexity is moved to the constraints $u \in \{0,1\}$, which are usually referred to as binary constraints.

Optimization problems with binary constraints are usually very hard to solve due to the existence of local minimum. One common way of relaxing binary constraints like $u \in \{0,1\}$ is to use $0 \le u \le 1$ instead. This leads to the *frame based segmentation model* of [**71**] given as follows:

$$\min_{0\le u\le 1;c^1,c^2\in\mathbb{R}} \|\lambda\cdot Wu\|_1 + \mu\langle\zeta(c^1,c^2),u\rangle. \tag{5.5}$$

Note that (5.5) is still a non-convex problem; however, with fixed c^1 and c^2 it is a convex problem.

When c^1 and c^2 are fixed, an optimal solution $u^\star$ of (5.5) is generally not a solution of (5.4), unless $u^\star$ happens to be binary itself. When $u^\star$ is not binary, a segmentation of image f can be obtained by taking $\Omega_\alpha := \{u^\star \ge \alpha\}$ for some $\alpha \in [0,1]$. Now the question is, for which α, is the function χ_{Ω_α} a solution to (5.4)? For the case $W = \nabla$, this question is answered by [**45, 12**], where they showed that for almost every $\alpha \in [0,1]$, χ_{Ω_α} is a solution to (5.4) (when $W = \nabla$, the model (5.4) is known as the piecewise Mumford-Shah model first considered by [**43**]). Although there is no corresponding theory for general tight frames where $W \ne \nabla$, our numerical simulations indicate that a similar result holds. Note that for different α, the set Ω_α is generally different. This raises the question of which choice of α results in Ω_α providing the best segmentation result. Recall from the theory of linear programming that an optimizer of a linear program must lie on one of the extreme points of the feasible set (see e.g. [**11**]). Therefore, when μ

is comparably large, the problem (5.5) is close to a linear program, and thus $u^\star$ is close to a binary array (which is confirmed by numerical experiments). As a consequence, choices of α are not crucial and we choose $\alpha = 0.5$.

1.3. Scene reconstruction from range data

High-quality 3D scene modeling has long been an important research topic in computer vision, robotic navigation, computer graphics and animation. The 3D geometrical model of a scene is usually reconstructed from pre-acquired range data sets of the scene, which are usually very noisy, corrupted with a lot of outliers, and sparsely sampled with large gaps. In the past, there have been extensive studies on reconstructing objects or scenes using range data from a single view or from multiple views. Interested readers are referred to a recent survey [**100**, Chapter 4] for more details. In Section 4, we focus on how to reconstruct the scene model using 3D range data of a single view. Since range data from a single view can only provide 3D information for the visible surfaces of the scene, a piecewise smooth explicit surface model is usually adequate to describe visible surfaces of a scene. It is noted that discontinuities on the reconstructed piecewise smooth surface provide very important information for many applications (e.g. robotics), because surface discontinuities represent either the boundaries of objects in the scene or sharp geometrical changes of individual objects. Thus, in contrast to what most traditional surface-fitting methods tend to do, a desired reconstruction algorithm should well preserve surface discontinuities.

Reconstructing a piecewise smooth surface using range data can be formulated as a function reconstruction problem. Assume that we are given a set of scattered data sites:

$$\Xi = \{x[1], x[2], \ldots, x[n]\} \subset \mathbb{R}^2$$

and associated function values

$$f|_\Xi = \{f[1], f[2], \ldots, f[n]\},$$

where f_i is the function value of an unknown data function $f(x)$ at $x[i]$ and possibly contains noise, i.e.,

$$f[i] = f(x[i]) + \epsilon[i].$$

Our goal is then to reconstruct the data function f under the assumption that f is a piecewise smooth function. It is emphasized that the discontinuities of f need to be well preserved in the reconstruction, because they encode important information about boundaries of objects on which many high-level tasks depend. On the other hand, the input data sites Ξ can be scattered such that they are non-uniformly sampled, with large gaps. Furthermore, the obtained function values $f|_\Xi$ are usually corrupted by different types of noise, e.g. Gaussian or impulsive noise. All of these challenges make reconstruction a rather difficult problem.

The frame based model of [**112**] is motivated by work in [**107, 117, 169**] which use a simple principal shift invariant space and its associated wavelet transform to fit scattered data, [**118**] proposed an efficient algorithm to approximate the solution of a smoothing spline in a principal shift invariant space.

Let $\Omega \subset \mathbb{R}^2$ be a bounded domain of interest that contains all data, and let ϕ be a carefully chosen compactly supported continuous function (e.g. uniform B-splines, box splines, nodal basis functions). We look for fitting functions in the space spanned by 2^{-L}-dilates and 2^{-L}-shifts of ϕ whose support intersects Ω, i.e.,

$$S^L(\phi,\Omega)=\left\{\sum_{k\in\mathbb{Z}^2}c(k)\phi(2^L\cdot-k):c(k)=0\text{ whenever }\operatorname{supp}\phi(2^L\cdot-k)\cap\Omega=\emptyset\right\},$$

where $L\in\mathbb{Z}^+$ is a scaling parameter that controls the refinement of the space. Then, any fitting function $s(x)\in S^L(\phi,\Omega)$ can be written as a finite expansion:

$$s(x)=\sum_{k\in\mathcal{I}}u[k]\phi(2^Lx-k), \tag{5.6}$$

where $\mathcal{I}:=\{k\in\mathbb{Z}^2:\operatorname{supp}\phi(2^L\cdot-k)\cap\Omega\neq\emptyset\}$. Let $\boldsymbol{u}$ and $\boldsymbol{f}$ denote the column vector $\{u[k]\}_{k\in\mathcal{I}}$ and $\{f[i]\}_{1\le i\le n}$ respectively. Our goal here is to find $\boldsymbol{u}$ such that $s(x)$ approximates the unknown function $f(x)$ based on the observed data $\boldsymbol{f}$. The function $f(x)$ that we are approximating is generally piecewise smooth and can be sparsely approximated by tight frame systems. Therefore, the key idea of [**112**] is to reconstruct $s(x)$ such that $s(x)$ can be sparsely approximated by certain tight frame systems, which naturally leads to the penalization of the ℓ_1-norm of tight frame coefficients of $s(x)$ as shown in the previous lecture. Therefore, we will obtain a desired $\boldsymbol{u}$ by solving the following problem

$$\min_{\boldsymbol{u}}\ \frac{1}{2}\|A\boldsymbol{u}-\boldsymbol{f}\|_2^2+\|\operatorname{diag}(\lambda)W\boldsymbol{u}\|_1,$$

where

$$A(i,k)=\phi(2^Lx[i]-k),\quad i=1,2,\ldots,n,\ k\in\mathcal{I},$$

for some given $L\in\mathbb{Z}^+$. We shall now leave the details to Section 4.

2. Frame based blind deconvolution

As we see from (5.1), blind deblurring is an under-constrained problem with many possible solutions. Thus we need to enforce extra constraints on the image and the kernel to overcome the ambiguity. In past years, there have been extensive studies showing that most natural images usually have a sparse approximation under some tight frame systems, which include, for example, shift-invariant orthonormal wavelets, Gabor transform, framelet, curvelet, etc. This inspires the authors in [**18**] to use the sparseness of the image under the suitable tight frame system as the constraint on the image. The same sparseness constraint is also applicable to the kernel if we understand the kernel as a special image.

2.1. Problem formulation and analysis

Given a blurred image f, our goal is to recover the clear image u and the blur kernel p from (5.1):

$$f=u*p+\eta.$$

Let W_1 and W_2 denote the fast framelet decomposition corresponding to two possibly different tight frame systems. Denote α and β the framelet coefficients of the clear image u with $u=W_1^\top\alpha$ and the blur kernel p with $p=W_2^\top\beta$. Then,

$$f=({W_1}^\top\alpha)*({W_2}^\top\beta)+\eta.$$

Thus, we can formulate the blind motion deblurring problem as:

$$E(\alpha,\beta)=\min_{\alpha,\beta}\|\alpha\|_1+\lambda\|\beta\|_1 \tag{5.7}$$

subject to

$$\|(W_1{}^\top\alpha) * (W_2{}^\top\beta) - f\|_2 \leqslant \sigma, \tag{5.8}$$

where λ is the parameter which balances the sparseness between the kernel and the image, σ is the noise parameter of the observed image f.

It is known that the ℓ_1 norm is a fairly good measurement of the sparseness of vectors. Thus, the first and second term in (5.7) penalize the sparseness of framelet coefficients of the image u and kernel p under two tight frame systems. Since u and p are usually two different types of images, the tight frame systems that can provide an ideally sparse approximation to them should be different. The constraint in (5.8) is the fidelity constraint between the observed and computed image. The minimization problem (5.7-5.8) is a challenging one. In the next section, we present an alternating minimization approach to solve this minimization problem efficiently. The key idea is to adopt a modified version of the linearized Bregman algorithm.

2.2. Numerical algorithm and analysis

The minimization (5.7-5.8) is a joint minimization problem on α and β. It may not always yield a physically meaningful solution because (5.7-5.8) does not have a unique solution in general. In order to obtain a physically meaningful solution, we chose to impose the following physical conditions:

$$\begin{cases} p = W_2{}^\top\beta \geq 0, \quad \textit{and} \quad \sum p = 1; \\ u = W_1{}^\top\alpha \geq 0. \end{cases} \tag{5.9}$$

Even with the constraint (5.9), the problem (5.7)-(5.8) is not easy to solve. Here, we take the traditional heuristic alternative approach. The outline of the alternative algorithm for solving (5.7)-(5.8) is given as follows:

ALGORITHM 5.1. *Let $\alpha_0 = W_1 f$ and $\beta_0 = W_2\delta_c$ be initial guess, where f is the observed image and δ_c takes value one at the center pixel of the image and zero elsewhere. Iterate on k until convergence.*

(1) *Fixing β_k, solve (5.7-5.8) with respect to α, i.e., set $\alpha_{k+\frac{1}{2}}$ be a solution of*

$$\begin{aligned} \min_{\alpha} \quad & \|\alpha\|_1 \quad \text{subject to} \\ & \|(W_2{}^\top\beta_k) * (W_1{}^\top\alpha) - f\|_2 \leqslant \sigma. \end{aligned} \tag{5.10}$$

Then impose

$$\alpha_{k+1} = W_1 u_{k+1},$$

where

$$u_{k+1}[j] = \begin{cases} W_1{}^\top\alpha_{k+\frac{1}{2}}[j], & \textit{if } W_1{}^\top\alpha_{k+\frac{1}{2}}[j] \geq 0, \\ 0, & \textit{otherwise.} \end{cases}$$

(2) *Fixing α_{k+1}, solve (5.7-5.8) with respect to β, i.e., set $\beta_{k+\frac{1}{2}}$, be a solution of*

$$\begin{aligned} \min_{\alpha} \quad & \|\beta\|_1 \quad \text{subject to} \\ & \|(W_1{}^\top\alpha_{k+1}) * (W_2{}^\top\beta) - f\|_2 \leqslant \sigma. \end{aligned} \tag{5.11}$$

Then impose

$$\beta_{k+1} = W_2 h_{k+1},$$

where

$$h_{k+1}[j] = \begin{cases} W_2^\top \beta_{k+\frac{1}{2}}[j], & \text{if } W_2^\top \beta_{k+\frac{1}{2}}[j] \geq 0, \\ 0, & \text{otherwise,} \end{cases}$$

followed by normalization $h_{k+1} = \frac{h_{k+1}}{\|h_{k+1}\|_1}$.

In each step of the alternative algorithm, there are two minimization problems (5.10) and (5.11), which in fact take exactly the same form. In the following, we present an efficient algorithm to solve such minimization problems, which is derived from the linearized Bregman algorithm (Algorithm 4.6). For (5.10), let $[p]_*$ denote the matrix form of the convolution operator corresponding to the kernel p. Define

$$w_0 = x_0 = 0.$$

Then we generate a sequence of w_i and x_i as follows

$$\begin{cases} w_{i+1} = \nu \mathcal{T}_\mu(x_i), \\ x_{i+1} = x_i - Z_k W_1 [W_2^\top \beta_k]_*^\top ([W_2^\top \beta_k]_* (W_1^\top w_{i+1} - f)), \end{cases} \tag{5.12}$$

where $\mathcal{T}_\mu$ is the soft-thresholding operator defined by (4.15) and (4.16), and $Z^{(k)}$ is a preconditioning matrix to accelerate the convergence of the iteration, which is chosen in the implementation as

$$Z_k = \left([W_2^\top \beta_k]_*^\top ([W_2^\top \beta_k]_*) + \lambda_i \Delta\right)^{-1},$$

where Δ is the discrete Laplacian. The convergence of the sequence $w^{(i)}$ is guaranteed by the following proposition which is a direct consequence of Theorem 4.7. It is proved in [**22**] that the sequence $w^{(i)}$ generated by (5.12) actually converges to an approximated solution of (5.10) when $\nu\mu \to \infty$.

PROPOSITION 5.1. *The sequence w_i generated via (5.12) with a proper ν converges to the unique solution of*

$$\begin{cases} \min_\alpha \|\alpha\|_1 + \frac{1}{2\nu\mu}\|\alpha\|_2^2, \\ \text{s.t. } (W_2^\top \beta_k) * (W_1^\top \alpha) = f, \end{cases} \tag{5.13}$$

if there exists at least one solution of $(W_2^\top \beta_k) * (W_1^\top \alpha) = f$.

PROOF. It is a direct consequence of Theorem 4.7 by noticing that the constraint $(W_2^\top \beta_k) * (W_1^\top \alpha) = f$ is equivalent to $Z_k^{\frac{1}{2}}(W_2^\top \beta_k) * (W_1^\top \alpha) = Z_k^{\frac{1}{2}} f$, and Z_k commutes with convolution operators. ∎

The method (5.12) is extremely efficient. Usually it takes only a few iterations for (5.12) to get an approximate solution of (5.10) (cf. [**22**]), and the accuracy of the approximation is adequate during the alternative iterations in the main algorithm.

For (5.11), the same method can be applied to find the solution: start with $z_i = x_i = 0$, and generate the sequence

$$\begin{cases} z_{i+1} = \nu \mathcal{T}_\mu(x_i), \\ x_{i+1} = x_i - Z_k W_2 [W_1^\top \alpha_k]_*^\top ([W_1^\top \alpha_k]_* (W_2^\top z_{i+1} - f)). \end{cases} \tag{5.14}$$

Then after a few iterations, z_i becomes a good approximated solution to (5.11).

2.3. Numerical experiments

Here, we present deblurring result using Algorithm 5.1 for one image with synthetic motion blur and one real image that was taken by a digital camera using a long shutter speed under a weak light condition. The synthetic motion blur is generated by MATLAB function "fspecial('motion',11,120)" and noise with $\sigma = 3$ is also added to the image. For simplicity, we choose $W_1 = W_2$ as the decomposition matrix corresponding to piecewise linear quasi-affine tight frame system. Numerical results of the two images are given in Figures 2 and 3. For more numerical experiments, one should consult [**18**].

FIGURE 2. Deblurring results for the synthetic motion blur. Columns 1-3 are: the original image, observed blurry and noisy image, and deblurring result of Algorithm 5.1. Rows 2 and 3 show two different close-ups of the corresponding images shown in row 1.

3. Frame based image segmentation

In this section, we review the frame based segmentation model proposed by [**71**]. First, we introduce a few notations that will be used throughout this section.

An observed s-dimensional image f is an array in $\mathcal{S}_s := \mathbb{R}^{N_1 \times \cdots \times N_s}$. Framelet decomposition and reconstruction operators are denoted as W and $W^\top$ which are

FIGURE 3. Deblurring results for the real image. For each row, the left image is the observed blurry and noisy image, and the right image is the deblurred image using Algorithm 5.1. Row 2 shows a close-up of the corresponding images shown in row 1.

given by Algorithm 2.3 and satisfy $W^\top W = I$ with I the identity operator. Here, we use the tensor product quasi-affine tight frame systems discussed in Lecture 2. Suppose there are r univariate tight framelets constructed from UEP. The total number of s-dimensional tight framelets obtained by tensor product is $r_s := (r+1)^s - 1$.

For simplicity of notation, we suppose the level of framelet decomposition is 1. We denote W_0 as the operator that corresponds to the decomposition with respect to the refinable function; and denote W_i, with $1 \le i \le r_s$, as the operator that corresponds to the decomposition with respect to the i-th framelet. Under this notation, we denote $Wu \in \mathcal{S}_s \times \mathbb{R}^{r_s+1}$ as

$$Wu := \big[W_0u, W_1u, W_2u, \ldots, W_{r_s}u\big].$$

3.1. Frame based segmentation model

For a given image $f \in \mathcal{S}_s$, we consider the following optimization problem

$$\min_{0\le u\le 1;\, c^1, c^2\in\mathbb{R}} \|\lambda \cdot Wu\|_1 + \mu\langle \zeta(c^1, c^2), u\rangle, \tag{5.15}$$

where $\zeta(c^1, c^2)$ is defined as

$$\zeta(c^1, c^2) := (c^1 - f)^2 - (c^2 - f)^2$$

with c^1 and c^2 real constants. Here $\lambda \in \mathcal{S}_s \times \mathbb{R}^{r_s+1}$ is a weight function that weights W_iu differently for different i in general. However, we choose the following λ for simplicity

$$\lambda[\cdot, i] = \begin{cases} 0, & i = 0; \\ \frac{1}{1+\sigma\sum_{j=1}^{r_s}|W_jf|^2}, & i = 1, 2, \ldots, r_s. \end{cases}$$

Notice that λ can be regarded as the so-called edge indicator function (see e.g. [**28, 45, 12**]) under the framelet transform W.

To solve (5.15), one can alternatively optimize variables u and c^i, $i = 1, 2$. Since when u is fixed, the optimal values c^i can be easily determined. Therefore, the key step is to optimize (5.15) with c^i fixed. Here, we adopt the idea of the split Bregman algorithm that was discussed in Section 3.

Let $d = Wu$ and write (5.15) equivalently as the following problem

$$\min_{d=Wu} \|\lambda \cdot d\|_1 + \mu\langle\zeta(c^1, c^2), u\rangle + \mathcal{C}_{0\leq u\leq 1},$$

where $\mathcal{C}_{0\leq u\leq 1}$ is the indicator function w.r.t. the set $\{0 \leq u \leq 1\}$. Then, from (4.52) in Section 3, we have the following iterative scheme

$$\begin{aligned} u_{k+1} &= \mathrm{argmin}_u \quad \mu\langle\zeta(c^1, c^2), u\rangle + \mathcal{C}_{0\leq u\leq 1} + \frac{\nu}{2}\|Wu - d_k + b_k\|_2^2, \\ d_{k+1} &= \mathrm{argmin}_d \quad \|\lambda \cdot d\|_1 + \frac{\nu}{2}\|Wu_{k+1} - d + b_k\|_2^2, \\ b_{k+1} &= b_k + \big(Wu_{k+1} - d_{k+1}\big). \end{aligned} \tag{5.16}$$

This, together with the optimization of c^1 and c^2, leads to the following algorithm for (5.15):

ALGORITHM 5.2. *Given image f.*

(1) *Start with $u_0 = 0$, $c^1 = \min f$, and $c^2 = \max f$.*

(2) *For a given $\alpha \in [0, 1]$, iterate until convergence:*

$$\begin{aligned} u_{k+\frac{1}{2}} &= W^\top(d_k - b_k) - \frac{\mu}{\nu}\zeta(c_k^1, c_k^2), \\ u_{k+1} &= \max\{\min\{u_{k+\frac{1}{2}}, 1\}, 0\}, \\ d_{k+1} &= \mathcal{T}_{\lambda/\nu}\left(Wu_{k+1} + b_k\right), \\ b_{k+1} &= b_k + \big(Wu_{k+1} - d_{k+1}\big), \\ c_{k+1}^1 &= M(f, \Omega_{k+1}), \quad c_{k+1}^2 = M(f, (\Omega_{k+1})^c), \quad \Omega_{k+1} = \{u_{k+1} > \alpha\}, \end{aligned} \tag{5.17}$$

where $M(f, \Omega)$ returns the mean value of f within domain Ω.

(3) *Let $u^\star$ be the output of step (2). Then the segmentation of f is given by $\Omega^\star := \{u^\star \geq \alpha\}$.*

Note that Algorithm 5.2 is very efficient. For each iteration k, the most time consuming operation is performing the fast framelet decomposition and reconstruction, which are of the same complexity as fast Fourier transform (FFT) (see Lecture 2, Section 1.4 and 3). Furthermore, numerical experiments show that we usually only need a few hundred iterations until the algorithm converges to a given tolerance.

3.2. Numerical results

In this section, we compare the frame based segmentation model (5.15) with the TV-based segmentation model [**45, 12**] using a 2D-slice of an MRI image and a 3D CT angiography (CTA) image. For more results of 3D CTA image segmentation, please see [**71**].

In our implementation, we adopt the stopping criterion: $\|b^{k+1}\| < 10^{-3}$. Based on this stopping criterion, the number of iterations for 2D and 3D cases varies from 100 to 500. Within each iteration, the comparably expensive operation is the framelet decomposition and reconstruction, i.e. W and $W^\top$. Although the complexity of applying W and $W^\top$ is of the same order as FFT, by applying the fast algorithm of [**65**], in practice the constants really matter. We note, however, that this constant is not big, and hence the framelet decomposition and reconstruction can be done rather efficiently. For example, for a 3D image of size $50 \times 50 \times 50$, the computational time for one level of framelet decomposition and reconstruction is approximately 5–6 times slower than the forward and inverse FFT. This comparison is done using MATLAB2007. Throughout this section, the parameter α in (5.17) is chosen to be 0.5, and the level of framelet decomposition L is chosen to be 1 for the 2D image and 2 for the 3D image for the best quality of segmentations.

We will solve the TV-based model using the split Bregman algorithm as given in [**95**]. Basically if one replace the update for $u^{k+\frac{1}{2}}$ in the first line of (5.17) by

$$\Delta u_{k+\frac{1}{2}} = \nabla \cdot (d_k - b_k) - \frac{\mu}{\nu}\zeta(c_k^1, c_k^2),$$

replace all W by ∇ and λ by g, then we obtain the split Bregman algorithm for TV-based segmentation model (see [**95**] for details). Note that the Laplace equation above is solved by FFT, instead of Gauss-Seidel relaxation as proposed in [**95**].

Since the decomposition operator W for tight frames corresponds to higher order difference operators, the frame based segmentation models should be more sensitive to low contrast boundaries and yet still robust to noise. To test this, we take a clean and high contrast image (first image in Figure 4) and then lower the contrast by 80%, add blurring and noise. We use the corrupted image (second image in Figure 4) as the input for both the TV-based and the frame based segmentation models.

In order to truly show the improvement of using tight frame systems, we pick the same set of parameters (μ, ν) and use the same λ for both the TV-based segmentation model and the frame based segmentation models (5.15). One can see from both Figures 4 and 5 that by using tight frame systems we can capture more features from the images and obtain better segmentations, especially for low contrast and blurry images.

4. Scene reconstruction from range data

4.1. Formulation and algorithm

For a given set of scattered sites $\{x[i]\}_{i=1}^n \subset \Omega \subset \mathbb{R}^2$ and the corresponding data $\{f[i]\}_{i=1}^n$, our task is to approximate the unknown surface $f(x)$ using $s(x) \in S^L(\phi, \Omega)$ by finding an appropriate vector $\boldsymbol{u}$ such that

$$s(x) = \sum_{k \in \mathcal{I}} u[k]\phi(2^L x - k),$$

where ϕ is the 2D tensor product of some uniform B-spline function and

$$\mathcal{I} := \{k \in \mathbb{Z}^2 : \operatorname{supp} \phi(2^L \cdot -k) \cap \Omega \neq \emptyset\}.$$

Since the function f that we want to recover can be sparsely approximated by tight frame systems, the desired $\boldsymbol{u}$ can be obtained by solving the following problem

$$\min_{\boldsymbol{u}} \ \frac{1}{2}\|A\boldsymbol{u} - \boldsymbol{f}\|_2^2 + \|\operatorname{diag}(\lambda)W\boldsymbol{u}\|_1, \tag{5.18}$$

where

$$A(i,k) = \phi(2^L x[i] - k), \quad i = 1, 2, \ldots, n, \ k \in \mathcal{I}.$$

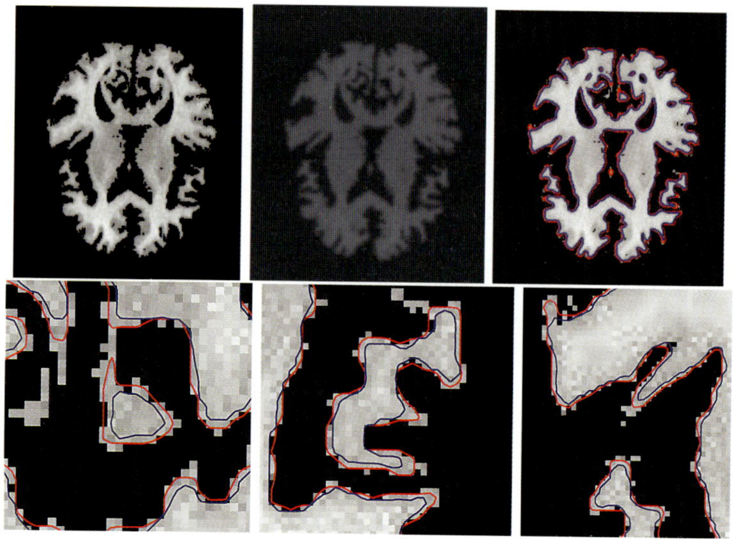

FIGURE 4. Comparisons of TV-based segmentation model with frame based segmentation model (5.15). First row: images from left to right are: original image; observed low contrast blurry image; segmentation results of the TV-based (blue) and the frame based (red) segmentations. Second row: zoom-in views of segmentation results. The set of parameters used is $(\mu, \nu) = (500, 5)$.

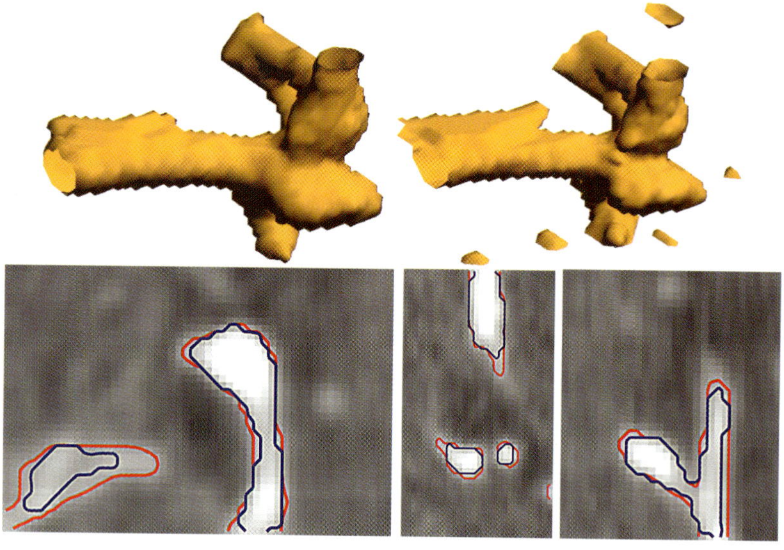

FIGURE 5. First row: comparisons of TV-based segmentation model (left) with frame based segmentation model (5.15) (right). Second row: axial, sagittal and coronal views of the results of the TV-based (blue) and the frame based (red) segmentation. The set of parameters used is $(\mu, \nu) = (100, 0.5)$.

The matrix W is the fast wavelet frame transform associated with the bivariate B-spline tight frame. Note that $W\boldsymbol{u}$ is the canonical frame coefficient sequence of $s(x)$, hence the regularization term $\|\text{diag}(\lambda)W\boldsymbol{u}\|_1$ penalizes the roughness of the fitting function $s(x)$, and also encourages preserving sharp edges as it is biased toward sparse solutions of $s(x)$. We note that the matrix W is constructed via the Neumann boundary condition, which implies that the symmetric boundary condition is used to extend $\boldsymbol{u}$ across the boundaries of Ω.

The key part of the scene reconstruction model is to solve (5.18) efficiently, which can be achieved using split Bregman algorithm given in Section 3 of Lecture 4. For the readers' convenience, we recall the algorithm for problem (5.18). Given initializations $\boldsymbol{u}^0 = 0, \boldsymbol{d}^0 = \boldsymbol{b}^0 = 0$, iterate the following steps until convergence

$$\begin{aligned} \boldsymbol{u}_{k+1} &= (A^\top A + \mu I)^{-1}(A^\top \boldsymbol{f} + \mu W(\boldsymbol{d}_k - \boldsymbol{b}_k)), \\ \boldsymbol{d}_{k+1} &= \mathcal{T}_{\lambda/\mu}(W\boldsymbol{u}_{k+1} + \boldsymbol{b}_k), \\ \boldsymbol{b}_{k+1} &= \boldsymbol{b}_k + \delta(W\boldsymbol{u}_{k+1} - \boldsymbol{d}_{k+1}). \end{aligned} \tag{5.19}$$

The stopping criterion for the iteration (5.19) is chosen as $\|\boldsymbol{d}_k - W\boldsymbol{u}_k\| \leq \epsilon$ with ϵ being some given tolerance. One aspect of split Bregman iteration that contributes to its efficiency is that it is unnecessary to exactly solve the first equation of (5.19). Therefore, we approximate the solution $\boldsymbol{u}_{k+1}$ of the first equation of (5.19) by a few steps of the conjugate gradient method.

In practice, there are usually a lot of outliers in the observed data (See e.g. [**177**]) and it is well-known that an ℓ_2 fidelity term is very sensitive to outliers. When there are outliers in the data, we use the ℓ_1 fidelity term instead:

$$\min_{\boldsymbol{u}} \quad \|A\boldsymbol{u} - \boldsymbol{f}\|_1 + \|\text{diag}(\lambda)W\boldsymbol{u}\|_1, \tag{5.20}$$

The implications of the ℓ_1 data fidelity term have been well studied in the literature (e.g. [**120, 111, 39, 144**]). In particular, [**144**] justified the use of the ℓ_1 data fitting term for processing data corrupted with outliers in both theoretical analysis and numerical experiments.

In terms of Bayesian statistical estimation (see e.g. [**8, 111**]), the models (5.18) and (5.20) can be interpreted as the maximization of a *posteriori* estimator:

$$\max_{\boldsymbol{u}} \text{Exp}\left(-\frac{1}{p}\|A\boldsymbol{u} - \boldsymbol{f}\|_p^p\right) \text{Exp}\left(-\|\text{diag}(\lambda)W\boldsymbol{u}\|_1\right), \tag{5.21}$$

where the first factor is the likelihood that models the noise, and the second factor is the prior of $\boldsymbol{u}$. The prior is usually chosen based on the a priori knowledge we have for the unknown $\boldsymbol{u}$. For example, we know that the surface we need to recover from range image data is piecewise smooth and hence can be sparsely approximated by tight wavelet frames. Therefore, the prior given by (5.21) is reasonable. As for the likelihood, when the noise $\boldsymbol{\eta} := A\boldsymbol{u} - f$ is i.i.d. Gaussian with zero mean, one should use $p = 2$ in order to obtain a statistically correct likelihood. When the noise $\boldsymbol{\eta}$ is impulsive noise, i.e. outliers, one may choose $p = 1$, because impulsive noise is sparse and can be well modeled by the ℓ_1-norm (i.e. we believe that outliers are i.i.d. Laplace distribution). Notice that for $0 < p < 1$, the objective function of (5.21) is not concave. Therefore, although the ℓ_p-norm with $0 < p < 1$ may model impulsive noise better than the ℓ_1-norm, we usually use the ℓ_1-norm in practice for computational concerns.

In contrast to the ℓ_2 fitting term, the ℓ_1 fitting term is not smooth. In our numerical simulations, we approximate the non-differentiable ℓ_1 fitting term by a smoothed version

$$\sum_{i=1}^{n} \sqrt{(A\boldsymbol{u}[i] - \boldsymbol{f}[i])^2 + s} \tag{5.22}$$

to approximate $\|A\boldsymbol{u} - f\|_1$, where s is a small positive number.

To solve the ℓ_1-norm fidelity model (5.20), we again use the split Bregman algorithm. Given initializations $\boldsymbol{u}^0 = 0, \boldsymbol{d}^0 = \boldsymbol{b}^0 = 0$, iterate the following steps until convergence

$$\begin{aligned} \boldsymbol{u}_{k+1} &= \arg\min_{\boldsymbol{u}} \sum_{i=1}^{n} \sqrt{(A\boldsymbol{u}[i] - \boldsymbol{f}[i])^2 + s} + \frac{\mu}{2}\|W\boldsymbol{u} - \boldsymbol{d}^k + \boldsymbol{b}^k\|_2^2, \\ \boldsymbol{d}_{k+1} &= \mathcal{T}_{\lambda/\mu}(W\boldsymbol{u}_{k+1} + \boldsymbol{b}_k), \\ \boldsymbol{b}_{k+1} &= \boldsymbol{b}_k + \delta(W\boldsymbol{u}_{k+1} - \boldsymbol{d}_{k+1}). \end{aligned} \tag{5.23}$$

For the first minimization subproblem of (5.23), since its objective functional is differentiable, it can be solved by the standard gradient descent method. One of the merits of the split Bregman algorithm that contributes to its efficiency is that it is unnecessary to solve the first subproblem of (5.23) to the full convergence. Instead, only a small number of inner iterations will be adequate at each split Bregman algorithm.

We remark that although we are using a smoothed version of the ℓ_1 data fidelity term in the above robust algorithm (5.23), the split Bregman algorithm actually can also be used to minimize an energy functional with both an ℓ_1 data fidelity term and an ℓ_1 regularization term without any smoothing (e.g. L1-TV minimization problem). Interested readers are referred to [**73, 85**] for more details.

4.2. Numerical experiments

The main purpose of this section is to briefly evaluate the performance of the tight frame based model (5.18) and (5.20) on reconstructing a piecewise smooth surface from sparse range data corrupted by Gaussian or impulse noise. More experiments and comparisons with some other existing methods can be found in [**112**]. Range image data are obtained from OSU (MSU/WSU) Range Image Database (http://sampl.ece.ohio-state.edu/data/3DDB/RID/index.htm). Figures 6 and 7 contain the reconstruction results of two scenes, where noise that is added to the original data is Gaussian and impulse plus small Gaussian noise respectively. Note that the input range data are scattered point sets, which are visualized as surfaces.

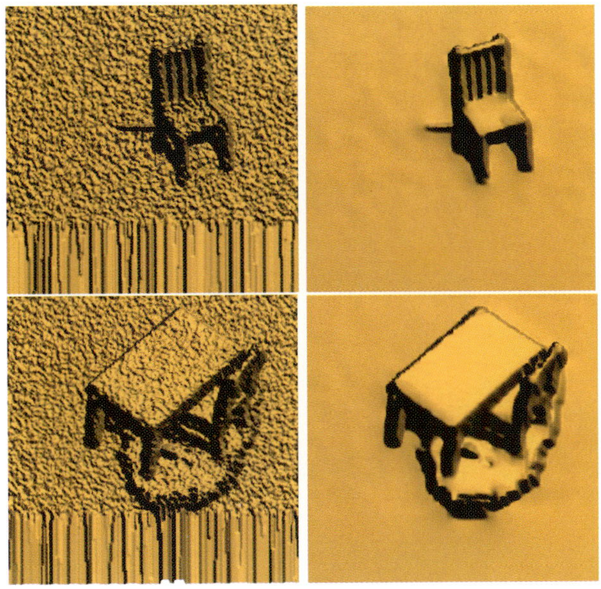

FIGURE 6. Reconstruction of scenes from range image data corrupted by Gaussian noise.

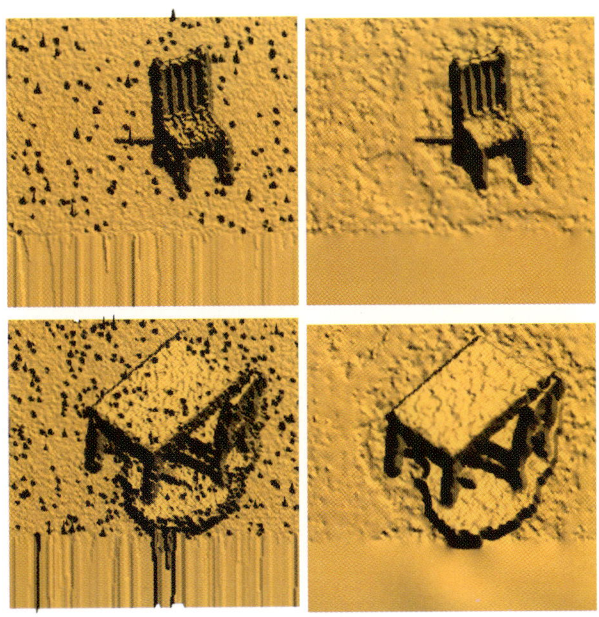

FIGURE 7. Reconstruction of scenes from range image data corrupted by impulse noise.

Bibliography

1. S. Arivazhagan and L. Ganesan, *Texture segmentation using wavelet transform*, Pattern Recognition Letters **24** (2003), no. 16, 3197–3203.
2. L. Bar, B. Berkels, M. Rumpf, and G. Sapiro, *A variational framework for simultaneous motion estimation and restoration of motion-blurred video*, International Conference on Computer Vision, 1–8, 2007.
3. B. Bascle, A. Blake, and A. Zisserman, *Motion deblurring and super-resolution from an image sequence*, Computer Vision — ECCV'96 (1996), 571–582.
4. A. Beck and M. Teboulle, *A fast iterative shrinkage-thresholding algorithm for linear inverse problems*, SIAM Journal on Imaging Sciences **2** (2009), no. 1, 183–202.
5. M. Ben-Ezra and S.K. Nayar, *Motion-based motion deblurring*, IEEE Transactions on Pattern Analysis and Machine Intelligence **26** (2004), no. 6, 689.
6. J.J. Benedetto and S. Li, *The Theory of Multiresolution Analysis Frames and Applications to Filter Banks*, Applied and Computational Harmonic Analysis **5** (1998), no. 4, 389–427.
7. M. Bertalmio, G. Sapiro, V. Caselles, and C. Ballester, *Image inpainting*, Proceedings of the 27th annual conference on Computer graphics and interactive techniques, ACM Press/Addison-Wesley Publishing Co., 2000, pp. 417–424.
8. J. Besag, *Towards Bayesian image analysis*, Journal of Applied Statistics **20** (1993), no. 5, 107–119.
9. L. Borup, R. Gribonval, and M. Nielsen, *Bi-framelet systems with few vanishing moments characterize Besov spaces* 1*, Applied and Computational Harmonic Analysis **17** (2004), no. 1, 3–28.
10. ———, *Tight wavelet frames in lebesque and sobolev spaces*, Journal of Function Spaces and Applications **2** (2004), no. 3, 227–252.
11. S.P. Boyd and L. Vandenberghe, *Convex optimization*, Cambridge Univ Pr, 2004.
12. X. Bresson, S. Esedoglu, P. Vandergheynst, J.P. Thiran, and S. Osher, *Fast global minimization of the active contour/snake model*, Journal of Mathematical Imaging and Vision **28** (2007), no. 2, 151–167.
13. J.F. Cai, E.J. Candès, and Z. Shen, *A Singular Value Thresholding Algorithm for Matrix Completion*, SIAM Journal on Optimization **20(4)** (2010), 1956–1982.
14. J.F. Cai, R. Chan, L. Shen, and Z. Shen, *Restoration of chopped and nodded images by framelets*, SIAM J. Sci. Comput **30** (2008), no. 3, 1205–1227.
15. J.F. Cai, R.H. Chan, L. Shen, and Z. Shen, *Convergence analysis of tight framelet approach for missing data recovery*, Advances in Computational Mathematics (2008), 1–27.
16. J.F. Cai, R.H. Chan, and Z. Shen, *A framelet-based image inpainting algorithm*, Applied and Computational Harmonic Analysis **24** (2008), no. 2, 131–149.
17. ———, *Simultaneous cartoon and texture inpainting*, Inverse Problems and Imaging (IPI) **4** (2010), no. 3, 379–395.
18. J.F. Cai, H. Ji, C. Liu, and Z. Shen, *Blind motion deblurring from a single image using sparse approximation*, Computer Vision and Pattern Recognition, 2009. CVPR 2009. IEEE Conference on, IEEE, 2009, pp. 104–111.
19. ———, *Blind motion deblurring using multiple images*, Journal of Computational Physics **228** (2009), no. 14, 5057–5071.
20. ———, *High-quality curvelet-based motion deblurring from an image pair*, Computer Vision and Pattern Recognition, 2009. CVPR 2009. IEEE Conference on, IEEE, 2009, pp. 1566–1573.
21. J.F. Cai, S. Osher, and Z. Shen, *Convergence of the linearized Bregman iteration for ℓ_1-norm minimization*, Mathematics of Computation **78** (2009), 2127–2136.

22. ———, *Linearized Bregman iterations for compressed sensing*, Math. Comp **78** (2009), 1515–1536.
23. ———, *Linearized Bregman iterations for frame-based image deblurring*, SIAM J. Imaging Sci **2** (2009), no. 1, 226–252.
24. ———, *Split Bregman methods and frame based image restoration*, Multiscale Modeling and Simulation: A SIAM Interdisciplinary Journal **8** (2009), no. 2, 337–369.
25. J.F. Cai and Z. Shen, *Framelet based deconvolution*, J. Comp. Math **28** (2010), no. 3, 289–308.
26. E.J. Candes and D.L. Donoho, *New tight frames of curvelets and optimal representations of objects with C2 singularities*, Comm. Pure Appl. Math **56** (2004), 219–266.
27. ———, *Continuous curvelet transform:: II. Discretization and frames*, Applied and Computational Harmonic Analysis **19** (2005), no. 2, 198–222.
28. Vicent Caselles, R. Kimmel, and G. Sapiro, *Geodesic active contours*, Int. J. Comput. Vision **22** (1997), no. 1, 61–79 (English).
29. A.S. Cavaretta, W. Dahmen, and C.A. Micchelli, *Stationary subdivision*, Amer. Math. Soc., 1991.
30. A.E. Cetin, *Reconstruction of signals from Fourier transform samples*, Signal Processing **16** (1989), no. 2, 129–148.
31. AE Cetin, *An iterative algorithm for signal reconstruction from bispectrum*, IEEE Transactions on Signal Processing **39** (1991), no. 12, 2621–2628.
32. A. Chai and Z. Shen, *Deconvolution: A wavelet frame approach*, Numerische Mathematik **106** (2007), no. 4, 529–587.
33. A. Chambolle, *An algorithm for total variation minimization and applications*, Journal of Mathematical Imaging and Vision **20** (2004), no. 1, 89–97.
34. R.H. Chan, T.F. Chan, L. Shen, and Z. Shen, *Wavelet algorithms for high-resolution image reconstruction*, SIAM Journal on Scientific Computing **24** (2003), no. 4, 1408–1432.
35. R.H. Chan, S.D. Riemenschneider, L. Shen, and Z. Shen, *High-resolution image reconstruction with displacement errors: A framelet approach*, International Journal of Imaging Systems and Technology **14** (2004), no. 3, 91–104.
36. ———, *Tight frame: an efficient way for high-resolution image reconstruction*, Applied and Computational Harmonic Analysis **17** (2004), no. 1, 91–115.
37. R.H. Chan, L. Shen, and Z. Shen, *A framelet-based approach for image inpainting*, Research Report **4** (2005), 325.
38. R.H. Chan, Z. Shen, and T. Xia, *A framelet algorithm for enhancing video stills*, Applied and Computational Harmonic Analysis **23** (2007), no. 2, 153–170.
39. T.F. Chan and S. Esedoglu, *Aspects of Total Variation Regularized L-Function Approximation*, SIAM Journal on Applied Mathematics **65** (2005), no. 5, 1817.
40. T.F. Chan, G.H. Golub, and P. Mulet, *A nonlinear primal-dual method for total variation-based image restoration*, SIAM Journal on Scientific Computing **20** (1999), no. 6, 1964–1977.
41. T.F. Chan and J. Shen, *Image processing and analysis: variational, pde, wavelet, and stochastic methods*, Society for Industrial Mathematics, 2005.
42. ———, *Variational image inpainting*, Commun. Pure Appl. Math **58** (2005), 579–619.
43. T.F. Chan and L.A. Vese, *Active contours without edges*, IEEE Transactions on image processing **10** (2001), no. 2, 266–277.
44. T.F. Chan and C.K. Wong, *Total variation blind deconvolution*, IEEE Transactions on Image Processing **7** (1998), no. 3, 370–375.
45. Tony F. Chan, Selim Esedoglu, and Mila Nikolova, *Algorithms for finding global minimizers of image segmentation and denoising models*, Siam. J. Appl. Math. **66** (2006), no. 5, 1632–1648 (English).
46. J. Chen, L. Yuan, C.K. Tang, and L. Quan, *Robust dual motion deblurring*, Proc. Conf. Comp. Vision and Pattern Recognition, 2008, pp. 1–8.
47. S.S. Chen, D.L. Donoho, and M.A. Saunders, *Atomic decomposition by basis pursuit*, SIAM review (2001), 129–159.
48. Y. Chen, H.D. Tagare, S. Thiruvenkadam, F. Huang, D. Wilson, K.S. Gopinath, R.W. Briggs, and E.A. Geiser, *Using prior shapes in geometric active contours in a variational framework*, International Journal of Computer Vision **50** (2002), no. 3, 315–328.

49. S. Cho, Y. Matsushita, S. Lee, and P. POSTECH, *Removing non-uniform motion blur from images*, IEEE 11th International Conference on Computer Vision, 2007. ICCV 2007, 2007, pp. 1–8.
50. O. Christensen, *Frames and bases: An introductory course*, Birkhäuser, 2008.
51. C.K. Chui and W. He, *Compactly supported tight frames associated with refinable functions*, Applied and Computational Harmonic Analysis **8** (2000), no. 3, 293–319.
52. C.K. Chui, W. He, and J. Stöckler, *Compactly supported tight and sibling frames with maximum vanishing moments*, Applied and computational harmonic analysis **13** (2002), no. 3, 224–262.
53. A. Cohen, *Biorthogonal wavelets*, Wavelets: A Tutorial in Theory and Applications, Academic Press Professional, Inc., 1993, pp. 123–152.
54. A. Cohen and J.P. Conze, *Régularité des bases d'ondelettes et mesures ergodiques*, Rev. Mat. Iberoamericana **8** (1992), no. 3, 351–365.
55. A. Cohen and I. Daubechies, *A stability criterion for biorthogonal wavelet bases and their related subband coding scheme*, Duke Math. J **68** (1992), no. 2, 313–335.
56. A. Cohen, I. Daubechies, and J.C. Feauveau, *Biorthogonal Bases of Compactly Supported Wavelets*, Communications on Pure and Applied Mathematics **45** (1992), no. 5, 485–560.
57. R.R. Coifman and D.L. Donoho, *Translation-invariant de-noising*, Lecture Notes in Statistics, New York, Springer-Verlag (1995), 125–125.
58. P.L. Combettes and V.R. Wajs, *Signal recovery by proximal forward-backward splitting*, Multiscale Modeling and Simulation **4** (2006), no. 4, 1168–1200.
59. J. Darbon and S. Osher, *Fast discrete optimizations for sparse approximations and deconvolutions*, preprint 2007.
60. I. Daubechies, *Orthonormal bases of compactly supported wavelets*, Commun. Pure Appl. Math. **41** (1988), no. 7, 909–996.
61. ———, *Ten lectures on wavelets*, vol. CBMS-NSF Lecture Notes, SIAM, nr. 61, Society for Industrial Mathematics, 1992.
62. I. Daubechies, M. Defrise, and C. De Mol, *An iterative thresholding algorithm for linear inverse problems with a sparsity constraint*, Communications on Pure and Applied Mathematics **57** (2004), no. 11, 1413–1457.
63. I. Daubechies, A. Grossmann, and Y. Meyer, *Painless nonorthogonal expansions*, Journal of Mathematical Physics **27** (1986), 1271.
64. I. Daubechies and B. Han, *Pairs of dual wavelet frames from any two refinable functions*, Constructive approximation **20** (2004), no. 3, 325–352.
65. I. Daubechies, B. Han, A. Ron, and Z. Shen, *Framelets: MRA-based constructions of wavelet frames*, Applied and Computational Harmonic Analysis **14** (2003), no. 1, 1–46 (English).
66. I. Daubechies, G. Teschke, and L. Vese, *Iteratively solving linear inverse problems under general convex constraints*, Inverse Problems and Imaging **1** (2007), no. 1, 29.
67. C. De Boor, *A practical guide to splines*, Springer-Verlag, 2001.
68. C. De Boor, R.A. DeVore, and A. Ron, *On the construction of multivariate (pre) wavelets*, Constructive approximation **9** (1993), no. 2, 123–166.
69. ———, *Approximation from shift-invariant subspaces of* $L_2(\mathbb{R}^d)$, Transactions of the American Mathematical Society **341** (1994), no. 2, 787–806.
70. ———, *The structure of finitely generated shift-invariant spaces in* $L_2(\mathbb{R}^d)$, J. Funct. Anal **119** (1994), no. 1, 37–78.
71. B. Dong, A. Chien, and Z. Shen, *Frame based segmentation for medical images*, Communications in Mathematical Sciences **9** (2011), no. 2, 551–559.
72. B. Dong, N. Dyn, and K. Hormann, *Properties of dual pseudo-splines*, Applied and Computational Harmonic Analysis **29** (2010), no. 1, 104–110.
73. B. Dong, E. Savitsky, and S. Osher, *A Novel method for enhanced needle localization using ultrasound-guidance*, Advances in Visual Computing (2009), 914–923.
74. B. Dong and Z. Shen, *Linear independence of pseudo-splines*, Proceedings of the American Mathematical Society **134** (2006), no. 9, 2685–2694.
75. ———, *Pseudo-splines, wavelets and framelets*, Applied and Computational Harmonic Analysis **22** (2007), no. 1, 78–104.
76. D.L. Donoho, *Interpolating wavelet transforms*, Preprint, Department of Statistics, Stanford University (1992).

77. ———, *De-noising by soft-thresholding*, IEEE transactions on information theory **41** (1995), no. 3, 613–627.
78. ———, *Ridge functions and orthonormal ridgelets*, Journal of Approximation Theory **111** (2001), no. 2, 143–179.
79. S. Dubuc, *Interpolation through an iterative scheme*, Journal of mathematical analysis and applications **114** (1986), no. 1, 185–204.
80. R.J. Duffin and A.C. Schaeffer, *A class of nonharmonic Fourier series*, Transactions of the American Mathematical Society **72** (1952), no. 2, 341–366.
81. N. Dyn, K. Hormann, M.A. Sabin, and Z. Shen, *Polynomial reproduction by symmetric subdivision schemes*, Journal of Approximation Theory **155** (2008), no. 1, 28–42.
82. I. Ekeland and R. Temam, *Convex analysis and variational problems*, Society for Industrial Mathematics, 1999.
83. M. Elad, P. Milanfar, and R. Rubinstein, *Analysis versus synthesis in signal priors*, Inverse Problems **23** (2007), 947.
84. M. Elad, J.L. Starck, P. Querre, and D.L. Donoho, *Simultaneous cartoon and texture image inpainting using morphological component analysis (MCA)*, Applied and Computational Harmonic Analysis **19** (2005), no. 3, 340–358.
85. E. Esser, *Applications of lagrangian-based alternating direction methods and connections to split Bregman*, CAM report **9** (2009), 31.
86. M.J. Fadili and J.L. Starck, *Sparse representations and bayesian image inpainting*, Proc. SPARS **5** (2005).
87. M.J. Fadili, J.L. Starck, and F. Murtagh, *Inpainting and zooming using sparse representations*, The Computer Journal **52** (2009), no. 1, 64.
88. W. Feller, *An introduction to probability theory and its applications. Vol. 1*, Wiley New York, 1968.
89. R. Fergus, B. Singh, A. Hertzmann, S.T. Roweis, and W.T. Freeman, *Removing camera shake from a single photograph*, ACM Transactions on Graphics (TOG) **25** (2006), no. 3, 787–794.
90. M.A.T. Figueiredo and R.D. Nowak, *An EM algorithm for wavelet-based image restoration*, IEEE Transactions on Image Processing **12** (2003), no. 8, 906–916.
91. ———, *A bound optimization approach to wavelet-based image deconvolution*, Image Processing, 2005. ICIP 2005. IEEE International Conference on, vol. 2, IEEE, 2005, pp. II–782.
92. N. Flasque, M. Desvignes, J.M. Constans, and M. Revenu, *Acquisition, segmentation and tracking of the cerebral vascular tree on 3D magnetic resonance angiography images*, Medical Image Analysis **5** (2001), no. 3, 173–183.
93. M. Frazier, G. Garrigós, K. Wang, and G. Weiss, *A characterization of functions that generate wavelet and related expansion*, Journal of Fourier Analysis and Applications **3** (1997), 883–906.
94. R. Glowinski and P. Le Tallec, *Augmented Lagrangian and operator-splitting methods in nonlinear mechanics*, Society for Industrial Mathematics, 1989.
95. T. Goldstein, X. Bresson, and S. Osher, *Geometric Applications of the Split Bregman Method: Segmentation and Surface Reconstruction*, UCLA CAM Report (2009), 09–06.
96. T. Goldstein and S. Osher, *The split Bregman algorithm for L1 regularized problems*, SIAM Journal on Imaging Sciences **2** (2009), no. 2, 323–343.
97. R. Gribonval and M. Nielsen, *On approximation with spline generated framelets*, Constructive Approximation **20** (2004), no. 2, 207–232.
98. G. Gripenberg, *A necessary and sufficient condition for the existence of a father wavelet*, Studia Math **114** (1995), no. 3, 207–226.
99. K. Gröchenig, *Foundations of time-frequency analysis*, Birkhäuser, 2001.
100. M. Gross and H. Pfister, *Point-based graphics*, Morgan Kaufmann Pub, 2007.
101. B. Han, *On Dual Wavelet Tight Frames*, Applied and Computational Harmonic Analysis **4** (1997), no. 4, 380–413.
102. B. Han, G. Kutyniok, and Z. Shen, *A Unitary Extension Principle for Shearlet Systems*, Arxiv preprint arXiv:0912.4529 (2009).
103. B. Han and Z. Shen, *Wavelets with short support*, SIAM Journal on Mathematical Analysis **38** (2007), no. 2, 530–556.
104. ———, *Compactly supported symmetric C-wavelets with spectral approximation order*, SIAM J. Math. Anal **40** (2008), 905–938.

105. ______, *Characterization of Sobolev spaces of arbitrary smoothness using nonstationary tight wavelet frames*, Israel Journal of Mathematics **172** (2009), no. 1, 371–398.
106. ______, *Dual wavelet frames and Riesz bases in Sobolev spaces*, Constructive Approximation **29** (2009), no. 3, 369–406.
107. X. He, L. Shen, and Z. Shen, *A data-adaptive knot selection scheme for fitting splines*, IEEE Signal Processing Letters **8** (2001), no. 5, 137.
108. E. Hernández and G. Weiss, *A first course on wavelets*, Studies in Advanced Mathematics, CRC Press, 1996.
109. J.B. Hiriart-Urruty and C. Lemaréchal, *Convex analysis and minimization algorithms: Fundamentals*, Springer, 1993.
110. Y. Hur and A. Ron, *CAPlets: wavelet representations without wavelets*, preprint (2005).
111. A. Hyvärinen, J. Hurri, and P.O. Hoyer, *Natural Image Statistics: A Probabilistic Approach to Early Computational Vision*, Springer-Verlag, New York, Inc., 2009.
112. H. Ji, Z. Shen, and Y. Xu, *Wavelet frame based scene reconstruction from range data*, Journal of Computational Physics **229** (2010), no. 6, 2093–2108.
113. J. Jia, *Single image motion deblurring using transparency*, IEEE Conference on Computer Vision and Pattern Recognition, 2007. CVPR'07, 2007, pp. 1–8.
114. R.Q. Jia and C.A. Micchelli, *Using the refinement equations for the construction of pre-wavelets II: Powers of two*, Curves and surfaces (1991), 209–246.
115. R.Q. Jia and Z. Shen, *Multiresolution and wavelets*, Proc. Edinb. Math. Soc., II. Ser. **37** (1994), no. 2, 271–300.
116. Q. Jiang and Z. Shen, *On existence and weak stability of matrix refinable functions*, Constructive Approximation **15** (1999), no. 3, 337–353.
117. M.J. Johnson, *Scattered Date Interpolation from Principal Shift-Invariant Spaces*, Journal of Approximation Theory **113** (2001), no. 2, 172–188.
118. M.J. Johnson, Z. Shen, and Y. Xu, *Scattered data reconstruction by regularization in B-spline and associated wavelet spaces*, Journal of Approximation Theory **159** (2009), no. 2, 197–223.
119. N. Joshi, R. Szeliski, and D.J. Kriegman, *PSF estimation using sharp edge prediction*, IEEE Conference on Computer Vision and Pattern Recognition, 2008. CVPR 2008, 2008, pp. 1–8.
120. S.A. Kassam and J.B. Thomas, *Signal detection in non-Gaussian noise*, Springer-Verlag, New York, NY, 1988.
121. A. Khodakovsky, P. Schröder, and W. Sweldens, *Progressive geometry compression*, Proceedings of the 27th annual conference on Computer graphics and interactive techniques, Citeseer, 2000, pp. 271–278.
122. C. Kirbas and F. Quek, *A review of vessel extraction techniques and algorithms*, ACM Computing Surveys **36** (2004), no. 2, 81–121.
123. G. Kutyniok and D. Labate, *Resolution of the wavefront set using continuous shearlets*, Amer. Math. Soc. **361** (2009), no. 5, 2719–2754.
124. W.M. Lawton, *Tight frames of compactly supported affine wavelets*, Journal of Mathematical Physics **31** (1990), 1898.
125. W.M. Lawton, S.L. Lee, and Z. Shen, *Convergence of multidimensional cascade algorithm*, Numerische Mathematik **78** (1998), no. 3, 427–438.
126. M. Leventon, W.E.L. Grimson, and O. Faugeras, *Statistical shape influence in geodesic active contours*, IEEE Computer Society Conference on Computer Vision and Pattern Recognition, vol. 1, IEEE Computer Society; 1999, 2000.
127. A. Levin, *Blind motion deblurring using image statistics*, Advances in Neural Information Processing Systems **19** (2007), 841.
128. C. Li, C.Y. Kao, J.C. Gore, and Z. Ding, *Minimization of region-scalable fitting energy for image segmentation*, IEEE Transactions on Image Processing **17** (2008), no. 10, 1940–1949.
129. M. Li, B. Hao, and X.C. Feng, *Iterative regularization and nonlinear inverse scale space based on translation invariant wavelet shrinkage*, International Journal of Wavelets Multiresolution and Information Processing **6** (2008), no. 1, 83.
130. C.G. Looney, *Convergence of minimizing sequences*, Journal of Mathematical Analysis and Applications **61** (1977), no. 3, 835–840.
131. S.G. Mallat, *Multiresolution approximations and wavelet orthonormal bases of $L^2(\mathbf{R})$*, Transactions of the American Mathematical Society **315** (1989), no. 1, 69–87.

132. ———, *A wavelet tour of signal processing*, vol. 2nd ed. New York: Academic, Academic press, 1999.
133. T. McInerney and D. Terzopoulos, *Deformable models in medical image analysis: a survey*, Medical image analysis **1** (1996), no. 2, 91–108.
134. Y. Meyer, *Wavelets and operators. Translated by D.H. Salinger*, Cambridge Studies in Advanced Mathematics, 1992.
135. ———, *Oscillating patterns in image processing and nonlinear evolution equations: the fifteenth Dean Jacqueline B. Lewis memorial lectures*, Amer Mathematical Society, 2001.
136. Q. Mo, Y. Shen, and S. Li, *A new proof of some polynomial inequalities related to pseudo-splines*, Applied and Computational Harmonic Analysis **23** (2007), no. 3, 415–418.
137. J.J. Moreau, *Fonctions convexes duales et points proximaux dans un espace hilbertien*, C.R. Acad. Sci. Paris Sér. A Math **255** (1962), 2897–2899.
138. ———, *Proximité et dualité dans un espace hilbertien*, Bull. Soc. Math. France **93** (1965), no. 2, 273–299.
139. D. Nain, A. Yezzi, and G. Turk, *Vessel segmentation using a shape driven flow*, Lecture Notes in Computer Science (2004), 51–59.
140. A.S. Nemirovskij and D.B. Yudin, *Problem complexity and method efficiency in optimization. transl. from the Russian by E.R. Dawson*, Wiley-Interscience Series in Discrete Mathematics, 1983.
141. Y. Nesterov, *A method of solving a convex programming problem with convergence rate* $o(1/k^2)$, Soviet Mathematics Doklady, vol. 27, 1983, pp. 372–376.
142. ———, *On an approach to the construction of optimal methods for minimizing smooth convex functions*, Ehkon. Mat. Metody **24** (1988), no. 3, 509–517.
143. ———, *Smooth minimization of non-smooth functions*, Mathematical Programming **103** (2005), no. 1, 127–152.
144. M. Nikolova, *A variational approach to remove outliers and impulse noise*, Journal of Mathematical Imaging and Vision **20** (2004), no. 1, 99–120.
145. S. Osher, M. Burger, D. Goldfarb, J. Xu, and W. Yin, *An iterative regularization method for total variation based image restoration*, Multiscale Model. Simul **4** (2005), no. 2, 460–489.
146. S. Osher and R.P. Fedkiw, *Level set methods and dynamic implicit surfaces*, Springer, 2003.
147. S. Osher, Y. Mao, B. Dong, and W. Yin, *Fast linearized bregman iteration for compressive sensing and sparse denoising*, Communications in Mathematical Sciences **8** (2010), no. 2, 93–111.
148. S. Osher and J. Sethian, *Fronts propagating with curvature-dependent speed- Algorithms based on Hamilton-Jacobi formulations*, Journal of Computational Physics **79** (1988), 12–49.
149. G. Pavlovic and AM Tekalp, *Maximum likelihood parametric blur identification based on a continuous spatial domain model*, IEEE Transactions on image processing **1** (1992), no. 4, 496–504.
150. S.M. Pizer, P.T. Fletcher, S. Joshi, A. Thall, J.Z. Chen, Y. Fridman, D.S. Fritsch, A.G. Gash, J.M. Glotzer, M.R. Jiroutek, et al., *Deformable m-reps for 3d medical image segmentation*, International Journal of Computer Vision **55** (2003), no. 2, 85–106.
151. E. Polak, *Computational methods in optimization*, Academic Press, New York, 1971.
152. G. Pólya and G. Szegö, *Aufgaben und Lehrsätze aus der Analysis*, Springer, 1971.
153. R. Raskar, A. Agrawal, and J. Tumblin, *Coded exposure photography: motion deblurring using fluttered shutter*, ACM Transactions on Graphics (TOG) **25** (2006), no. 3, 795–804.
154. R.T. Rockafellar and J.B.W. Roger, *Variational analysis*, Springer-Verlag, 1997.
155. A. Ron, *Factorization theorems for univariate splines on regular grids*, Israel Journal of Mathematics **70** (1990), no. 1, 48–68.
156. A. Ron and Z. Shen, *Frames and Stable Bases for Shift-Invariant Subspaces of* $L_2(\mathbb{R}^d)$, Canadian Journal of Mathematics **47** (1995), no. 5, 1051–1094.
157. ———, *Affine systems in* $L_2(\mathbb{R}^d)$ *II: dual systems*, Journal of Fourier Analysis and Applications **3** (1997), no. 5, 617–638.
158. ———, *Affine systems in* $L_2(\mathbb{R}^d)$*: The analysis of the analysis operator*, Journal of Functional Analysis **148** (1997), no. 2, 408–447.
159. ———, *Weyl-Heisenberg frames and Riesz bases in* $L_2(\mathbb{R}^d)$, Duke Mathematical Journal **89** (1997), no. 2, 237–282.

160. ______, *Compactly supported tight affine spline frames in* $L_2(\mathbb{R}^d)$, Mathematics of Computation **67** (1998), no. 221, 191–207.
161. ______, *The Sobolev regularity of refinable functions*, Journal of Approximation Theory **106** (2000), 185–225.
162. ______, *Generalized shift-invariant systems*, Constructive Approximation **22** (2005), no. 1, 1–45.
163. W. Rudin, *Real and complex analysis*, Tata McGraw-Hill, 2006.
164. G. Sapiro, *Geometric partial differential equations and image analysis*, Cambridge Univ Pr, 2001.
165. I.W. Selesnick, *Smooth Wavelet Tight Frames with Zero Moments* 1*, Applied and Computational Harmonic Analysis **10** (2001), no. 2, 163–181.
166. Q. Shan, J. Jia, and A. Agarwala, *High-quality motion deblurring from a single image*, ACM Transactions on Graphics-TOG **27** (2008), no. 3, 73–73.
167. Z. Shen, *Wavelet frames and image restorations*, Proceedings of the International Congress of Mathematicians, vol. 4, 2010, pp. 2834–2863.
168. Z. Shen, K. C. Toh, and S. Yun, *An accelerated proximal gradient algorithm for frame based image restorations via the balanced approach*, SIAM Journal on Imaging Sciences **4(2)** (2011), 573–596.
169. Z. Shen and S. Waldron, *Scattered data interpolation by box splines*, Study in Advanced Mathematics, AMS/IP **42** (2008), no. 2, 749.
170. J.L. Starck, M. Elad, and D.L. Donoho, *Image decomposition via the combination of sparse representations and a variational approach*, IEEE transactions on image processing **14** (2005), no. 10, 1570–1582.
171. X.C. Tai and C. Wu, *Augmented Lagrangian method, dual methods and split Bregman iteration for ROF model*, Scale Space and Variational Methods in Computer Vision (2009), 502–513.
172. Y.W. Tai, H. Du, M.S. Brown, and S. Lin, *Image/video deblurring using a hybrid camera*, Computer Vision and Pattern Recognition, 2008. CVPR 2008. IEEE Conference on, IEEE, 2008, pp. 1–8.
173. K.C. Toh and S. Yun, *An accelerated proximal gradient algorithm for nuclear norm regularized linear least squares problems*, Pacific Journal of Optimization **6** (2010), no. 20, 615–640.
174. A. Tsai, A. Yezzi, W. Wells, C. Tempany, D. Tucker, A. Fan, W.E. Grimson, and A. Willsky, *A shape-based approach to the segmentation of medical imagery using level sets*, IEEE Transactions on Medical Imaging **22** (2003), no. 2, 137–154.
175. P. Tseng, *On accelerated proximal gradient methods for convex-concave optimization*, submitted to SIAM Journal on Optimization (2008).
176. Z. Tu, K.L. Narr, P. Dollar, I. Dinov, P.M. Thompson, and A.W. Toga, *Brain anatomical structure segmentation by hybrid discriminative/generative models*, IEEE Transactions on Medical Imaging **27** (2008), no. 4, 495–508.
177. M. Umasuthan and AM Wallace, *Outlier removal and discontinuity preserving smoothing of rangedata*, IEEE Proceedings-Vision, Image and Signal Processing **143** (1996), no. 3, 191–200.
178. M. Unser, *Texture classification and segmentation using wavelet frames*, IEEE Transactions on image processing **4** (1995), no. 11, 1549–1560.
179. L.A. Vese and T.F. Chan, *A multiphase level set framework for image segmentation using the Mumford and Shah model*, International Journal of Computer Vision **50** (2002), no. 3, 271–293.
180. C.R. Vogel, M.E. Oman, et al., *Fast, robust total variation-based reconstruction of noisy, blurred images*, IEEE Transactions on Image Processing **7** (1998), no. 6, 813–824.
181. N. Wiener, *The Fourier integral and certain of its applications*, Cambridge Univ Pr, 1988.
182. J. Xu and S. Osher, *Iterative regularization and nonlinear inverse scale space applied to wavelet-based denoising*, IEEE Transactions on Image Processing **16** (2007), no. 2, 534.
183. J. Yang, L.H. Staib, and J.S. Duncan, *Neighbor-constrained segmentation with 3d deformable models*, Lecture Notes in Computer Science (2003), 198–209.

184. W. Yin, S. Osher, D. Goldfarb, and J. Darbon, *Bregman iterative algorithms for l 1-minimization with applications to compressed sensing*, SIAM J. Imaging Sci **1** (2008), no. 1, 143–168.
185. X. Zhang, M. Burger, X. Bresson, and S. Osher, *Bregmanized nonlocal regularization for deconvolution and sparse reconstruction*, SIAM Journal on Imaging Sciences **3** (2010), no. 3, 253–276.

Five Lectures on Sparse and Redundant Representations Modelling of Images

Michael Elad

IAS/Park City Mathematics Series
Volume 19, 2010

Five Lectures on Sparse and Redundant Representations Modelling of Images

Michael Elad

Preface

The field of sparse and redundant representations has evolved tremendously over the last decade or so. In this short series of lectures, we intend to cover this progress through its various aspects: theoretical claims, numerical problems (and solutions), and lastly, applications, focusing on tasks in image processing.

Our story begins with a search for a proper model, and of ways to use it. We are surrounded by huge masses of data, coming from various sources, such as voice signals, still images, radar images, CT images, traffic data and heart signals just to name a few. Despite their diversity, these sources of data all have something in common: each and everyone of these signals has some kind of internal structure, which we wish to exploit for processing it better.

Consider the simplest problem of removing noise from data. Given a set of noisy samples (e.g., value as a function of time), we wish to recover the true (noise-free) samples with as high accuracy as possible. Suppose that we know the exact statistical characteristics of the noise. Could we perform denoising of the given data? The answer is negative: Without knowing anything about the properties of the signal in mind, this is an impossible task. However, if we could assume, for example, that the signal is piecewise-linear, the noise removal task becomes feasible. The same can be said for almost any data processing problem, such as deblurring, super-resolution, inpainting, compression, anomaly detection, detection, recognition, and more—relying on a good model is in the base of any useful solution.

When approaching a problem in image processing, there is a wide range of models to choose from, such as Principal Component Analysis (PCA), Anisotropic diffusions, Markov Random Fields, Total-Variation, Besov spaces and many more. There are two important (and sometimes somewhat contradictory) requirements for a model to be useful. On one hand, it should be reliable, in the sense that it should be able to represent the data well. On the other hand, it has to be simple enough to be applied in practice to the various tasks at hand.

A simple example of a model used in a common task can be found in the JPEG algorithm. At its core, this algorithm relies on the claim that when transforming patches of 8×8 pixels in a typical image using the Discrete Cosine Transform (DCT), only the top-left (low frequency) components are dominant, while the rest are close

The Computer Science Department, the Technion, Haifa 32000, Israel
E-mail address: elad@cs.technion.ac.il
The author wishes to thank Dr. Matan Protter for the extensive help in writing these lecture-notes.

to zero. This model is used for compression by discarding the non-dominant components, thus using much fewer numbers for representing the image. While simple and relatively effective, one immediately sees potential failures of this model, and ways to further improve it.

We therefore turn our attention to the model of "Sparse and Redundant Representations" and example-based methods, which will be the focus of this booklet (and corresponding course). This model has many "fathers", drawing from various areas of research, such as signal processing (wavelet theory, multi-scale analysis, signal transforms), machine learning, mathematics (approximation theory, linear algebra, optimization theory) and more. We will see that it can be applied to various applications, such as blind source separation, compression, denoising, inpainting, demosaicing, super-resolution and more.

What is this model all about? Let us focus on the task of representing signals, which are 8×8 patches of images, as done above. We shall assume that a *dictionary* of such image patches is given to us, containing 256 *atom-images*, each of which is also 8×8. The *Sparseland* model assumption is that *every* image patch from the family of interest could be constructed as a linear combination of a *few* atoms from the dictionary.

To represent a signal of length 64, this model uses 256 numbers, and therefore it is redundant. However, of these 256 all but a few are zero, and therefore the model is sparse. Effectively, if there are 3 atoms involved in the found combination, only 6 numbers are needed to accurately represent the signal using this model—the three locations of the non-zeros, and their values.

While this model sounds near-trivial, it raises several very difficult questions. Given an image patch, can we find what atoms it was built from? Given a set of patches, how should we get the dictionary for them? Is this model applicable to various signals? The first question is problematic, as crunching the numbers indicates that a search through every possible combination is unlikely to finish any time soon, even for low-dimensional problems. However, several approximation algorithms have been proposed, with surprising guarantees on their performance.

The first two lectures will focus on such algorithms and their theoretical analysis. The third lecture will discuss how this model and theoretical foundations are brought to the realm of image processing. The second question can be addressed by learning the dictionary from a large collection of signals. Such algorithms have been developed in the past several years, and will be discussed in the fourth lecture. The last question is yet to be answered from a theoretical point of view, but empirically, this model is extremely effective in representing different sources of signals and in various problems, leading to state of the art results, which will be demonstrated in the fifth and final lecture.

Having answered the above three important questions positively, we are now ready to dive into the wonderful world of sparse and redundant representation modelling of images, or *Sparseland*. For extensive coverage of this area, the reader is encouraged to turn to my book:

> M. Elad, *Sparse and Redundant Representations: From Theory to Applications in Signal and Image Processing*, Springer, New York, 2010, ISBN: 978-1-4419-7010-7 (http://www.springer.com/978-1-4419-7010-7).

I would like to take this opportunity and thank Springer for their permission to publish these lecture notes with PCMI. It should be noted that the copyright of some of the material here (e.g. the figures in Chapters 4 and 5) belongs to Springer, and are used with kind regards from Springer Science+Business Media B.V. Finally, the reader will find it useful to read these lecture notes while looking at the slides: http://www.cs.technion.ac.il/~elad/talks/2010/Summer-School-2010-PCMI.rar.

LECTURE 1

Introduction to sparse approximations - algorithms

1. Motivation and the sparse-coding problem

We start our journey with a pure linear algebra point of view, and specifically by considering a linear under-determined set of equations,

$$\mathbf{D}_{n\times k}\alpha = \mathbf{x}\ ,\ \ k > n. \tag{1.1}$$

In order to simplify the following developments, we will assume that the columns of $\mathbf{D}$ are (ℓ_2) normalized.

Now, as for terminology used: We shall refer hereafter to $\mathbf{x}$ as a *signal* to be processed, and α will stand for its *representation.* The matrix $\mathbf{D}$ will be referred to as the *dictionary*, and its columns will be called *atoms*.

Assuming $\mathbf{D}$ is full rank, there is an infinite number of solutions to this equation. In engineering, such a situation is unacceptable, since we aim to get a single solution to our problems. In order to be able to choose the "best" single solution, we require some quality measure $J(\alpha)$ that basically ranks the solutions based on some measure, this way choosing the solution that solves the problem

$$\hat{\alpha} = \arg\min_{\alpha} J(\alpha)\ s.t.\ \mathbf{D}\alpha = \mathbf{x}. \tag{1.2}$$

Of course, the natural question is how to select J properly. In engineering, the most common answer is some quadratic term $J(\alpha) = \|\mathbf{B}\alpha\|_2^2$, as it is easy to minimize, having a closed form solution. When chosen as a non-quadratic term, J is preferred to be convex in order to ensure the existence of a unique global minimum to the above problem. Of course, in many cases such choices are not aligned with our needs, as we shall see next.

In these lectures we are interested in cases when α is sparse. In order to gain a mathematical intuition, we consider the general ℓ_p-norm, $J(\alpha) = \|\alpha\|_p^p$. This expression sums the absolute entries of the vector α after being raised to a power p. Note that when p drops below 1 it is no longer a norm, but we shall disregard this technicality.

When p goes below 1 and approaches zero, the function $|\alpha_i|^p$ resembles the indicator function, being 1 for every $\alpha_i \neq 0$, and 0 for $\alpha_i = 0$. Effectively, we define the ℓ_0-norm, as a function counting the number of non-zeros in α and denote it by $\|\alpha\|_0^0$. While ℓ_0 is definitely not a norm, we will abuse this notation for simplicity. We also define the term *Support* as the locations of non-zeros in α. We therefore write the minimization problem (P_0):

$$(P_0)\ \hat{\alpha} = \arg\min_{\alpha} \|\alpha\|_0^0\ s.t.\ \mathbf{D}\alpha = \mathbf{x}. \tag{1.3}$$

In many cases our system might contain noise, and we cannot expect a perfect solution to the linear system $\mathbf{D}\alpha = \mathbf{x}$. Instead, we require some proximity between $\mathbf{D}\hat{\alpha}$ and $\mathbf{x}$, arriving at the following alternative problem:

$$(1.4) \qquad (P_0^\epsilon)\ \hat{\alpha} = \arg\min_\alpha \|\alpha\|_0^0 \ s.t.\ ||\mathbf{D}\alpha - \mathbf{x}||_2 \leq \epsilon,$$

where ϵ is closely related to the properties of the noise. We will be seeing this minimization problem many times during the course.

In order to understand the complexity of the above defined problem, let us propose a method to perform this minimization in a direct way. Set $L = 1$, and consider all supports $\{S_i\}_{i=1}^I$ of cardinality L (there are $I = \binom{k}{L}$ such supports). For each support, we attempt to represent $\mathbf{x}$ using only columns from $\mathbf{D}$ in the support. This can be done by solving a simple Least-Squares (LS) problem,

$$\min_\alpha ||\mathbf{D}\alpha - \mathbf{x}||_2 \ s.t.\ Sup(\alpha) = S_j.$$

If the result LS error is below the threshold ϵ, than we have found a solution and can terminate. If none of the supports yields a solution with an error below the threshold, L should be incremented and the process should be repeated.

Taking a nominal case where $k = 2000$, and a known support size $L = 15$ (so searching the value of L is not needed), and requiring a mere 1 nano-second for each LS operation, solving just one problem will take about $7.5 \cdot 10^{20}$ years. This is due to the exponential nature of the problem. Indeed, the problem posed in Equation (1.4) is known to be NP-Hard [**41**]. This limitation leads us to seek approximation algorithms for these two problems, (P_0) and (P_0^ϵ).

There are two main approaches to approximate the solution to the problem posed in Equation (1.4), which we will explore in this lecture. The first approach is the greedy family of methods, where we attempt to build the solution one non-zero at a time. The second approach is the relaxation methods, which attempt to solve the problem by smoothing the ℓ_0-norm and using continuous optimization techniques. All these techniques that aim to approximate the solution of the problem posed in Equation (1.4) are commonly referred to as pursuit algorithms.

While this lecture will be devoted to the various approximation algorithms, the next lecture will discuss in depth the quality of the estimations they provide. The second lecture will also discuss conditions that should be met in order to ensure that (P_0) has a unique solution, and (P_0^ϵ) has a stable solution (as we shall see, we will not be able to require uniqueness due to the noise).

2. Greedy algorithms

Our first journey into approximation algorithms will deal with the family of greedy algorithms. These algorithms attempt to build the support of α one non-zero element at a time. There is a variety of greedy techniques – see [**22, 3, 9, 38, 40, 42**] for more details.

The most basic of algorithm is the Matching Pursuit (MP). This is an iterative algorithm that starts by finding the *one* atom that best describes the input signal, i.e., finding the index $\hat{i} = \arg\min_i \min_c ||\mathbf{x} - c \cdot \mathbf{d_i}||_2^2$. Once found, we compute the signal residual as $\mathbf{x} - c \cdot \mathbf{d_{\hat{i}}}$ with the optimally found constant c and the atom $\mathbf{d}_{\hat{i}}$.

Then, at each iteration, the *one* atom that best describes the residual is added to the support, with the appropriate coefficient contributing to reducing the residual the most. This process is repeated until the norm of the threshold drops below the given threshold. Note that the same atom may be added to the support several times – in such cases, its corresponding coefficient is aggregated.

The Orthogonal Matching Pursuit (OMP) is an improvement of the MP. The OMP re-computes the set of coefficients for the entire support each time an atom is added to the support. This stage speeds up the convergence rate, as well as insures that once an atom has been added to the support, it will not be selected again. This is a result of the LS stage, making the residual orthogonal to all atoms in the chosen support.

A further improvement to the OMP observes that the atom selection stage does not completely correspond to the objective of reducing the residual (because of the LS stage). Therefore, this version suggests to select the atom that will best reduce the residual after the LS stage. This method is accordingly known as LS-OMP. While LS-OMP seems to require more computations, there are some numerical short-cuts to employ, which bring the efficiency of this method very close to that of the OMP.

While improving the results of the MP, both the OMP and the LS-OMP require more complex implementations and more calculations. The other direction, of simplifying the MP, has been attempted too, with the goal of reducing the amount of computations needed at each stage. The Weak-MP is born from the observation that in the MP, it is probably not necessary to find the atom that most reduces the residual, but an atom that can come close to this performance might do as well. As a result, the atoms are scanned in order, and as soon as an atom is found to reduce the residual by some threshold (or beyond), it is added to the support and the iteration ends.

A further simplification can be obtained by avoiding the search at each stage completely, and instead determine the order in which atoms are to be added in advance. Such an algorithm is called the Thresholding algorithm. In this method, the inner product between the signal and all the atoms are computed, and then sorted in descending order of magnitude. This order determines the order in which atoms will be added to the support. At each iteration, the next atom is added to the support, and LS over the support is carried out to best represent the signal (this can be done using recursive least-squares, which avoids a direct matrix inversion). As soon as the norm of the residual drops below the given threshold, the process terminates.

3. Relaxation algorithms

In the previous section we reviewed the greedy algorithms, which attempt to construct the support one atom at a time. Now it is time to turn to relaxation methods, that smooth the ℓ_0-norm and solve the alternative obtained problem using classical methods from continuous optimization [**22, 3**].

One natural way to do this is to replace the $\|\alpha\|_0^0$ penalty with the expression $\sum_{i=1}^{K}\left[1-\exp\left(-\beta\alpha_i^2\right)\right]$, and observe that as $\beta \to \infty$, this term better approximates the ℓ_0 norm. Therefore, it is possible to approximate the ℓ_0 solution by optimizing for a small value of β, and increase the value of β every few iterations.

An alternative is to "convexise" the ℓ_0-norm, which brings us to the Basis Pursuit (BP) algorithm [**7**]. In this algorithm, the ℓ_0-norm is simply replaced by the ℓ_1-norm, leading to the optimization problem

$$\min_{\alpha} \|\alpha\|_1 \ s.t. \ \|\mathbf{D}\alpha - \mathbf{x}\|_2 \le \epsilon, \tag{1.5}$$

with the hope that the solution of this optimization problem is close to the solution of the original one. This problem can be formulated as linear programming for the exact case ($\epsilon = 0$), and quadratic programming for the general case. For this task, several types of efficient solvers can be employed, such as Interior Point methods [**33**], Least-Angle-Regression (LARS) [**20**] and Iterative Shrinkage [**8, 21, 24, 25, 26, 27**].

Another interesting option for handling the problem formed above (either (1.4) or (1.5)), is the Iterative Reweighted Least-Squares (IRLS). Before explaining this algorithm, we slightly change the formulation of the problem. Instead of enforcing the constraint $\|\mathbf{D}\alpha - \mathbf{x}\|_2 \le \epsilon$, we now insert it into the optimization problem:

$$\min_{\alpha} \ \lambda \cdot \|\alpha\|_p^p + \|\mathbf{D}\alpha - \mathbf{x}\|_2^2, \tag{1.6}$$

with λ balancing the effects of both terms. A proper choice of λ makes the two formulations equivalent. Notice that our formulation uses the ℓ_p-norm, thus covering both cases posed in Equations (1.4) or (1.5).

Now, let us re-write the first term:

$$\|\alpha\|_p^p = \sum_{i=1}^{k} |\alpha_i|^p = \sum_{i=1}^{k} |\alpha_i|^{p-2} \cdot |\alpha_i| = \alpha^T \mathbf{A}(\alpha)\, \alpha. \tag{1.7}$$

Based on this, we can suggest an iterative optimization scheme in which we freeze $\mathbf{A}(\alpha)$, and minimize for α:

$$\hat{\alpha}_{j+1} = \arg\min_{\alpha} \lambda \alpha^T \mathbf{A}(\alpha_j)\, \alpha + \frac{1}{2}\|\mathbf{D}\alpha - \mathbf{x}\|_2^2. \tag{1.8}$$

This is a quadratic term, and as such, it is easy to minimize. In the following step, we freeze α_{j+1} and compute $\mathbf{A}(\alpha_{j+1})$, and so on, until the solution converges. More on the IRLS (sometimes called FOCUSS) can be found in [**28**].

In the previously described algorithm, our only demand of the residual is that its norm is bounded. However, in the general case, we expect the residual not only to be bounded, but we also expect it to look like noise. In other words, we expect the residual to show low correlation with any of the atoms in the dictionary. This observation leads to the formulation of the Dantzig Selector (DS) [**6**]:

$$\min_{\alpha} \|\alpha\|_1 \ s.t. \ \|\mathbf{D}^T(\mathbf{D}\alpha - \mathbf{x})\|_\infty \le T. \tag{1.9}$$

Interestingly, this is a linear programming task even in the noisy case (unlike the BP, which leads to a quadratic programming problem in the noisy case), and therefore a wide range of tools can be used for solving it.

Simulations comparing the performance of the DS with those of the BP (see [**22**]) show that these two algorithm perform (on average) at the same level. It is well known that in the unitary case, the two are perfectly equivalent.

4. A closer look at the unitary case

What happens in the special case where the dictionary $\mathbf{D}$ is unitary, i.e., $\mathbf{D}^T\mathbf{D} = \mathbf{I}$ (the dictionary is of course not redundant in such a case)? Starting from the problems posed in Equations (1.4) and (1.5), let us write a more general optimization goal,

$$f(\alpha) = \frac{1}{2}\|\mathbf{D}\alpha - \mathbf{x}\|_2^2 + \lambda \cdot \rho(\alpha) \tag{1.10}$$

where we use a general function $\rho(\alpha) = \sum_{j=1}^{k} \rho(\alpha_j)$ measuring the "quality" of the solution. This could be the ℓ_0 pseudo-norm, the ℓ_1-norm or any other separable choice of penalty. Exploiting the fact that $\mathbf{D}$ is unitary, we get

$$\begin{aligned} f(\alpha) &= \frac{1}{2}\|\mathbf{D}\alpha - \mathbf{x}\|_2^2 + \lambda \cdot \rho(\alpha) = \frac{1}{2}\|\mathbf{D}\alpha - \mathbf{D}\mathbf{D}^T\mathbf{x}\|_2^2 + \lambda \cdot \rho(\alpha) \\ &= \frac{1}{2}\|\mathbf{D}(\alpha - \beta)\|_2^2 + \lambda \cdot \rho(\alpha) = \frac{1}{2}\|\alpha - \beta\|_2^2 + \lambda \cdot \rho(\alpha). \end{aligned} \tag{1.11}$$

The first transition employed the unitary property, $\mathbf{D}\mathbf{D}^T = \mathbf{I}$. The second is done by denoting $\beta = \mathbf{D}^T\mathbf{x}$. The last transition is due to the fact that the ℓ_2-norm is invariant to a unitary transformation. Since $\rho()$ works on each entry of α independently, we can finally write

$$f(\alpha) = \sum_{j=1}^{k} \left[\frac{1}{2}(\alpha_j - \beta_j) + \lambda \cdot \rho(\alpha_j)\right], \tag{1.12}$$

which indicates that this problem is separated into k easy to solve scalar optimization problems.

As the problem becomes a set of independent 1-D problems, it is possible to compute the output $\hat{\alpha}$ for any input value of β, effectively forming a lookup table. Plotting this graph generally looks like a shrinkage curve[1] that nulls the small entries and the large ones intact. Of course, such a graph can be created for any penalty function ρ and value λ.

4.1. ℓ_0 and greedy algorithms for unitary dictionaries

When working with the ℓ_0, the function ρ is the indicator function, $\rho(\alpha_j) = |\alpha_j|^0$. Therefore, the independent optimization problem to solve is

$$\alpha_j^{opt} = \arg\min_{\alpha_j} \left[\frac{1}{2}(\alpha_j - \beta_j)^2 + \lambda \cdot |\alpha_j|^0\right]. \tag{1.13}$$

This is a very simple task - there are only two logical options for the value of α_j. Either $\alpha_j = \beta_i$, paying λ for the second term, or $\alpha_j = 0$, and paying $\frac{1}{2}\beta_j^2$ in the first term. Any other value will result in a necessarily larger penalty.

Therefore, the shrinkage curve for this case is very simple. For $|\beta| < \sqrt{2}\lambda$, we choose $\alpha = 0$, while for $|\beta| \geq \sqrt{2}\lambda$, we choose $\alpha = \beta$. Therefore, this method is known as Hard-Thresholding, as any coefficient below the threshold $\sqrt{2} \cdot \lambda$ is set to zero, while every coefficient above this threshold remains untouched. This solution is the *global* minimizer of the problem we tackled.

It is possible to show that when the dictionary is unitary, all the greedy algorithms we have seen (thresholding, OMP, LS-OMP, weak-OMP, MP) obtain the exact solution, which is the one obtained by the hard-thresholding algorithm described here. Just to illustrate this, the thresholding algorithm will compute $\mathbf{D}^T\mathbf{x}$, and work on the coefficients in descending order of magnitude. The amount of atoms it will choose is exactly those that the hard-thresholding will leave intact, i.e., the amount of atoms above the threshold. The OMP does the same, simply

[1] The name comes from the fact that the resulting value of $\hat{\alpha}$ is always smaller or equal to the input β.

choosing one atom at a time. Since the dictionary is unitary, adding another atom to the support does not effect the inner products of the residual with the remaining atoms.

4.2. Relaxation algorithms for unitary dictionaries

The basis pursuit uses the ℓ_1 -norm as a penalty on α. In the unitary case, this leads to separate optimization problem of the form

$$\alpha_j^{opt} = \arg\min_{\alpha_j} \left[\frac{1}{2} (\alpha_j - \beta_j)^2 + \lambda \cdot |\alpha_j| \right]. \tag{1.14}$$

Analyzing this problem, it turns out that the solution becomes

$$\alpha_j^{opt} = \begin{cases} \beta_j - \lambda & \beta_j > \lambda, \\ 0 & |\beta_j| \le \lambda, \\ \beta_j + \lambda & \beta_j < -\lambda. \end{cases} \tag{1.15}$$

The curve this solution implies is known as soft-thresholding, as when the coefficient is above the threshold it does not remain the same (as in the hard-thresholding), but is made smaller instead.

The Dantzig Selector tries to solve an alternative problem (1.9). When we use the unitary property, it becomes

$$\begin{aligned} \hat{\alpha} &= \min_{\alpha} \|\alpha\|_1 \ s.t. \ \|\mathbf{D}^T (\mathbf{D}\alpha - \mathbf{x})\|_\infty = \min_{\alpha} \|\alpha\|_1 \ s.t. \ \|\alpha - \mathbf{D}^T\mathbf{x}\|_\infty \\ &= \min_{\alpha} \|\alpha\|_1 \ s.t. \ \|\alpha - \beta\|_\infty, \end{aligned} \tag{1.16}$$

where in the first transition we used $\mathbf{D}^T\mathbf{D} = \mathbf{I}$, which is the unitarity property, and in the second transition simply re-introduced the notation of β. The last step results in K independent optimization problems of the form:

$$\arg\min_{\alpha_i} |\alpha_j| \ s.t. \ |\alpha_j - \beta_j| \le T. \tag{1.17}$$

It is not difficult to see that the solution of this optimization problem is exactly identical to the solution offered for the BP algorithm in (1.15). The conclusion is that BP and DS are exactly identical when the dictionary is unitary, and both of them can be solved using a closed-form formula for each of the coefficients (such that they do not require a complex optimization procedure).

4.3. Unitary dictionaries - summary

We have seen that when the dictionary is unitary, both families of algorithms enjoy a simple closed-form solution. Given the signal $\mathbf{x}$, multiply it by the dictionary to obtain a vector of coefficients $\beta = \mathbf{D}^T\mathbf{x}$. On each of the coefficients, apply independently a Look-Up Table operation with a shape that depends on ρ and λ, and this results in the vector of coefficients $\hat{\alpha}$. This simple procedure yields the *global* optimum of $f(\alpha)$, even if f itself is not convex. This leads to the obvious question – can this be used to gain an insight to the general case, when the dictionary is non-unitary? Can we use it to construct simpler, more effective algorithms for such cases? In the 4th lecture we will see a family of algorithms doing exactly that.

LECTURE 2

Introduction to sparse approximations - theory

This lecture deals with the theory of sparse approximations. We will start by defining several properties of dictionaries - the Spark, the Mutual-Coherence, and the Restricted Isometry Property (RIP). We will then show theoretical guarantees for the uniqueness of the solution for (P_0), and show the equivalence of MP and BP for the exact case. We will then continue with the discussion of (P_0^ϵ), and show theoretical guarantees on the stability of the solution. Finally, we will mention briefly results related to near-oracle performance of pursuit techniques in the noisy case.

1. Dictionary properties

The purpose of this section is finding ways to characterize the dictionary $\mathbf{D}$, in order for us to be able to make claims about the conditions under which pursuit algorithms are guaranteed to succeed. Ideally, we would like these bounds to be tight, and the characterizations easy to compute. As we will see, this is not necessarily the case.

1.1. The matrix spark

The Spark was defined by Donoho & Elad [**14**] (similar definitions have been made in tensor decomposition by Kruskal [**34, 35**] as well as in coding theory):

DEFINITION 2.1 (Spark). Given a matrix $\mathbf{D}$, $\sigma = \mathit{Spark}(\mathbf{D})$ is the smallest number of atoms from $\mathbf{D}$ that are linearly dependent.

While this definition is likely to remind the reader of the definition of the rank of a matrix, it is very different. There are two differences between the definitions - while the rank is the *largest* number of *independent* atoms, the Spark is the *smallest* number of *dependent* atoms. For an example, consider the matrix

$$\mathbf{D} = \begin{pmatrix} 1 & 0 & 0 & 0 & 1 \\ 0 & 1 & 0 & 0 & 1 \\ 0 & 0 & 1 & 0 & 0 \\ 0 & 0 & 0 & 1 & 0 \end{pmatrix}.$$

The rank of this matrix is 4, the maximal number of independent columns (e.g., the first four columns). The Spark of this matrix is 3: one column could only be linearly dependent if it is the zero vector; Any two columns in this dictionary are linearly independent; There is a group of three columns - the first, second and last - that are linearly dependent. Therefore, $\sigma(\mathbf{D}) = 3$.

It is easy to observe that for a dictionary $\mathbf{D}_{n\times k}$, the Spark satisfies $1 \le \sigma \le n+1$. Also, if for some non-trivial vector $\mathbf{v}$ it exists that $\mathbf{Dv} = \mathbf{0}$, than is is certain that $\|\mathbf{v}\|_0^0 \ge \sigma$, since the support of $\mathbf{v}$ constitutes a linearly dependent set of atoms. Since the smallest possible such set is σ, the support of $\mathbf{v}$ also contains at least σ atoms.

1.2. The mutual coherence

DEFINITION 2.2 (Mutual Coherence). For a matrix $\mathbf{D}$, the Mutual-Coherence M is the largest (in magnitude) off-diagonal element of the Gram matrix of $\mathbf{D}$, i.e., $\mathbf{G} = \mathbf{D}^T\mathbf{D}$.

For this definition, we assume that the columns of $\mathbf{D}$ are normalized. The Mutual-Coherence measures the maximal similarity (or anti-similarity) between two columns on the dictionary. Therefore, its value ranges between 0 (an orthonormal basis) and 1 (when there are two identical atoms). As we will see later, a large mutual coherence is bad news for our algorithms, as the algorithms are likely to confuse between the two atoms.

1.3. The relationship between the spark and the mutual coherence

Our next task is to understand the relationship between the Spark and the Mutual Coherence measures. In order to do that, we need the following well-known theorem:

THEOREM 2.1 (Gershgorin's Disk Theorem). *Consider a general square matrix* $\mathbf{A}$, *and define the k'th disk as* $disk_k = \left\{x \left| |x - a_{kk}| \le \sum_{j\neq k} |a_{jk}| \right.\right\}$. *Then, the eigenvalues of* $\mathbf{A}$ *all lie in the union of all disks,* $\{\lambda_1, \lambda_2, .., \lambda_n\} \subset \cup_{k=1..n} disk_k$.

Alternatively, we can write $\lambda_{\min} \ge \arg\min_{1\le k\le n}\left\{a_{kk} - \sum_{j\neq k}|a_{jk}|\right\}$ and $\lambda_{\max} \le \arg\max_{1\le k\le n}\left\{a_{kk} + \sum_{j\neq k}|a_{jk}|\right\}$. One of the implications of the theorem is that if $\mathbf{A}$ is diagonally dominant (assuming real entries), than $\mathbf{A}$ is necessarily positive definite. This is because the origin is not in any of the disks, and therefore all eigenvalues are positive. Using this theorem, we are now ready to make the following claim:

THEOREM 2.2 (Relation between Spark and Mutual Coherence). $\sigma \ge 1 + \frac{1}{M}$

PROOF. Let us consider a group of columns S from $\mathbf{D}$, and compute its sub-Gram matrix. We know that these columns are independent, if this matrix is positive definite. From the Gershgorin's disk theorem, we know that the matrix will be positive definite if it is diagonally dominant.

The Mutual-Coherence bounds any off-diagonal element by M, while the diagonal elements are all 1. Therefore, the matrix is diagonally dominant if $(S-1)M < 1$. As long as $S < 1 + 1/M$, any group of S columns are independent. Therefore, the minimal number of columns that could be dependent is $S = 1 + 1/M$, and this is exactly the claimed result. □

1.3.1. *Grassmanian Frames.* We now turn to deal with the following question: For a general matrix $\mathbf{D}_{n\times k}$, with $k > n$, what is the best (smallest) Mutual-Coherence we can hope for? Obviously, it can't be zero, as that would mean that there are k columns that are orthogonal, and that is not possible when their length is n. It turns out that

$$M_{\min} = \sqrt{\frac{k-n}{n(k-1)}} = O\left(\frac{1}{\sqrt{n}}\right), \tag{2.1}$$

with the second step using the substitution $k = \rho n$, and the approximation is true for small values of ρ (e.g., 2,3). We will omit the proof, and only state that this question is closely related to questions about packing vectors in the n-dimensional space in the most spacious ways. As an example, for a matrix with $n = 50$ rows and $k = 100$ columns, the minimal possible coherence is roughly 0.1.

Can matrices achieving this bound be found in practice? The answer is yes, and this family of matrices is known as Grassmanian Frames [**49**]. The prevailing property of these frames is that the off-diagonal elements in their Gram-matrices all have the same magnitude. They are difficult to built, and a numerical algorithm by Tropp has been developed for their construction [**52**]. It is important to note, however, that for some pairs of n and k such frames do not exist.

1.4. The restricted isometry property

DEFINITION 2.3 (Restricted Isometry Property). Given a matrix $\mathbf{D}$, consider all s-sparse vectors α, and find the smallest value δ_s that satisfies

$$(1-\delta_s)\|\alpha\|_2^2 \leq \|\mathbf{D}\alpha\|_2^2 \leq (1+\delta_s)\|\alpha\|_2^2.$$

$\mathbf{D}$ is then said to have an s-RIP with a constant δ_s.

What is the meaning of this definition? $\mathbf{D}\alpha$ combines linearly atoms from $\mathbf{D}$. The RIP suggests that every such group of s atoms behaves like an isometry, i.e., does not considerably change the length of the vector it operates on.

As an example, when $\mathbf{D}$ is unitary, the length does not change for any s. Therefore, $\delta_s(\mathbf{D}) = 0$, for every s. In this sense, the RIP asks how close is $\mathbf{D}$ to a unitary matrix, when considering multiplications by s-sparse vectors.

1.5. The RIP and the mutual coherence

Let us denote $\mathbf{D}_S$ as the sub-matrix of $\mathbf{D}$ containing only the columns in the support S, and similarly, α_S as containing only the non-zeros in α, so $\mathbf{D}_S\alpha_S = \mathbf{D}\alpha$. Using this terminology, the RIP property states that

$$(1-\delta_s)\|\alpha_S\|_2^2 \leq \alpha_S^T\mathbf{D}_S^T\mathbf{D}_S\alpha_S \leq (1+\delta_s)\|\alpha_S\|_2^2.$$

On the other hand, from an eigenvalue perspective, we know that

$$\|\alpha_S\|_2^2 \cdot \lambda_{\min}\left(\mathbf{D}_S^T\mathbf{D}_S\right) \leq \alpha_S^T\mathbf{D}_S^T\mathbf{D}_S\alpha_S \leq \|\alpha_S\|_2^2 \cdot \lambda_{\max}\left(\mathbf{D}_S^T\mathbf{D}_S\right).$$

From the Gershgorin's theorem we already know that

$$\begin{aligned}\lambda_{\min}\left(\mathbf{D}_S^T\mathbf{D}_S\right) &\geq 1-(s-1)M\\ \lambda_{\max}\left(\mathbf{D}_S^T\mathbf{D}_S\right) &\leq 1+(s-1)M,\end{aligned}$$

and finally, we see that

$$\delta_s \leq (s-1)M. \tag{2.2}$$

1.6. The RIP and the spark

Suppose that the Spark of $\mathbf{D}$ is known to be σ. Then, for $s = \sigma$, there is at least one vector α of cardinality s such that $\mathbf{D}\alpha = \mathbf{0}$, therefore indicating that $\delta_\sigma = 1$. Furthermore, for any value of $s < \sigma$, it follows that $\delta_s < 1$.

1.7. Computing the spark, the MC and the RIP

Summarizing the discussion so far, we consider the question of computing the above described properties. The Spark requires going through each and every support and checking for linear dependence. Thus, it is impossible to compute in general as it is combinatorial. There is an exception, however, as a random iid matrix has a full Spark of $(n+1)$.

The RIP similarly has to scan all supports, and is therefore also impossible to compute in practice. There is also an exception here as well, in the case of a random Gaussian matrix, the RIP can be bounded (with high probability) by $\delta_s \le C \cdot \sqrt{\frac{s}{n} \cdot \log\left(\frac{k}{s}\right)}$. The C is a coefficient controlling the probability – as the required confidence grows, so does the constant C, making the bound less tight.

Fortunately, the Mutual Coherence is very easy to compute, as we have seen - it only requires computing the Gram matrix, and scanning all elements in it. Unfortunately, as this measure only quantifies pairs of atoms, it is also the weakest of the three measures.

Lastly, it is important to note that all of the measures introduced are of worst-case nature, and as such lead to worse-case performance analysis of algorithms. For example, it is enough that the dictionary contains a duplicate atom to set the Mutual Coherence to 1. The rest of the dictionary can behave perfectly, but that will be unknown to the coherence. The Spark and the RIP will also suffer in such a case in a similar fashion. Different types of measures are needed for average-case performance analysis, but such measures are still very much missing.

2. Theoretical guarantees - uniqueness for P_0

After the introduction of some tools to be used, we now turn to the first problem of interest. Suppose Alice holds a dictionary $\mathbf{D}$, and generates some sparse vector of coefficients α. Multiplying the two, Alice generates the signal $\mathbf{x} = \mathbf{D}\alpha$. Then, Alice contacts her friend, Bob, and gives him the dictionary $\mathbf{D}$ and the vector $\mathbf{x}$, and asks him to find the vector α she had used.

Bob knows that α is sparse, and so he sets out to seek the sparsest possible solution of the system $\mathbf{D}\alpha = \mathbf{x}$, namely, solve the optimization problem

$$\hat{\alpha} = \arg\min_{\alpha} \|\alpha\|_0^0 \ s.t. \ \mathbf{D}\alpha = \mathbf{x}.$$

For the time being, we shall assume that Bob can actually solve this problem exactly (we have seen in the first lecture that this is generally NP-hard, and therefore impossible in practice). The question we want to ask is if (or when) can we expect that $\hat{\alpha} = \alpha$. It might turn our that $\hat{\alpha}$ is even sparser than the original α. This is essentially the question of uniqueness – what properties must α and the dictionary $\mathbf{D}$ have so that α is the sparsest vector to create $\mathbf{x}$? The answer turns out to be quite simple [**14**]:

THEOREM 2.3 (Uniqueness Requirements of P_0). *The following two claims define the uniqueness condition and implication:*

(1) *Suppose we have found a sparse solution $\hat{\alpha}$ that satisfies $\mathbf{D}\hat{\alpha} = \mathbf{x}$. If this solution satisfies $\|\hat{\alpha}\|_0^0 < \sigma/2$, then this this solution is unique (i.e., it is the sparsest solution possible).*
(2) *If Alice generated a signal with $\|\alpha\|_0^0 < \sigma/2$, than Bob is able to recover α exactly: $\hat{\alpha} = \alpha$.*

PROOF. Suppose we have found two candidate solutions α_1 and α_2 to the linear system of equations $\mathbf{D}\alpha = \mathbf{x}$, leading to $\mathbf{D}\alpha_1 = \mathbf{D}\alpha_2$ and $\mathbf{D}(\alpha_1 - \alpha_2) = 0$. From the Spark property, the cardinality of any vector $\mathbf{v}$ that satisfied $\mathbf{D}\mathbf{v} = 0$ is at least the Spark, indicating that

$$\|\alpha_1 - \alpha_2\|_0^0 \ge \sigma. \tag{2.3}$$

We can also observe that

$$\|\alpha_1\|_0^0 + \|\alpha_2\|_0^0 \geq \|\alpha_1 - \alpha_2\|_0^0. \tag{2.4}$$

This property can be understood by analyzing the supports of both vectors. If there is no overlap in their supports, then the equality holds. If there is an overlap, and two terms in the overlap happen to cancel each other exactly, then the resulting support is smaller than the sum of sizes of the two supports. We can refer to this property as some sort of triangle inequality of the ℓ_0, even though it is not actually a norm.

Combining the two inequalities, it turns out that the sum of cardinalities of both vectors is at least the Spark,

$$\|\alpha_1\|_0^0 + \|\alpha_2\|_0^0 \geq \sigma. \tag{2.5}$$

This can be viewed as an uncertainty principle – two solutions cannot be jointly very sparse. If one solution is indeed sparse, the other must be non-sparse. The uncertainty principle implies that if we have found a solution α_1 with cardinality less than half the Spark, i.e., $\alpha_1 < \sigma/2$, than *every* other solution α_2 must have a cardinality of more than half the Spark, in order to fulfill the requirement that $\|\alpha_1\|_0^0 + \|\alpha_2\|_0^0 \geq \sigma$. Therefore, α is sparser than any other solution, and indeed, the sparsest solution.

It directly follows that if Alice chose α such that $\|\alpha\|_0^0 < \sigma/2$, than it is the sparsest solution to $\mathbf{D}\alpha = \mathbf{x}$, and therefore Bob will recover it exactly. □

3. Equivalence of the MP and BP for the exact case

Now we turn to walk in Bob's shoes. We have seen that Bob's task is impossible to solve directly, and therefore he has to turn to approximation algorithms, such as Orthogonal Matching Pursuit (OMP)—, Basis Pursuit (BP), etc. The question we ask now is: Is there any hope that any of these algorithms will actually recover the exact solution? Surprisingly, we will see that under some conditions, these algorithms are guaranteed to succeed.

3.1. Analysis of the OMP

Recall the OMP algorithm, introduced in lecture 1. To make our life easier, we assume that the s non-zero locations in α are in the first s locations. Furthermore, we assume that these coefficients are sorted in decreasing order of magnitude, with $|\alpha_1| \geq |\alpha_2| \geq .. \geq |\alpha_s| > |\alpha_{s+1}| = 0$. Thus, our signal can be written as $\mathbf{x} = \sum_{i=1}^{s} \alpha_i \mathbf{d}_i$. This assumption does not effect the generality of our claims, as we can achieve this structure by reordering the columns of $\mathbf{D}$ and the locations of the non-zeros in α. The re-structured problem remains equivalent to the original one.

We now ask ourselves what conditions are needed for the first OMP step to succeed, i.e., that the first atom added to the support is an atom in the true support (one of the first s atoms)? Since the OMP selects the atom with the largest magnitude inner-product, the first step succeeds if

$$\left|\mathbf{x}^T \mathbf{d}_1\right| > \max_{j>s} \left|\mathbf{x}^T \mathbf{d}_j\right|. \tag{2.6}$$

This condition guarantees that the atom selection will prefer the first atom to any atom atom outside the support, and thus guarantees that an atom in the true support is chosen. It may happen that another atom in the true support is chosen instead of the first, but this is also considered as a success. Therefore, the question we want to answer is when does the condition (2.6) hold.

In order to find out under what conditions this inequality is satisfied, we lower bound the left-hand-side (LHS), and upper bound the right-hand-side(RHS). We will then find the conditions for the bound on the LHS still being greater than the upper bound of the RHS, thus fulfilling the original inequality as well.

We start with an upper-bound of the Right-Hand-Side (RHS):

$$\max_{j>s}\left|\mathbf{x}^T\mathbf{d_j}\right| = \max_{j>s}\left|\sum_{i=1}^{s}\alpha_i\mathbf{d}_i^T\mathbf{d}_j\right| \le \max_{j>s}\sum_{i=1}^{s}|\alpha_i|\cdot\left|\mathbf{d}_i^T\mathbf{d}_j\right| \le |\alpha_1|\cdot s\cdot M. \tag{2.7}$$

The first transition assigns the expression $\mathbf{x}=\sum_{i=1}^{s}\alpha_i\mathbf{d}_i$, the second uses the property $|a+b|\le|a|+|b|$, and the third uses the definition of the Mutual-Coherence as an upper bound on $\left|\mathbf{d}_i^T\mathbf{d}_j\right|$, as well as the monotonicity of the coefficients, implying that $|\alpha_k|\le|\alpha_1|$ for $k=2,..,s$. We now turn to lower-bound the Left-Hand-Side(LHS) in inequality (2.6):

$$\begin{aligned}\left|\mathbf{x}^T\mathbf{d}_1\right| &= \left|\sum_{i=1}^{s}\alpha_i\mathbf{d}_i^T\mathbf{d}_1\right| = \left|\alpha_1+\sum_{i=2}^{s}\alpha_i\mathbf{d}_i^T\mathbf{d}_1\right| \ge |\alpha_1| - \left|\sum_{i=2}^{s}\alpha_i\mathbf{d}_i^T\mathbf{d}_1\right| \\ &\ge |\alpha_1| - \sum_{i=2}^{s}|\alpha_i|\cdot\left|\mathbf{d}_i^T\mathbf{d}_1\right| \ge |\alpha_1|\cdot[1-(s-1)\cdot M].\end{aligned} \tag{2.8}$$

The transitions in the first line are made by first substituting $\mathbf{x}=\sum_{i=1}^{s}\alpha_i\mathbf{d}_i$ and then using the fact that the atoms are normalized, i.e., $\mathbf{d}_1^T\mathbf{d}_1=1$. Then, the bounding starts by the reversed triangle inequality $|a+b|\ge|a|-|b|$, and then repeating the steps performed in bounding the RHS (note that since the second term is with a negative sign, increasing its magnitude reduces the magnitude of the overall expression).

Using the bounds in (2.7) and (2.8) and plugging them into (2.6), we get that a condition that guarantees that the first OMP step succeeds is

$$\left|\mathbf{x}^T\mathbf{d}_1\right| \ge |\alpha_1|\cdot[1-(s-1)\cdot M] > |\alpha_1|\cdot s\cdot M \ge \max_{j>s}\left|\mathbf{x}^T\mathbf{d_j}\right|,$$

which leads to the following bound on the cardinality of the representation,

$$s < \frac{1}{2}\left[1+\frac{1}{M}\right]. \tag{2.9}$$

Assuming this condition is met, the first atom chosen is indeed an atom from the true support, with index i^*. Then, a LS step is performed, and a coefficient value α_1^* is assigned to this atom. Now, we look at the residual as our new signal, being

$$\mathbf{r}^1 = \mathbf{x}-\alpha_1^*\mathbf{d}_{i^*} = \sum_{i=1}^{s}\alpha_i\mathbf{d}_i - \alpha_1^*\mathbf{d}_{i^*}.$$

This signal has the same structure we the one we started with – a linear combination of the s first atoms. This means that the exact same condition is needed for the second step to succeed, and in fact, for all other steps to succeed as well.

This means, that if $s<\frac{1}{2}\left[1+\frac{1}{M}\right]$ the OMP is guaranteed to find the correct sparse representation. This is done in exactly s iterations, as in each iteration one atom in the support is chosen, and due to the LS stage, the residual is always orthogonal to the atoms already chosen.

Using this result, and the bound we have found in the previous section regarding the Spark, we come to the following conclusion:

THEOREM 2.4 (OMP and MP Equivalence). *Given a vector* $\mathbf{x}$ *with a representation* $\mathbf{D}\alpha = \mathbf{x}$, *and assuming that* $\|\alpha\|_0^0 < 0.5\,(1 + 1/M)$, *then (i)* α *is the sparsest possible solution to the system* $\mathbf{D}\alpha = \mathbf{x}$, *and (ii) OMP (and MP) are guaranteed to find it exactly* [**50, 16**].

3.2. Analysis of the BP

We now turn to perform the same analysis for the BP, checking under which conditions are we guaranteed to succeed in recovering the sparsest solution. We remind ourselves that the BP aims to approximate the solution of the problem $\hat{\alpha} = \arg\min_\alpha \|\alpha\|_0^0$ *s.t.* $\mathbf{D}\alpha = \mathbf{x}$, by solving instead $\hat{\alpha} = \arg\min_\alpha \|\alpha\|_1$ *s.t.* $\mathbf{D}\alpha = \mathbf{x}$. Therefore, our aim is now to show under which condition is the sparse vector α used to generate the signal $\mathbf{x}$ also the shortest one with respect to the ℓ_1 norm.

The strategy for this analysis starts by first defining the set of all solutions for the linear set of equations $\mathbf{D}\beta = \mathbf{x}$ (excluding the one we started with), such that their ℓ_1 length is shorter (or equal) to the length of the original vector α,

$$C = \{\beta \neq \alpha \,|\mathbf{x} = \mathbf{D}\alpha = \mathbf{D}\beta \text{ and } \|\beta\|_1 \leq \|\alpha\|_1\}. \tag{2.10}$$

We will gradually inflate this set, while simplifying it, eventually showing that it is still empty. When done, this will imply that the initial set is also empty, and therefore α is the result of the ℓ_1 minimization, leading to α being the vector with the shortest ℓ_1 norm, therefore recovered by the BP.

Our first step starts with defining the vector $\mathbf{e} = \beta - \alpha$, or in other words, $\mathbf{e} + \alpha = \beta$. Our set remains unchanged, and is simply rewritten as

$$C_{\mathbf{e}} = \{\mathbf{e} \neq \mathbf{0} \,|\mathbf{0} = \mathbf{D}\mathbf{e} \text{ and } \|\alpha + \mathbf{e}\|_1 - \|\alpha\|_1 \leq 0\}. \tag{2.11}$$

Next, we simplify the linear constraint $\mathbf{0} = \mathbf{D}\mathbf{e}$,

$$\begin{aligned}\{\mathbf{e}\,|\mathbf{0} = \mathbf{D}\mathbf{e}\} &= \left\{\mathbf{e}\,\middle|\mathbf{0} = \mathbf{D}^T\mathbf{D}\mathbf{e}\right\} = \left\{\mathbf{e}\,\middle|-\mathbf{e} = \left(\mathbf{D}^T\mathbf{D} - \mathbf{I}\right)\mathbf{e}\right\} \\ &\subseteq \left\{\mathbf{e}\,\middle||\mathbf{e}| = \left|\left(\mathbf{D}^T\mathbf{D} - \mathbf{I}\right)\mathbf{e}\right|\right\} \\ &\subseteq \left\{\mathbf{e}\,\middle||\mathbf{e}| \leq \left|\mathbf{D}^T\mathbf{D} - \mathbf{I}\right| \cdot |\mathbf{e}|\right\} \\ &\subseteq \left\{\mathbf{e}\,\middle||\mathbf{e}| \leq M \cdot \left(\mathbf{1}\mathbf{1}^T - \mathbf{I}\right) \cdot |\mathbf{e}|\right\} \\ &= \{\mathbf{e}\,|(1 + M) \cdot |\mathbf{e}| \leq M\|\mathbf{e}\|_1\}.\end{aligned}$$

The transition from the first row to the second is done by taking absolute value of both sides (an absolute value of a vector is done element-wise). Obviously, some new $\mathbf{e}$'s have been added to the set, as we are no longer forcing the sign equality, implying that the set becomes larger. The next step stems from the triangle inequality. The transition between the third and fourth lines is slightly more complex. Observe that the matrix $\mathbf{D}^T\mathbf{D}$ is effectively the Gram Matrix, and all off-diagonal elements of it are bounded by M. Furthermore, the matrix $\left(\mathbf{D}^T\mathbf{D} - \mathbf{I}\right)$ has zeros along the main diagonal, and all remaining elements are still bounded by M. Observing that $\mathbf{1}\mathbf{1}^T$ is a matrix of the proper size, all containing ones, than $M\left(\mathbf{1}\mathbf{1}^T - \mathbf{I}\right)$ is a matrix containing zeros on the diagonal, and M on the off-diagonal terms. Thus $\left|\mathbf{D}^T\mathbf{D} - \mathbf{I}\right| \leq M \cdot \left(\mathbf{1}\mathbf{1}^T - \mathbf{I}\right)$ element-wise. This way, we no longer need $\mathbf{D}$ directly, instead characterizing it using only its Mutual-Coherence.

The last step re-orders the elements, and uses $\mathbf{1}^T|\mathbf{e}| = \|\mathbf{e}\|_1$. Replacing the linear constraint in (2.11) with the one derived above, we obtain an inflated set

$$C_{\mathbf{e}} \subseteq \left\{ \mathbf{e} \neq \mathbf{0} \,\middle|\, |\mathbf{e}| \leq \frac{M}{M+1} \cdot \|\mathbf{e}\|_1 \text{ and } \|\alpha + \mathbf{e}\|_1 - \|\alpha\|_1 \leq 0 \right\}. \tag{2.12}$$

Now we turn to handle the second constraint, $\|\alpha + \mathbf{e}\|_1 - \|\alpha\|_1 \leq 0$,

$$\begin{aligned}
\{\mathbf{e} \,|\, \|\alpha + \mathbf{e}\|_1 - \|\alpha\|_1 \leq 0\} &= \left\{ \mathbf{e} \,\middle|\, \sum_{i=1}^{K} (|\alpha_i + e_i| - |\alpha_i|) \leq 0 \right\} \\
&= \left\{ \mathbf{e} \,\middle|\, \sum_{i=1}^{s} (|\alpha_i + e_i| - |\alpha_i|) + \sum_{i>s} |e_i| \leq 0 \right\} \\
&\subseteq \left\{ \mathbf{e} \,\middle|\, \sum_{i>s} |e_i| - \sum_{i=1}^{s} |e_i| \leq 0 \right\} \\
&= \left\{ \mathbf{e} \,|\, \|\mathbf{e}\|_1 - 2 \cdot \mathbf{1}_s^T |\mathbf{e}| \leq 0 \right\}.
\end{aligned}$$

The first step simply writes the ℓ_1 norm as a sum of elements, and the second step uses the knowledge that only the first s coefficients in α are non-zeros (as we have assumed earlier). The third term uses a version of the triangle inequality, $|x + y| - |x| \geq -|y|$. Using this replacement, each term in the sum is made smaller, thus inserting more possible $\mathbf{e}$'s into the set. The last step merely rewrites the expression, since $\sum_{i>s} |e_i| - \sum_{i=1}^{s} |e_i| = \sum_{i=1}^{K} |e_i| - 2 \cdot \sum_{i=1}^{s} |e_i| = \|\mathbf{e}\|_1 - 2 \cdot \mathbf{1}_s^T |\mathbf{e}|$, with $\mathbf{1}_s$ denoting a vector of length k with the first s elements being 1-es and the remaining elements being 0-es.

Plugging this inflated group into our definition of the group C_e in (2.12), we arrive at

$$C_{\mathbf{e}} \subseteq \left\{ \mathbf{e} \neq \mathbf{0} \,\middle|\, |\mathbf{e}| \leq \frac{M}{M+1} \cdot \|\mathbf{e}\|_1 \text{ and } \|\mathbf{e}\|_1 - 2 \cdot \mathbf{1}_s^T |\mathbf{e}| \leq 0 \right\}. \tag{2.13}$$

The attractive part about this set is that it no longer depends directly on $\mathbf{D}$, but only on its Mutual-Coherence, and furthermore, it does not depend on α any more, allowing us to obtain a general bound applicable to all vectors.

Our next simplification uses the observation that if we have found any vector $\mathbf{e} \in C_e$, then for every real number z, $z \cdot \mathbf{e}$ is also in C_e. Since we are only interested in the question weather C_e is empty or not, we choose to intersect this set with the ℓ_1-ball, i.e., all the vectors $\mathbf{e}$ that have an ℓ_1-norm of 1,

$$C_{\mathbf{e}} \subseteq \left\{ \mathbf{e} \neq \mathbf{0} \,\middle|\, |\mathbf{e}| \leq \frac{M}{M+1} \text{ and } 1 - 2 \cdot \mathbf{1}_s^T |\mathbf{e}| \leq 0 \text{ and } \|\mathbf{e}\|_1 = 1 \right\}. \tag{2.14}$$

Our purpose now is to derive conditions for this set to be empty. The first constraint is that every element of $\mathbf{e}$ is smaller than $M/(1 + M)$. In order to satisfy the second constraint, the energy of the vector $\mathbf{e}$ should be concentrated in the elements in the support. Using the bound on the size of each element, for the second constraint to be satisfied (and the set being not empty) it is required that

$$1 - 2 \cdot s \cdot \frac{M}{1+M} \leq 0 \Rightarrow s \geq \frac{M+1}{2M}.$$

This means that as long as $s < \frac{M+1}{2M}$, the set is empty, and therefore C_e is also empty. This implies that under this condition, a vector β different from α, satisfying

$\mathbf{D}\beta = \mathbf{x}$ and with $\|\beta\|_1 \leq \|\alpha\|_1$ does not exist. This leads to the BP actually arriving at α as the result of its minimization, which means that when $s < \frac{M+1}{2M}$, the basis pursuit is guaranteed to recover α exactly.

Stating this formally, we obtain the equivalence theorem for the Basis Pursuit:

THEOREM 2.5 (BP Equivalence). *Given a vector* $\mathbf{x}$ *with a representation* $\mathbf{D}\alpha = \mathbf{x}$, *and assuming that* $\|\alpha\|_0^0 < 0.5\,(1 + 1/M)$, *then BP is guaranteed to find the* sparsest *solution* [**14, 29**].

Obviously, this claim is exactly identical to the equivalence claim for the OMP, meaning that the bound to guarantee success of the two algorithms is identical. However, it does not mean that the two algorithms are perform the same, as they are different in general (specifically, when this bound is not met, or in the noisy case). It is hard to tell which is better.

Furthermore, the bounds presented here are worst-case performance guarantees. When analyzing actual performance, these bounds are usually too pessimistic. Analysis of average performance has also been done. In a series of papers [**5, 12, 13, 18, 19, 53**], bounds are developed from a probabilistic point of view, showing that up to some bounds, the actual vector may be recovered with a probability close to 1. These bounds are proportional to $O(1/M^2)$, instead of $O(1/M)$.

4. Theoretical guarantees - stability for (P_0^ϵ)

4.1. Stability versus uniqueness

Consider the following problem, which is very similar to the one we dealt with above: Alice chooses a sparse vector α, multiplies it by $\mathbf{D}$, and obtains $\mathbf{x} = \mathbf{D}\alpha$. Alice wants to send the signal $\mathbf{x}$ to Bob, but unfortunately, unlike before, the transmission line is imperfect. Therefore, noise is added to the signal, and Bob receives the signal $\mathbf{y} = \mathbf{x} + \mathbf{v}$, where we know that the norm of $\mathbf{v}$ is bounded, $\|\mathbf{v}\|_2 \leq \epsilon$.

Bob would like to recover α. Since $\mathbf{y}$ is noisy, there is no sense in solving the exact equation $\mathbf{y} = \mathbf{D}\alpha$, and thus Bob would attempt to solve

$$\hat{\alpha} = \arg\min_{\alpha} \|\alpha\|_0 \ s.t. \ \|\mathbf{D}\alpha - \mathbf{x}\|_2 \leq \epsilon.$$

Due to the noise being added, there might be somewhere in the sphere around $\mathbf{x}$ a vector with a sparser representation than α. If not, α itself will be chosen. Therefore, we know that $\|\hat{\alpha}\|_0 \leq \|\alpha\|_0$. Can we hope that $\hat{\alpha} = \alpha$? If not, can we at least hope that they are similar or close-by?

It turns out (we will not show it hear, but the slides present it through a simple and illustrative example) that exact equality in the recovery, $\hat{\alpha} = \alpha$, is impossible to expect in general, and thus we will turn to discuss *stability*, which is a claim about the proximity between Alice's $\mathbf{a}$, and Bob's proposed solution $\hat{\alpha}$.

Suppose we created the true signal with an s-sparse vector of coefficients α_1, and the solution we got is an s-sparse (or sparser) vector α_2. These vectors satisfy $\|\mathbf{D}\alpha_1 - \mathbf{y}\|_2 \leq \epsilon$ and $\|\mathbf{D}\alpha_2 - \mathbf{y}\|_2 \leq \epsilon$, which means that both of these solutions are inside a sphere around $\mathbf{y}$ with radius ϵ. As these two points are inside the sphere, the maximal distance between the two is twice the radius, and therefore

$$\|\mathbf{D}\,(\alpha_1 - \alpha_2)\,\|_2 \leq 2\epsilon. \tag{2.15}$$

On the other hand, we can use the RIP property. The vector $\alpha_1 - \alpha_2$ is $2s$-sparse at most (if the vectors have different supports), and we have the RIP property

$$(1-\delta_{2s})\|\alpha_1-\alpha_2\|_2^2 \le \|\mathbf{D}(\alpha_1-\alpha_2)\|_2^2. \tag{2.16}$$

Combining (2.15) (by squaring both sides) with (2.16) leads to the relation

$$(1-\delta_{2s})\|\alpha_1-\alpha_2\|_2^2 \le 4\epsilon^2 \tag{2.17}$$

Rearranging the terms leads to the following bound on the distance between the two solutions

$$\|\alpha_1-\alpha_2\|_2^2 \le \frac{4\epsilon^2}{1-\delta_{2s}} \le \frac{4\epsilon^2}{1-(2s-1)M}, \tag{2.18}$$

where in the last step we have used the relationship between the RIP and the Mutual-Coherence $\delta_s \le (s-1)M$. We have obtained a bound on the distance between the two representation vectors, and we can see that this bound is in the order of the noise. It is slightly higher, since the denominator is smaller than 1, but it is not far-off. This is what we strived to show – stability.

The above analysis assumes that the power of the noise is bounded, $\|\mathbf{v}\|_2 \le \epsilon$. However, there was no assumption made as to the shape of the noise, and the noise might have taken any shape to work against our analysis and algorithm. For this reason, the bounds we obtain is weak, showing no effect of noise attenuation.

An alternative approach could be to model the noise as a random, i.i.d Gaussian vector, the elements of which are randomly drawn as $v_i \sim N(0,\sigma^2)$. When such a model is assumed, we can expect to get much better stability bound: The norm of the error is expected to be close to (but not exactly) $s\cdot\sigma^2$ instead of $n\cdot\sigma^2$, which is a much better result. However, this claim will no longer be deterministic, but only probabilistic, with a probability very close to 1.

5. Near-oracle performance in the noisy case

So far we discussed the theoretical solution of (P_0^ϵ). However, we know that this task is impossible, and approximation algorithms are used, such as the OMP and the BP. Very similar analysis exists that shows stability of these algorithms – a successful recovery of the representation vector. We will not show this here, and we refer the readers to [**15, 16, 51**]

As above, there is a distinction between adversive noise, which leads to worst-case performance bounds similar to the one given in Equation (2.18). When random noise is assumed, these bounds improve dramatically. In that respect, it is customary to compare the performance of pursuit techniques to the one obtained by an oracle - a pursuit technique that *knows the exact support*. The oracle's error is typically $O(s\sigma^2)$ (as described above), and practical pursuit methods show bounds that are this expression multiplied by a constant (larger than 1) and a $log k$ factor [**1, 6**].

LECTURE 3

Sparse and redundant representation modelling

1. Modelling data with sparse and redundant representations

1.1. Inherent low-dimensionality of images

Let us think of the following virtual experiment: Consider images $\mathbf{x}$ of size $\sqrt{n}\times\sqrt{n}$ (e.g., $n = 400$ for 20×20 images). Each such image can be thought of as a point in the $\mathbb{R}^n$ space. Once we accumulate many such images, we will have many such points in $\mathbb{R}^n$. However, no matter how many images we accumulate, the space $\mathbb{R}^n$ is not going to be completely filled, but rather many empty areas will remain. Furthermore, the local density of the images from one place to another is expected to vary considerably.

The bottom line from these two observations is that we believe that the true dimension of the images is not n. Rather, we believe that the images form a low-dimensional manifold in $\mathbb{R}^n$, with spatially varying densities. It is our task to model this manifold reliably.

The tool with which we represent the manifold is the Probability Density Function (PDF) $Pr(\mathbf{x})$. This function tell us for each $\mathbf{x} \in \mathbb{R}^n$ how probable it is to be an image. Obviously, this function should be non-negative and when summed over the entire space it should sum to 1. In order to fulfill our assumption of the images forming a low-dimensional manifold, the PDF $Pr(\mathbf{x})$ should be zero for most of the cube $[0, 255]^n$.

The function $Pr(\mathbf{x})$ is called the *Prior* as it reflects our prior knowledge about what images should look like. Unfortunately, characterizing this function explicitly is probably an impossible task, which leads us to look for ways to simplify it.

1.2. Who needs a prior?

Let us consider one simple example in which we need a prior. Suppose we would like to obtain an image $\mathbf{x}$ (lying on the low-dimensional manifold). Unfortunately, we are only able to obtain its noisy version, $\mathbf{z} = \mathbf{x} + \mathbf{v}$, where $\mathbf{v}$ is a random noise vector with a limited magnitude $\|\mathbf{v}\|_2 \leq \epsilon$. Using the prior over the image $\mathbf{x}$, we can now use it to try and recover $\mathbf{x}$, by solving the problem

$$\hat{\mathbf{x}} = \arg\min_{\mathbf{x}} Pr\left(\mathbf{x}\right) \ \ s.t. \ \|\mathbf{x} - \mathbf{z}\|_2 \leq \epsilon.$$

In effect, we are looking for the most probable vector $\mathbf{x}$ (according to our prior), which is also close enough to $\mathbf{z}$ to explain it. This approach is a very known one – it is called the Maximum A'posteriori Probability (MAP) approach.

This simple example is just one of many that require a good model (prior). All inverse problems, such as denoising, deblurring, super-resolution, inpainting, demosaicing, tomographic reconstruction, single image scale-up and more, require a good image prior, in order to be able to differentiate between the true image content and the degradation effects. Compression algorithms require a model in order to be able to find the most compact representation of the image. Anomaly detection requires a model of the image, in order to pick up on anything not conforming to the model. The number of problems requiring a good image model is ever growing, which leads us to the natural question - what is this prior $Pr(\mathbf{x})$? Where can we get it from?

1.3. Evolution of priors

The topic of a proper choice of $Pr(\mathbf{x})$ for images has been at of the foundation of the research in image processing since its early days, and many models have been suggested. Since it is not possible to characterize $Pr(\mathbf{x})$ explicitly, some form of computing it is suggested. Usually, instead of suggesting $Pr(\mathbf{x})$ directly, its minus log is used, $G(\mathbf{x}) = -\log(Pr(\mathbf{x}))$, thereby giving a penalty of 0 to the "perfect" image, and some positive penalty to any other, a larger value as the probability of the image decreases.

There has been an evolution of choices for $G(\mathbf{x})$ through the years. It started with the ℓ_2-norm, and the very simple energy term, $G(\mathbf{x}) = \lambda\|\mathbf{x}\|_2^2$. It then progressed to smoothness terms, demanding low energy of some derivatives, $G(\mathbf{x}) = \lambda\|\mathbf{L}\mathbf{x}\|_2^2$, with $\mathbf{L}$ being the Laplacian operator. A further step was made by adaptively demanding this low energy, while allowing pixels detected as edges to contain high energy in the derivative, $G(\mathbf{x}) = \lambda\|\mathbf{L}\mathbf{x}\|_{\mathbf{W}}$, where $\mathbf{W}$ is a pre-computed diagonal matrix, assigning different weights to different pixels.

The next major step was the abandonment of the convenient ℓ_2 norm and the progression to robust statistics, $G(\mathbf{x}) = \lambda\rho\{\mathbf{L}\mathbf{x}\}$. One of the most known priors from this family is the total variation, $G(\mathbf{x}) = \lambda\| \, |\nabla\mathbf{x}| \, \|_1$, suggested in the early 1990's. [**44**].

Around the same time, the prior of Wavelet sparsity $G(\mathbf{x}) = \lambda\|\mathbf{W}\mathbf{x}\|_1$ started evolving in the approximation theory and the harmonic analysis communities [**11, 17, 38**]. The idea behind this prior is that looking at the wavelet coefficients of an image, they should be sparse - most of them should be zeros or very close to zero.

The prevailing model in the last decade, which is an evolution of all models described so far, is the one involving sparse and redundant representations. Similarly to the wavelets sparsity, this model assumes sparse coefficients, however, it no longer relies on fixed transforms. In the reminder of these course, we will focus on this model and show its applicability to image processing. Of course, this model is not the last in the evolution, and in the coming years new models will be suggested, and they will perform even better (hopefully).

2. The *Sparseland* prior

Our model (and any other model) represents a belief - how we believe an image should look like, or what are the rules it obeys. The sparse and redundant representation model believes that all images are generated by a "machine" that is able to create $\sqrt{n} \times \sqrt{n}$ images. Inside this machine there exists a dictionary $\mathbf{D}_{n\times k}$, and each of the columns of this dictionary is a prototype signal – an atom.

Each time this machine is required to generate a signal, a vector α is (randomly) chosen. Not any vector can be selected, however, as this vector must be sparse - it may only contain a small number L of non-zeros, at random locations and with random values, which we will assume are Gaussian. Then, the result signal is created by multiplying the dictionary by the sparse vector of coefficients: $\mathbf{x} = \mathbf{D}\alpha$.

We call this machine *Sparseland* and we assume that all our signals in the family of interest are created this way.

The *Sparseland* model has some interesting properties. It is simple, as any signal generated is a linear combination of only a small number of atoms. Instead of describing the signal using n numbers to represent the signal, we only need $2L$ values – the locations of the non-zeros and their values. These signals are thus very compressible.

On the other hand, this source is also very rich. We have many (combinatorial) ways to choose subsets of L atoms, and can therefore create a wealth of signals using this model. As a matter of fact, the signals generated are a union of many low-dimensional Gaussians.

The last reason, which is also the actual reason this model is indeed being used, is that it is in fact familiar and has been used before, in other contexts and perhaps in a simplified way, like wavelets and JPEG. This progress have been observed in several trends recently. For example, the transition from JPEG to JPEG2000 is essentially a move from the ℓ_2-norm and taking the leading coefficients, to the notion of sparsity of wavelet coefficients. The progress from Wiener filtering (Fourier) to robust estimation also promotes sparsity. Similar trend exists in the evolution of wavelets, from unitary transforms to over-complete transforms with concepts and tools such as frames, shift-invariance, steerable wavelets, contourlets, curvelets and more, all point to the direction of redundancy in the representation [**39, 10, 46, 45**]. The ICA (independent component analysis) transform was also shown to be tied to sparsity and independence between the coefficients.

Another field progressing towards an extensive usage of sparse and redundant representation modeling is that of machine learning. Methods such as LDA, PCA, SVM and feature-selection methods have all been extended or improved recently by the introduction of sparse and redundant representations.

All of these directions clearly point to *Sparseland*. It is here, and we are going to use it. In order to do that, however, we must define it clearly, analyze it, understand it, and find out how it should be applied to the various tasks at hand.

One more thing we must take into account in our model is the model noise. While we expect our signals to look like they have been created as $\mathbf{D}\alpha$ with a sparse α, we may want to allow for some perturbations from this exact model, with the understanding that, just like any other model, this model is not perfect or exact. For this purpose, we assume that signals are created as $\mathbf{y} = \mathbf{D}\alpha + \mathbf{e}$, with $\mathbf{e}$ the model noise with bounded energy $\|\mathbf{e}\|_2 \leq \epsilon_0$.

3. Processing *Sparseland* signals

The model we will be considering in this section is the exact model, i.e. $\mathbf{x} = \mathbf{D}\alpha$. We will assume from now on that there is no model noise, i.e., $\epsilon_0 = 0$. This does not mean that we work only on clean signals; It only means that true signals can indeed be created exactly by multiplying the dictionary by a sparse vector of coefficients. Now, we ask – How do we process such signals?

3.1. Transform

Suppose we have a *Sparseland* signal $\mathbf{y}$, and we want to design a transform for it, similarly to the wavelet transform or the Fourier transform. What would we like such a transform to do?

First, we would like it to be invertible - we do not want to lose information by applying the transformation. We would also like the transformation to be compact, i.e., for its energy to be concentrated in a small number of coefficients. Lastly, we would also want the coefficients to be independent of each other.

We can propose a very simple transformation that follows all these requirements – let us transform $\mathbf{y}$ into the sparsest vector α such that $\mathbf{D}\alpha = \mathbf{y}$,

$$\hat{\alpha} = \arg\min_{\alpha} \|\alpha\|_0^0 \; s.t. \; \mathbf{D}\alpha = \mathbf{y}.$$

We know that such an α indeed exists, making this transform possible, since it is our model's assumption that $\mathbf{y}$ was generated by a sparse vector multiplying the dictionary.

As the dictionaries used are redundant, α is usually longer than $\mathbf{y}$, making the transform suggested also redundant. It is also invertible, as multiplying $\mathbf{D}$ by α brings us back to our original signal. Lastly, we observe that the energy of the transform is concentrated in only a few non-zero coefficients, as α is sparse, and those non-zeros are independent.

3.2. Compression

Suppose we now try to compress the *Sparseland* signal $\mathbf{y}$. In order to allow compression, we allow some error ϵ between the compressed signal and the original one. Changing this value will create different rates of compression, thus creating the rate-distortion curve for the compression scheme. The compression can be done by solving

$$\hat{\alpha} = \arg\min_{\alpha} \|\alpha\|_0^0 \; s.t. \; \|\mathbf{D}\alpha - \mathbf{y}\|_2^2 \leq \epsilon^2,$$

which means searching for the smallest number of coefficients that can explain the signal $\mathbf{y}$ with an error of less than ϵ. Note that the final rate-distortion curve is obtained from taking into account the further error introduced by quantizing the non-zero coefficients in α.

3.3. Denoising

Given a noisy *Sparseland* signal $\mathbf{z} = \mathbf{y} + \mathbf{v}$, created by the addition of additive noise $\mathbf{v}$ (with known power $\|\mathbf{v}\|_2 = \epsilon$) to our original signal, we would like to recover $\mathbf{y}$ as accurately as possible.

Since we know the signal $\mathbf{y}$ is sparse over the dictionary $\mathbf{D}$ and we expect the noise to not be sparse over the same dictionary (since it is not a "real" image), the following task can be proposed:

$$\hat{\alpha} = \arg\min_{\alpha} \|\alpha\|_0^0 \; s.t. \; \|\mathbf{D}\alpha - \mathbf{z}\|_2^2 \leq \epsilon^2,$$

with the final solution being $\hat{\mathbf{y}} = \mathbf{D}\hat{\alpha}$. This denoising method works quite well in practice, as we will see later on in the lectures.

There is a subtle point to address regarding denoising. Our model assumes that the signals we operate on can be indeed represented exactly using a sparse representation over the dictionary. However, for real-world signals, this is not exactly the case, and there are some model mismatches. The denoising algorithm will remove these true details from the signal, along with the noise. This implies that if the noise is very weak, such that its power is in the order of the model mismatch, the denoising result might actually be worse than not denoising at all. This is because the price of removing the model mismatch may be greater than the price of not removing the noise. Usually, when the standard deviation of the noise is 5 grey levels or more (when the image is in the range $[0, 255]$), it is considered to be stronger than the model mismatches, and denoising the image indeed improves its quality.

3.4. General inverse problems

As a generalization of the denoising problem, we now assume that the *Sparseland* signal is given to us after a degradation operator $\mathbf{H}$ and an additive noise, $\mathbf{z} = \mathbf{H}\mathbf{y} + \mathbf{v}$. The operator $\mathbf{H}$ can be any linear operator that can be written as a matrix, representing blur, projection, down-scaling, masking and more.

We can try to recover $\mathbf{y}$ by looking for the sparsest α that when multiplied by the dictionary and then by $\mathbf{H}$ is close enough to the input signal $\mathbf{z}$. This vector is found by solving the minimization problem:

$$\hat{\alpha} = \arg\min_{\alpha} \|\alpha\|_0^0 \ s.t. \ \|\mathbf{H}\mathbf{D}\alpha - \mathbf{z}\|_2^2 \leq \epsilon^2,$$

and again, the final solution is obtained by $\hat{\mathbf{y}} = \mathbf{D}\hat{\alpha}$. The minimization is carried out in practice by defining the equivalent dictionary $\tilde{\mathbf{D}} = \mathbf{H}\mathbf{D}$, and solving the same minimization problem we have already encountered:

$$\hat{\alpha} = \arg\min_{\alpha} \|\alpha\|_0^0 \ s.t. \ \|\tilde{\mathbf{D}}\alpha - \mathbf{z}\|_2^2 \leq \epsilon^2,$$

There is a wide range of problems that can be solved using this mechanism: deblurring, inpainting, demosaicing, super-resolution, tomographic reconstruction, image-fusion and many more.

3.5. Compressed-sensing

Suppose that we want to sample the *Sparseland* signal $\mathbf{y}$, such that we use much less than n samples. There may be various reasons for this desire, such as the signal being too long, having limited time to sample it, each sample being expensive, and so on. Instead, we propose to sample noisy projections of it, using a known projection matrix $\mathbf{P}$, making our signal $\mathbf{z} = \mathbf{P}\mathbf{y} + \mathbf{v}$, with $\|\mathbf{v}\|_2 = \epsilon$ representing the sampling noise. Again, we suggest to recover the original signal by solving

$$\hat{\alpha} = \arg\min_{\alpha} \|\alpha\|_0^0 \ s.t. \ \|\mathbf{P}\mathbf{D}\alpha - \mathbf{z}\|_2^2 \leq \epsilon^2,$$

and multiplying the found vector of coefficients by the dictionary, $\hat{\mathbf{y}} = \mathbf{D}\hat{\alpha}$. This problem again is of the same form seen above, considering the effective dictionary as $\tilde{\mathbf{D}} = \mathbf{P}\mathbf{D}$. The Compressed Sensing field deals with questions like the minimal number of rows in $\mathbf{P}$ (the minimal number of projections) needed to recover the original signal $\mathbf{y}$ with sufficient accuracy, and what kind of $\mathbf{P}$ should be chosen.

3.6. Morphological-component-analysis

The MCA task [**47, 48**] deals with the case where a given signal $\mathbf{z} = \mathbf{y}_1 + \mathbf{y}_2 + \mathbf{v}$ is a mixture of two *Sparseland* signals $\mathbf{y}_1$ and $\mathbf{y}_2$, each created from a different dictionary $\mathbf{D}_1$ and $\mathbf{D}_2$ respectively, with noise also being added. The goal is to recover each of the original signals $\mathbf{y}_1$ and $\mathbf{y}_2$, assuming their dictionaries $\mathbf{D}_1$ and $\mathbf{D}_2$ are known.

Since both signals should have a sparse representations, it is suggested to solve the minimization problem

$$\hat{\alpha_1}, \hat{\alpha_2} = \arg\min_{\alpha_1, \alpha_1} \|\alpha_1\|_0^0 + \|\alpha_2\|_0^0 \ s.t. \ \|\mathbf{D}_1\alpha_1 + \mathbf{D}_2\alpha_2 - \mathbf{z}\|_2^2 \leq \epsilon^2,$$

and recover the original signals using $\hat{\mathbf{y}_1} = \mathbf{D}_1\hat{\alpha_1}$ and $\hat{\mathbf{y}_2} = \mathbf{D}_2\hat{\alpha_2}$.

By concatenating the vectors α_1 and α_2 vertically into a single vector α, and concatenating the two dictionaries by putting them side by side $\mathbf{D} = [\mathbf{D}_1\mathbf{D}_2]$, it turns out that we have already seen this task:

$$\hat{\alpha} = \arg\min_{\alpha} \|\alpha\|_0^0 \ s.t. \ \|\mathbf{D}\alpha - \mathbf{z}\|_2^2 \leq \epsilon^2.$$

3.7. Applications: Summary

We have gone over several applications that can be solved using the *Sparseland* model. There are many more that we can envision using this model – encryption, watermarking, scrambling, target detection, recognition, feature extraction and more. What is common to all these problems is the need to solve a variant of the problem (P_0^ϵ),

$$\hat{\alpha} = \arg\min_{\alpha} \|\alpha\|_0^0 \ s.t. \ \|\mathbf{D}\alpha - \mathbf{z}\|_2^2 \leq \epsilon^2.$$

This is a problem we have already seen and discussed - we know it is NP-hard, but we have seen several algorithms to approximate its solution and some guarantees on the quality of their solutions. The conclusion is that signal processing for *Sparseland* signals is reasonable and practical.

LECTURE 4

First steps in image processing

In this lecture we will start working on image processing applications that rely on the *Sparseland* model. We will consider image deblurring, image denoising, and image inpainting. Several other, more advanced applications, using the tools shown in the lecture, will be given in the following (and last) lecture. More on these applications can be found in [**22**].

1. Image deblurring via iterative-shrinkage algorithms

1.1. Defining the problem

The deblurring problem is one of the most fundamental problems in image processing: A high-quality image $\mathbf{x}$ undergoes blurring, be it due to atmospheric blur, camera blur, or any other possible source. We assume this blur is known (could be space variant). We measure a noisy (white iid Gaussian) version of the blurred image. The noise $\mathbf{v}$ has a known standard deviation. Thus, the image we actually obtain $\mathbf{y}$ is related to the ideal image $\mathbf{x}$ through

$$\mathbf{y} = \mathbf{H}\mathbf{x} + \mathbf{v}, \tag{4.1}$$

and our task is to recover $\mathbf{x}$ given $\mathbf{y}$. Since we are dealing with the *Sparseland* model, we assume that the image $\mathbf{x}$ has a sparse representation over a carefully chosen dictionary, and therefore we will try to solve

$$\hat{\alpha} = \arg\min_{\alpha} \frac{1}{2}\|\mathbf{H}\mathbf{D}\alpha - \mathbf{y}\|_2^2 + \lambda \cdot \rho(\alpha). \tag{4.2}$$

Here, the function ρ is sparsity promoting penalty, such as the ℓ_1-norm, or something similar (see the end of lecture 1). Since we are not interested in $\hat{\alpha}$ in itself but in the recovered image, we finish the processing with $\hat{\mathbf{x}} = \mathbf{D}\hat{\alpha}$.

Let us define the ingredients in Equation (4.2). The vector $\mathbf{y}$ is the given degraded image, and $\mathbf{H}$ is the known blur operator. The dictionary $\mathbf{D}$ should be chosen by us. In the experiments we report here, we follow the work by Figueiredo and Nowak [**26, 27**] and use the 2D un-decimated Haar wavelet transform with 2 resolution levels. This is a global dictionary with a redundancy factor of $7:1$. As for the function ρ, we use an approximation of the ℓ_1 norm, with a slight smoothing,

$$\rho(\alpha) = |\alpha| - S \cdot \log\left(1 + \frac{\alpha}{S}\right). \tag{4.3}$$

The function ρ is a scalar function, operating on each entry of α individually, and summing the result over all entries. The benefit of using this function rather than the ℓ_1 norm directly is that it is smooth and derivable in the origin (and everywhere else) and is therefore much more convenient to work with.

Lastly, the value of λ is set manually to get the best possible result. There are algorithms that can find its value for optimal performance automatically. However, we will not discuss those here and proceed with the manually chosen value.

There is a wide knowledge in the area of continuous optimization, and it seems like once we have written down our penalty function, we can use a wide variety of methods or even general purpose optimization software packages (e.g., Matlab, MOSEK) for its solution. However, such general purpose solvers typically perform very poorly for this specific optimization task. One possible reason for this poor performance is the fact that these solvers disregard the knowledge about the sparsity of the desired solution. A second possible reason can be understood from look at the gradient of our penalty function,

$$\nabla f(\alpha) = \mathbf{D}^T(\mathbf{D}\alpha - \mathbf{y}) + \lambda \cdot \rho'(\alpha)$$

and the Hessian

$$\nabla^2 f(\alpha) = \mathbf{D}^T\mathbf{D} + \lambda \cdot \rho''(\alpha).$$

We must remember that α will contain many zero elements, and few non-zero elements in *unknown locations*. In the locations of the zeros, the value of ρ'' will be very large, while for the non-zeros, these will be small numbers. Therefore, the Hessian's main diagonal will have a high contrast of values, causing this matrix to be of very high condition-number, hurting the performance of general-purpose solvers. Furthermore, as we do not know in advance the locations of these non-zeros (this is actually the goal of the optimization - recovering the support), we cannot even pre-condition the Hessian. This forces us to seek for tailored optimization methods that will take advantage of the expected sparsity in the solution.

How should we do that? In lecture 1 we have seen that when the matrix $\mathbf{D}$ is unitary, the solution is obtained by a simple scalar-wise shrinkage step $S_{\rho,\lambda}(\mathbf{D}^T\mathbf{y})$. When $\mathbf{D}$ is not unitary, we would like to propose a similarly simple solution. This idea leads to a family of algorithms known as "Iterative Shrinkage" [**25**]. There are several such algorithms, and here we will present one of them.

1.2. The parallel coordinate descent (PCD) method

An interesting iterative shrinkage algorithm emerges from the classic coordinate descent (CD) algorithm. Put generally, the CD operates as follows: Given a function to minimize with a vector as the unknown, all entries but one are fixed, and the remaining entry is optimized for. Then, this coordinate is fixed, and a different entry is optimized. This is done serially over all entries, and in several iterations, until convergence.

In our problem, solving for one entry at a time considers only the atom $\mathbf{d}_j$ and the coefficient α_j. Taking it out of the current support by setting $\alpha_j = 0$, we get that the current error in representation is $\mathbf{e}_j = \mathbf{y} - \mathbf{D}\alpha$. We now attempt to best reduce this error using only the atom $\mathbf{d}_j$, by choosing the optimal value for α_j, i.e., minimizing

$$f(\alpha_j) = \frac{1}{2}\|\alpha_j \cdot \mathbf{d}_j - \mathbf{e}_j\|_2^2 + \lambda\rho(\alpha_j).$$

This is a scalar optimization problem that can be solved using shrinkage on the scalar value $\mathbf{d}_j^T\mathbf{e}_j$. The shrinkage is applied by

$$\alpha_j^{OPT} = S_{\rho,\lambda/\|\mathbf{d}_j\|^2}\left(\mathbf{d}_j^T\mathbf{e}_j\right),$$

where S denotes the shrinkage operation – its shape is dictated by the choice of ρ and λ/c. Notice that the shrinkage is done differently to each entry, based on the norm of the j-th atom.

For low-dimensional problems ($k \approx 100$) it makes sense to solve our problem this way exactly. However, in the image deblurring problem, the unknown contains many thousands entries, and it is impossible to use the above algorithm, as it requires to extract one column at a time from the dictionary. The solution to this problem is the Parallel-Coordinate-Descent (PCD) algorithm. The idea is the following: Compute all the k descent trajectories $\{v_j\}_{j=1}^{k}$, each considering one coordinate optimization. Since all are descent directions, their sum is also a descent direction. The PCD takes this sum and uses this for the current iteration. The problem with this direction is that it is not known how far to go along it. Therefore, a 1D line-search is needed, and the overall iteration step is written as

$$\alpha_{k+1} = \alpha_k + \mu \cdot \left[S_{\rho,\mathbf{Q}\lambda}\left(\mathbf{Q}\mathbf{D}^T\left(\mathbf{y}\mathbf{D}\alpha_k\right) - \alpha_k\right)\right], \tag{4.4}$$

with $\mathbf{Q} = \operatorname{diag}^{-1}\left(\mathbf{D}^T\mathbf{D}\right)$. μ is the found step-size in the line-search step.

1.3. Results

Figure 1 shows the deblurring result obtained by the PCD algorithm after 10 iterations. More details on this experiment are given in the accompanying slides.

FIGURE 1. Debluring. Left: The original image, Middle: the measured (blurred and noisy) image, and Right: the PCD deblurring result.

2. Image denoising

The image denoising problem can be viewed as a special case of the image deblurring problem discussed above. For this to be evident, all we need to do is select $\mathbf{H} = \mathbf{I}$ (i.e., no blur at all). In this section we shall present two very different ways to handle this problem.

2.1. Global shrinkage

Embarking from the above-described deblurring algorithm, we can suggest the following simplified denoising method: Instead of the iterative mechanism, we can apply the thresholding algorithm, as follows

$$\hat{\mathbf{x}} = \mathbf{D}S_T\left(\mathbf{D}^T\mathbf{y}\right). \tag{4.5}$$

Of course, since we are using the redundant Haar dictionary, we should normalize its columns (since they are not ℓ_2 normalized) in order to have a single scalar threshold for all atoms. Put differently, this leads to

$$\hat{\mathbf{x}} = \mathbf{DW}S_T\left(\mathbf{W}^{-1}\mathbf{D}^T\mathbf{y}\right). \tag{4.6}$$

The only missing ingredient is determining the value of the threshold T. Experimenting with this algorithm, different values of T will have a different quality of denoising. As an example, on the well known image "barbara" with noise with standard deviation $\sigma = 20$, the optimal value of T is around 55. The denoised image using this method is more than 5 dB better in PSNR than the noisy image. This result is shown in Figure 2.

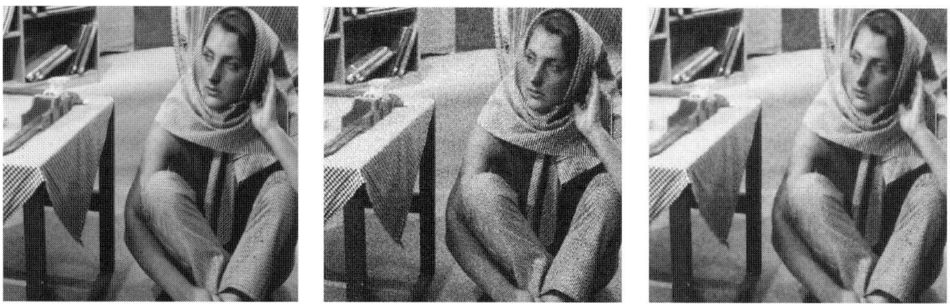

FIGURE 2. Global denoising. Left: The original image, Middle: the measured (noisy) image, and Right: the global denoising result.

2.2. From global to local processing

While the above result is good, it is definitely not the best that can be achieved. We will now turn to a much better method (and see an even better one in the next lecture).

When discussing algorithms for the solution of the (P_0^ϵ) problem, we saw that very large signals could pose a computational problem. Therefore, when we work on the image as a whole, we are forced to use only simple algorithms, such as (iterative) shrinkage. Also, when using dictionaries for an entire image, the result may be sparse when compared to the overall number of atoms, but we still require a large number of coefficients. These two observations lead us to attempt operating on small image patches instead.

When processing a large image, we claim that every $N \times N$ patch (e.g., 8×8) in it has a sparse representation over a dictionary. Furthermore, we use full overlaps between the patches. By operating this way, we in fact force shift-invariance sparsity. In order to write this formally, we define the operator $\mathbf{R}_{ij}$, which extracts an $N \times N$ patch from the image around pixel (i, j). Now, the minimization task to solve is [**30, 31, 23**]

$$\hat{\mathbf{x}} = \arg\min_{\mathbf{x},\{\alpha_{ij}\}_{ij}} \left[\frac{1}{2}\|\mathbf{x}-\mathbf{y}\|_2^2 + \mu\sum_{ij}\|\mathbf{R}_{ij}\mathbf{x} - \mathbf{D}\alpha_{ij}\|_2^2\right] \quad s.t. \; \|\alpha_{ij}\|_0 \le L. \tag{4.7}$$

The first term requires a proximity between the recovered image and the input image, a direct consequence of the log-likelihood. The second term requires that each patch in the reconstructed image can be represented well as a linear combination of atoms from the dictionary $\mathbf{D}$. The constraint forces all of these linear combinations to be sparse.

At this stage, the dictionary we will be using for this algorithm is the over-complete DCT dictionary with 256 atoms, each representing an 8×8 patch. This dictionary is separable, allowing it so be applied efficiently.

How can we perform the minimization in (4.7)? We will take a coordinate descent approach. First, we will fix $\mathbf{x} = \mathbf{y}$, and solve for $\{\alpha_{ij}\}_{ij}$. The first term disappears, and we are left with $M \cdot N$ different minimization tasks:

$$\hat{\alpha}_{ij} = \arg\min_{\alpha} \|\mathbf{R}_{ij}\mathbf{x} - \mathbf{D}\alpha\| \ s.t. \ \|\alpha\|_0 \leq L. \tag{4.8}$$

We have seen this problem before - this is the sparse coding problem, and can be solved by any pursuit algorithm (e.g., OMP). Once we have found $\{\alpha_{ij}\}_{ij}$, we fix them, and recover $\mathbf{x}$. This is a least-squares problem and a little bit of algebra leads to a closed-form solution

$$\hat{\mathbf{x}} = \left[\mathbf{I} + \mu \sum_{ij} \mathbf{R}_{ij}^T \mathbf{R}_i j\right]^{-1} \left[\mathbf{y} + \mu \sum_{ij} \mathbf{R}_{ij}^T \mathbf{D} \hat{\alpha}_i j\right]. \tag{4.9}$$

While this expression seems complicated, it is actually simple to implement. While the operator $\mathbf{R}_{ij}$ extracts a patch around the pixel (i, j), the operator $\mathbf{R}_{ij}^T$ returns a patch to the location (i, j). Therefore, the left matrix is in fact a diagonal one, indicating for each pixel the amount of patches it belongs in, serving as normalization.

Therefore, the entire denoising process is done by extracting each patch, performing sparse coding, and getting a denoised patch instead. The denoised patch is returned to its original position, with aggregation performed in the overlaps area. Then, each pixel is normalized by the number of patches it belongs too (which is generally N^2, apart from the borders of the image).

It turns out that this very simple method performs about 2.5dB better than the global thresholding show previously. Furthermore, this algorithm can be implemented very efficiently on parallel processors. We will see this local processing approach in several more applications. The result obtained is shown in Figure 3.

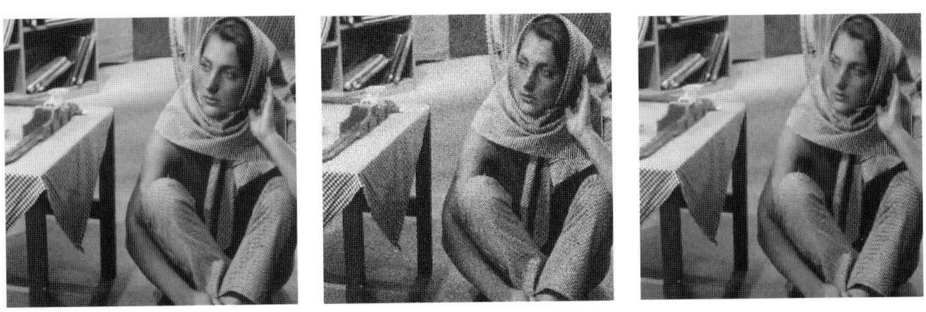

FIGURE 3. Local denoising. Left: The original image, Middle: the measured (noisy) image, and Right: the local denoising result.

3. Image inpainting

Assume that a signal has been created as $\mathbf{x} = \mathbf{D}\alpha_0$ using a very sparse vector of coefficients α_0. Unfortunately, we do not get $\mathbf{x}$ directly, but rather obtain $\tilde{\mathbf{x}}$, which is the same signal $\mathbf{x}$, but with some samples removed. We shall assume that we know which samples are missing. Each missing sample from $\mathbf{x}$ implies one lost equation in the system $\mathbf{x} = \mathbf{D}\alpha_0$. Therefore, we remove these corresponding rows from the dictionary $\mathbf{D}$, obtaining the dictionary $\tilde{\mathbf{D}}$. Now, we may try to solve

$$\hat{\alpha} = \arg\min_{\alpha} \|\alpha\|_0 \ s.t. \ \tilde{\mathbf{x}} = \tilde{\mathbf{D}}\alpha. \tag{4.10}$$

If α_0 was sparse enough, it will also be the solution of this modified problem. Then, by computing $\mathbf{D}\hat{\alpha}$ we will recover the original signal completely.

Using the above intuition, we now turn to extend the local denoising algorithm into an inpainting one. The minimization task here is very similar to the one we had for the denoising (4.7), with the needed modification to take into account the masking operator [**30, 31, 36**],

$$\hat{\mathbf{x}} = \arg \min_{\mathbf{x},\{\alpha_{ij}\}_{ij}} \left[\frac{1}{2}\|\mathbf{M}\mathbf{x} - \mathbf{y}\|_2^2 + \mu \sum_{ij} \|\mathbf{R}_{ij}\mathbf{x} - \mathbf{D}\alpha_{ij}\|_2^2\right] \ s.t. \ \|\alpha_{ij}\|_0 \le L, \tag{4.11}$$

where we now require that $\mathbf{x}$ is close to $\mathbf{y}$ only in the locations of un-masked pixels.

Again this minimization will be done using the coordinate descent approach. First, we compute $\mathbf{X} = \mathbf{M}^T\mathbf{y}$ (e.g., by zero filling), and do sparse coding for each patch, using the matching pursuit

$$\hat{\alpha}_{ij} = \arg\min_{\alpha} \|\mathbf{M}_{ij}\left(\mathbf{R}_{ij}\mathbf{x} - \mathbf{D}\alpha\right)\| \ s.t. \ \|\alpha\|_0 \le L, \tag{4.12}$$

with $\mathbf{M}_{ij}$ the appropriate masking for each patch. Once the sparse coding is finished, the result image is again reconstructed by aggregating the patches in the overlaps area, normalizing by the number of patches in each pixel,

$$\hat{\mathbf{x}} = \left[\mathbf{M}^T\mathbf{M} + \mu \sum_{ij} \mathbf{R}_{ij}^T\mathbf{R}_{ij}\right]^{-1} \left[\mathbf{M}^T\mathbf{y} + \mu \sum_{ij} \mathbf{R}_{ij}^T\mathbf{D}\hat{\alpha}_{ij}\right]. \tag{4.13}$$

Figure 4 presents inpainting results obtained using this algorithm.

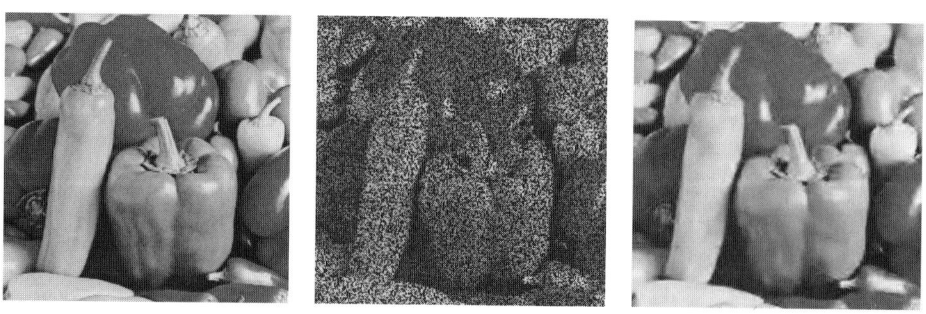

FIGURE 4. Local inpainting. Left: The original image, Middle: the measured image with 50% missing pixels, and Right: the inpainting result.

4. Dictionary learning

4.1. General dictionary learning

Our entire model and applications rely on having a suitable dictionary. A good dictionary would be one that sparsifies our data. The question we aim look at now is how to construct such dictionaries. We have already seen one possible solution – we can choose a dictionary from an existing family of transforms, such as DCT, curvelets, contourlets, wavelets and so on. This approach has its benefits, as usually these transforms can be applied rather efficiently. However, a dictionary selected this way is typically not the best fit for the actual data.

A better approach is to train a dictionary from examples. The core idea is to collect a set of representative examples of the images we work with, and train a dictionary so that the examples can be sparsely represented over it. Such a dictionary will be better suited for our data, therefore leading to better performance in various tasks.

Let us collect all the examples $\{\mathbf{x}_j\}$ into a matrix $\mathbf{X}$, where each column is one example. Then, we can write the decomposition as $\mathbf{X} \approx \mathbf{DA}$, where each column of $\mathbf{A}$ contains the coefficients for the representation of the corresponding example. A good dictionary is therefore one that is able to adequately represent each of the examples, while the representation for each example in indeed sparse, i.e., each column in $\mathbf{A}$ has only a small number of non-zeros. Formally, we would like to solve

$$\hat{\mathbf{D}}, \hat{\mathbf{A}} = \arg\min_{\mathbf{D},\mathbf{A}} \sum_{j=1}^{P} \|\mathbf{D}\alpha_j - \mathbf{x}_j\| \; s.t. \; \forall j, \; \|\alpha_j\|_0 \leq L. \tag{4.14}$$

This problem will be handled as follows: The first stage is sparse-coding, assuming $\mathbf{D}$ is fixed, and only the representation vectors are sought. In this stage we should solve for each example

$$\min_{\alpha} \|\mathbf{D}\alpha - \mathbf{x}_j\|_2^2 \; s.t. \; \|\alpha\|_0 \leq L. \tag{4.15}$$

We have seen in the first lecture several ways to perform this minimization, via pursuit algorithms. The second stage of dictionary training is the update of the dictionary, once the representation vectors $\mathbf{A}$ have been recovered. There are two main methods to do so: the Method of Optimal Directions (MOD) method [**32**] and the K-SVD method [**2**].

4.2. MOD dictionary update

One option is to try and update the dictionary at once. The goal is minimizing

$$\min_{\mathbf{D}} \|\mathbf{DA} - \mathbf{X}\|_F^2, \tag{4.16}$$

with $\mathbf{X}$ and $\mathbf{A}$ fixed. Note the use of the Frobenius norm, as both $\mathbf{DA}$ and $\mathbf{X}$ are matrices. Few simple linear algebra steps lead to

$$\mathbf{D} = \mathbf{XA}^T \left(\mathbf{AA}^T\right)^{-1} = \mathbf{XA}^{\dagger}. \tag{4.17}$$

Once the dictionary is found this way, its columns need to be re-normalized.

4.3. K-SVD dictionary learning

Another option to train the dictionary is imitating the K-means algorithm in the dictionary update stage. This means that we process one atom at a time and attempt to optimally replace it.

For processing the i-th atom, the first stage is collecting only those examples that use it into the matrix $\mathbf{X}^i$. Then, the dictionary $\mathbf{D}$ and coefficient matrix $\mathbf{A}$ are all fixed apart from the i-th column in $\mathbf{D}$ and the i-th row in $\mathbf{A}$. Denoting by $\mathbf{D}_i$ and $\mathbf{A}_i$ these matrices with the i-th column/row removed respectively, the following residual is computed

$$\mathbf{E}^i = \mathbf{X}^i - \mathbf{D}_i\mathbf{A}_i = \mathbf{X}^i - \mathbf{D}\mathbf{A} + \mathbf{d}_i\alpha_i.$$

This is the residual matrix for all the examples that use the i-th atom, without taking into account the i-th atom itself. The new atom is the one that is best able to reduce the mean residual, i.e., the result of the minimization problem

$$\min_{\mathbf{d}_i,\alpha_i} \|\alpha_i\mathbf{d}_i^T - \mathbf{E}^i\|_F^2. \tag{4.18}$$

The vector α_i is the optimal set of coefficients for this atom in each example. The resulting atom $\hat{\mathbf{d}}_i$ is inserted into the dictionary, and the new vector of coefficients α_k replaces the i-th row in $\mathbf{A}$ (only in the columns relevant to the examples using this atom).

The optimization task in (4.18) is in fact a rank-1 approximation, which can be solved directly using the SVD decomposition. As every atom is updated using an SVD operation, this algorithm is appropriately named K-SVD. Note, that in the MOD training stage, the matrix of coefficients $\mathbf{A}$ is held constant, and the dictionary $\mathbf{D}$ is adapted to it. In the K-SVD algorithm, on the other hand, the atom update stage also optimizes for the k'th row of $\mathbf{A}$ at the same time.

LECTURE 5

Image processing - more practice

In the previous lecture we saw how to denoise and inpaint an image based on local, patch-processing. However, in both this cases, we used a fixed dictionary – the local DCT. In this lecture we shall revisit these two applications, and incorporate into them the K-SVD, in an attempt to improve their performance. We shall also demonstrate the capabilities of the *Sparseland* model in handling image processing tasks by considering two new applications – an image scale-up process based on a pair of trained dictionaries, and a facial image compression algorithm that performs better than JPEG-2000. More details on the applications presented here can be found in [**22**, **3**].

1. Image denoising with a learned dictionary

When thinking about dictionary learning, both the MOD and the K-SVD are not able to process signals with large dimensions. The reason is obvious – a high dimensional signal implies a dictionary held as a very large explicit matrix, intolerable amounts of computations, both for the training and for using this dictionary, and a huge amount of examples to learn from, in order to avoid over-fitting. Thus, learned dictionaries are naturally used for low-dimensional signals.

We have already seen in the previous lecture that denoising (and inpainting) can be done very effectively by operating on small (e.g., 8×8) patches. The processing is then done on local blocks of images, using a dictionary that handles low-dimensional signals. Formally, the denoising we have proposed is written as

$$\hat{\mathbf{x}} = \arg \min_{\mathbf{x},\{\alpha_{ij}\}_{ij}} \left[\frac{1}{2}\|\mathbf{x}-\mathbf{y}\|_2^2 + \mu \sum_{ij} \|\mathbf{R}_{ij}\mathbf{x} - \mathbf{D}\alpha_{ij}\|_2^2 \right] \; s.t. \; \|\alpha_{ij}\|_0 \leq L. \tag{5.1}$$

The second term is the prior, expressing our belief that every patch in the image should have a sparse representation over the dictionary $\mathbf{D}$. Our objective is to introduce a learned dictionary into the above paradigm. One option is collecting a set of high-quality images, extract patches from them, and use these examples to train a dictionary. This dictionary will be a "universal dictionary", as it is trained on general content images. This option is useful, and brings reasonably good results - about 0.5-1dB below the state of the art.

An interesting alternative is to use patches from the corrupted image itself for the training process. Suppose an image of 1000×1000 is to be denoised. Such an image contains nearly 10^6 patches, which are more than enough to allow training the dictionary. Furthermore, since the dictionary training process contains some sort of averaging, the resulting dictionary will actually be (almost) noise free, and can thus be used for denoising.

This option works much better than using the "universal dictionary", and leads to state-of-the-art results [**23**]. In fact, going back to the penalty term in (5.1), we can bring about this option by minimizing over $\mathbf{D}$ as well. Again, a coordinate descent approach is used. First, the image $\mathbf{x} = \mathbf{y}$ is fixed, and the dictionary $\mathbf{D}$ is fixed on some initialization (e.g., the redundant DCT). Each patch in the image undergoes sparse coding. Freezing the representations, the dictionary is updated. All the patches then undergo sparse coding with the new dictionary, and so. Iterating several times between sparse coding and dictionary update is exactly the training procedure we have seen in the previous lecture, performed over the patches in the image. After several such training iterations, the final dictionary is used for one last sparse-coding stage. Then, the sparse representations and the dictionary are fixed, and the result image is recovered, using the same simple aggregation method shown in the previous lecture,

$$\hat{\mathbf{x}} = \left[\mathbf{I} + \mu \sum_{ij} \mathbf{R}_{ij}^T \mathbf{R}_i j\right]^{-1} \left[\mathbf{y} + \mu \sum_{ij} \mathbf{R}_{ij}^T \mathbf{D} \hat{\alpha}_i j\right]. \tag{5.2}$$

Figure 1 presents the result obtained using this algorithm, for the same test shown in Lecture 4. The obtained image has $PSNR = 31$dB, which is 1dB higher compared to the DCT-based outcome.

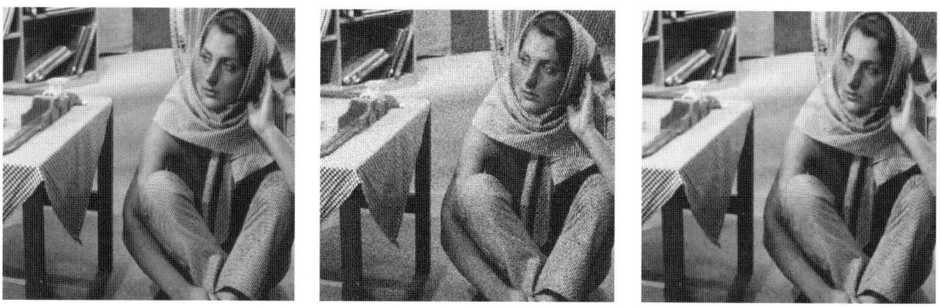

FIGURE 1. Local denoising with a trained dictionary. Left: The original image, Middle: the noisy image ($\sigma = 20$), and Right: the denoising result.

This denoising algorithm has been applied to many variation of the denoising problem. One is of course the denoising of gray-scale images. When published, this method led to state-of-the-art results. Later, it was extended by Mairal et. al. by using joint sparsity (i.e., sparse coding similar patches together) [**37**]. This method is currently the best performing denoising algorithm.

The denoising algorithm could also be applied to denoising of color images. Usually, there is a sensitivity in how the relationships between the color channels should be handled. In this algorithm, the solution is simple – the training is performed on $8 \times 8 \times 3$ patches, i.e., on colored patches extracted from the image. The dictionary contains color atoms, and the rest of the algorithm remains the same. This method is also the state-of-the-art in color image denoising [**36**].

Video sequences can also be handled by this method. When processing image sequences, the redundancy between frames can and should be used in order to achieve better denoising. While most video denoising algorithms require motion estimation (which is error-prone), the suggested method can simply train on space-time patches extracted from the video. Therefore, the dictionary is also constructed of "movielets", as each atom is a very short, very small (e.g., $8 \times 8 \times 5$) space-time patch. Furthermore, since consecutive images are very similar, it is not necessary to retrain a dictionary for each image. Instead, it can be propagated from one frame to the next, only requiring fine tuning (1 K-SVD iteration) for each one. This method's performance is equivalent to the state-of-the-art [**43**].

2. Image inpainting with dictionary learning

Turning to image inpainting, we would like to apply dictionary learning here as well, for handling the local patches. The process is very similar to the one done in denoising. The algorithm starts with dictionary training using the K-SVD, by performing alternating steps between sparse coding and dictionary update. During the sparse coding stage, only the un-masked (i.e., known) pixels in each patch are used. This is similar to the sparse coding undertaken in the fixed dictionary scenario. Once the sparse coding is done, each atom is updated using the SVD operation, just like the K-SVD method.

This dictionary training is repeated for several (e.g., 10) iterations. Then, the image is reconstructed from the final sparse coding using the same formula we have seen:

$$\hat{\mathbf{x}} = \arg \min_{\mathbf{x}, \{\alpha_{ij}\}_{ij}} \left[\frac{1}{2} \|\mathbf{M}\mathbf{x} - \mathbf{y}\|_2^2 + \mu \sum_{ij} \|\mathbf{R}_{ij}\mathbf{x} - \mathbf{D}\alpha_{ij}\|_2^2 \right] \quad s.t. \ \|\alpha_{ij}\|_0 \leq L, \tag{5.3}$$

Of course, just like in the denoising case, this method can be easily generalized to inpainting of color images and inpainting of video sequences using the same extensions as in the denoising case. Figure 2 presents the results obtained with this algorithm, showing the benefit of using learned dictionary.

This approach to inpainting works well when the missing pixels are in (near-random) locations, or organized in small groups. If the missing pixels form a large area to reconstruct, the local approach will not work, and an alternative, global, approach will be needed.

Another problem could arise if the missing pixels create a periodic mask, for example, when every other pixel is missing. In this case, the dictionary will not be able to learn the relationship between neighboring pixels, because it has no information about such neighbors. Such is the case, for example, in image single-image scale-up and demosaicing. The periodic nature of the mask pattern may cause the learned dictionary to absorb this pattern into its atoms. In order to prevent that, the dictionary should trained while allowing only a very small number of non-zeros, and furthermore, only few training iterations in order to avoid over-fitting. Such a solution leads to state of the art results, which can be seen in [**36**].

3. Image scale-up with a pair of dictionaries

We turn to describe an image scale-up algorithm that uses trained dictionaries. The method described here follows the work of Yang et. al. [**54**], with several important modifications that lead to a simpler and more efficient algorithm, with better results [**22**].

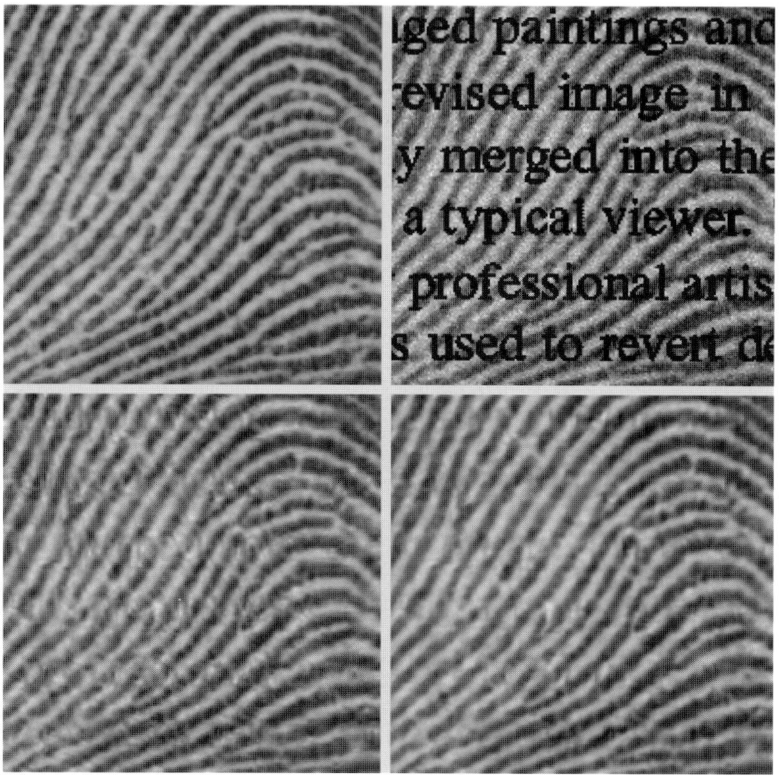

FIGURE 2. Image inpainting with the K-SVD. Top-Left: The original image, Top-Right: the measured image (masked with missing pixels), Bottom Left: DCT-based inpainting, and Bottom Right: the proposed K-SVD inpainting result.

3.1. Preliminaries

The image scale-up problem is defined as follows. A high-quality (high-resolution) image $\mathbf{y}_h$ is blurred, decimated, and contaminated by additive noise, obtaining the image $\mathbf{z}_l$,

$$\mathbf{z}_l = \mathbf{SH}\mathbf{y}_l + \mathbf{v}, \tag{5.4}$$

with $\mathbf{H}$ a blur operation, $\mathbf{S}$ the decimation operator, and $\mathbf{v}$ the noise. We assume that the decimation $\mathbf{S}$ and blur $\mathbf{H}$ operators, and the statistical properties of the noise $\mathbf{v}$, are known.

We would like to reverse this process and recover the high quality image from the given low quality one. This is a severely ill-posed inverse problem, and therefore we shall need an image model for its regularization – we shall use the *Sparseland* model.

In order to simplify notations, we will assume the the low resolution image has undergone simple interpolation (such as bicubic) to the same size as the target image

$$\mathbf{y}_l = \mathbf{Q}\mathbf{z}_l, \tag{5.5}$$

with $\mathbf{Q}$ being the interpolation operator. This initialization step will save us the technicalities of handling two scales in our formulation.

The algorithm we suggest relies on two core ideas. The first is operating locally, in the sense that each patch in the low-quality image will undergo resolution enhancement on its own. Then, the improved patches will be merged (using averaging in the overlap areas, just as in the previously described denoising and inpainting) in order to create the final output image.

How each patch is enhanced? This enhancement requires two dictionaries, $\mathbf{A}_l$ and $\mathbf{A}_h$ for the low and high quality patches respectively. Each atom in $\mathbf{A}_h$ should be the high-quality counterpart of the corresponding low quality atom in $\mathbf{A}_l$. In order to enhance a low-quality patch, its sparse representation over the low-quality dictionary $\mathbf{A}_l$ is found. This sparse representation is then used to reconstruct a patch from the high-quality dictionary, which is the enhancement result for the low-quality patch. We will now go into further details of this algorithm, and focus on learning these two dictionaries.

3.2. The core idea

As we have been practicing throughout this course, we assume that every $\sqrt{n} \times \sqrt{n}$ patch extracted from location k in the high-quality image through $\mathbf{p}_h^k = \mathbf{R}_k\mathbf{y_h}$, can be represented sparsely over a high-quality dictionary $\mathbf{A}_h$, i.e., for every patch $\mathbf{p}_h^k$ we assume that there exists a very sparse representation vector $\mathbf{q}^k$ such that $\mathbf{p}_h^k \cong \mathbf{A}_h\mathbf{q}^k$.

Next, we aim to find a connection between a high-quality patch in $\mathbf{y}_h$ and a patch in the same location in $\mathbf{y}_l$. We can use our knowledge of the creation process of $\mathbf{z}_l$ from $\mathbf{y}_h$ through blur, decimation and the addition of noise, with interpolation used to create $\mathbf{y}_l$. Therefore,

$$\mathbf{y}_l = \mathbf{Q}\left(\mathbf{DH}\mathbf{y}_h + \mathbf{v}\right). \tag{5.6}$$

Combining all operators into one $\mathbf{L}^{ALL} = \mathbf{QDH}$, we come to the relationship for each patch

$$\|\mathbf{L}\mathbf{p}_h^k - \mathbf{p}_l^k\|_2 \leq \epsilon, \tag{5.7}$$

with $\mathbf{L}$ a local portion of $\mathbf{L}^{ALL}$. This expression ties the low-quality patch and the high-quality patch in the same location. Using our assumption that the high quality patch can be sparsely represented over the high quality dictionary $\mathbf{p}_h^k \cong \mathbf{A}_h\mathbf{q}^k$, we obtain the relationship $\|\mathbf{L}\mathbf{A}_h\mathbf{q}^k - \mathbf{p}_l^k\|_2 \leq \epsilon$. Defining a new dictionary, $\mathbf{A}_l = \mathbf{L}\mathbf{A}_h$, we obtain

$$\|\mathbf{A}_l\mathbf{q}^k - \mathbf{p}_l^k\|_2 \leq \epsilon. \tag{5.8}$$

This observation leads to the conclusion that the sparse representation $\mathbf{q}^k$ is also the representation of the low quality patch $\mathbf{p}_l^k$ over the low quality dictionary $\mathbf{A}_l$.

3.3. Dictionary training

Since we want to use trained dictionaries, the obvious question is what do we train them on. One option is obtaining one (or more) high quality images. The high-quality image is then blurred and down-sampled it using the known (expected) operators. The low-resolution image is then interpolated back to the size of the high-quality image. Then, all the patches from each image are extracted, with patches in the same location in the original and interpolated image being paired.

Finally, each patch undergoes some pre-processing steps (different for the high-quality and low-quality patches), to form our final corpus for training. More on this pre-processing can be found in [**22**].

An alternative to using a pre-determined image for training, we can attempt training on the input image, as was done for denoising and inpainting and shown to improve results. However, we only have the low-quality image, so it seems something is missing. To our aid will come to observation that images tend to have a multi-resolution nature. If we consider an image pyramid, the characterizations of the local relationship between two consecutive levels in the pyramid is relatively the same in all levels.

From this observation comes the bootstrapping approach. The input image is blurred, decimated and then interpolated to create an even lower quality image. Now, we use the pair of low-quality and lowest-quality images as the basis for the training of the high and low quality dictionaries respectively. Once these two dictionaries are trained, the algorithm proceeds regardless of the training corpus.

Once the training patches have been collected and prepared, dictionary training can begin. The first step is training the low-quality dictionary from the set of low-quality patches. This training is done using the K-SVD algorithm. The next step is training the high-quality dictionary. We must remember that the two dictionaries should align with each other, such that a high-quality patch and its low-quality counterpart have the same representation over their dictionaries. This means that given a low quality patch $\mathbf{p}_l^k$ with a representation vector $\mathbf{q}^k$, it should also be the representation vector of the high-quality patch $\mathbf{p}_h^k$. Therefore, it makes sense to use the high-quality patches to train the high-quality dictionary such that this scale-up process gets as close as possible to its objective. Formally, we should minimize

$$\min_{\mathbf{A}_h} \sum_k \|\mathbf{p}_h^k - \mathbf{A}_l \mathbf{q}^k\|_2^2. \tag{5.9}$$

This is a simple least-squares problem to solve, and it results in a dictionary that is coupled with the low-quality dictionary that was already trained.

Once both dictionaries have been trained, we are now armed with all that we need to recover a high-resolution image given a low-resolution input. Note that this training algorithm may be done off-line on a high-quality image (which we expect to be similar to the image we would later enhance) or online, on the input image itself.

Figure 3 presents an example for the results obtained by this algorithm. The figure shows a high-resolution image and its scaled-down version. The scaling-up process aims to return to the original image – we provide two possible results, one obtained by the plain bicubic interpolation, and the second using the above-described algorithm.

4. Image compression using sparse representation

Compressing images in one of the most researched fields in image processing, because of its importance to the ability to transfer/store images easily while retaining as much of their quality as possible. Trained dictionaries that lead to the sparsest representations seem ideal for compression. However, there is a difficulty with training a global dictionary for all images, as it is expected to be less effective. The alternative, training a dictionary tailored for the specific given image, as practiced

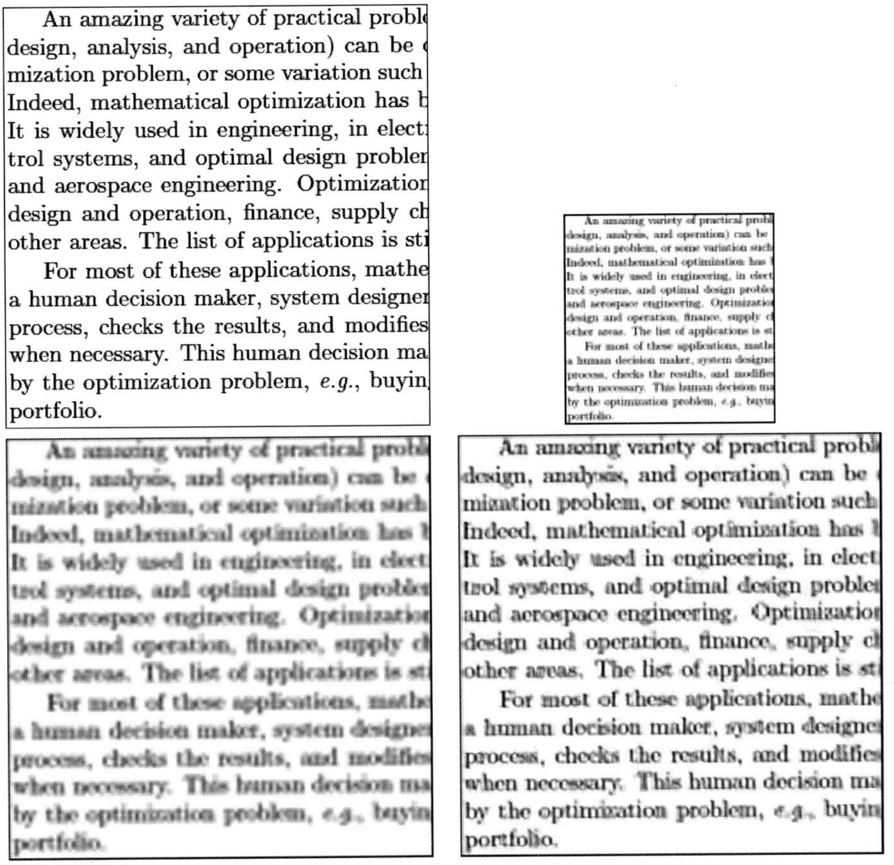

FIGURE 3. Image Scale-Up. Top-Left: The original image, Top-Right: the measured image (blurred, decimated, and noisy), Bottom Left: Bicubic scaled-up image, and Bottom Right: the proposed scale-up result.

in denoising and inpainting, is also difficult, as it requires allocated bits for the dictionary, which may compromise the success of the compression algorithm.

When turning to a specific family of images, the story changes, as it is now possible to tailor specific dictionaries to the task, while not actually needing to send them. One example for such a problem is the compression of high-quality (e.g., 500×400) facial ID images, to put on biometric identity cards or credit cards. It is possible to use general purpose compression algorithms, such as JPEG and JPEG2000 for this purpose, but a specifically trained algorithm is expected to do much better.

The proposed algorithm we describe here [**4**] has two stages – first, training appropriate dictionaries, and then, applying the trained dictionaries to compress (and decompress) the facial images.

4.1. Facial image compression algorithm - training

For training, a set of high-quality facial images are collected. This set is then used for training a dictionary for each portion of the image. The stages are as following:

(1) Key points in each image are detected (e.g., eyes, nose). These locations are used to transform each image (using a piecewise affine transform) such that these key points are in exactly the same coordinates in all images.
(2) Each image is divided into non-overlapping 15×15 patches. Overlaps are avoided as redundancy will only worsen the compression rate.
(3) All patches in the same location are collected into a set.
(4) The mean of each set of patches is computed, stored, and removed from each patch.
(5) For each set, a dictionary is trained, either using a linear approximation (i.e., PCA) or non-linear approximation (i.e., K-SVD).
(6) Each patch location is assigned a constant number of atoms to use, and quantization levels to use, both varying from location to location. This assignment is done by minimizing the mean error on the training set while achieving the desired compression rate.

Once the training procedure ends, for each patch location, its mean patch, dictionary, number of allotted atoms and quantization levels are stored both in the encoder and decoder.

4.2. Compressing and decompressing

When a new image is to be compressed, it undergoes the following stages (similar to the training stage):

(1) The same key points in the image are detected and the image is warped.
(2) The (warped) image is divided into non-overlapping 15×15 patches.
(3) For each patch, the mean of this location is removed.
(4) Sparse coding over this location's dictionary (or projection onto the PCA) is carried out, using the number of atoms allotted to this location.
(5) The values of the coefficients selected are quantized.
(6) Entropy coding is used to further compress the sent values.

Note that if K-SVD is used, when sending the atoms' coefficients their indices are sent as well, while in PCA, only the coefficients are sent. The recovery algorithm is the exact opposite. Entropy de-coding is performed on the sent data, and each patch is then reconstructed from the send coefficients (and indices if K-SVD is used). The inverse transform is applied to the image, returning to an image that is as close as possible to the original. Figure 4 presents the obtained results for 550 Bytes per image.

5. Summary

The field of sparse and redundant representation modeling offers a new model, and this can be brought to various problems in image processing, often leading to state-of-the-art results. We have seen in these lecture-notes a partial list of these success stories, and many more have been introduced to the scientific community in the past few years. More applications, and more details in general about this field and its impact to image processing, can be found in my book:

> M. Elad, *Sparse and Redundant Representations: From Theory to Applications in Signal and Image Processing*, Springer, New York, 2010.

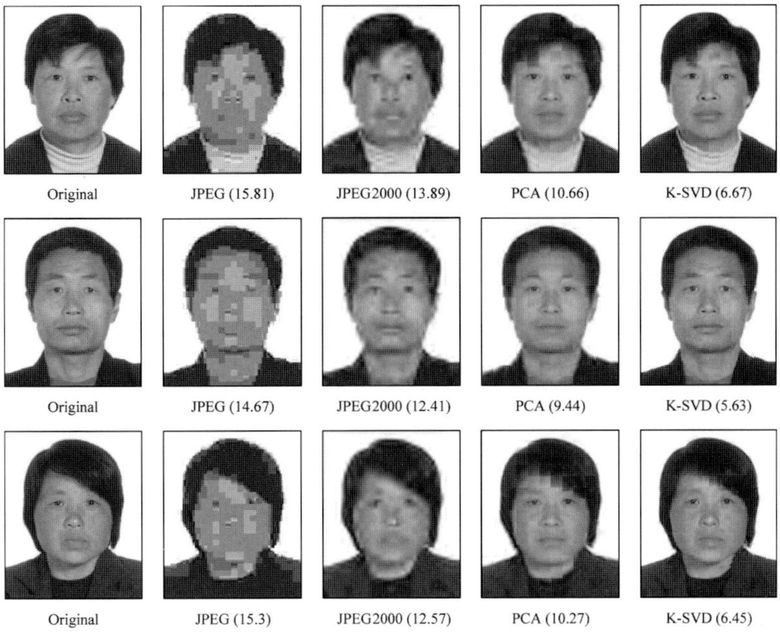

FIGURE 4. Facial image compression. From left to right: the original image, JPEG results, JPEG-2000 results, local-PCA, and K-SVD. The numbers represent the mean-squared-error per pixel after compression and decompression with 550 Bytes.

Bibliography

[1] Z. Ben-Haim, Y.C. Eldar, and M. Elad, Coherence-based performance guarantees for estimating a sparse vector under random noise, to appear in *IEEE Trans. on Signal Processing.*

[2] M. Aharon, M. Elad, and A.M. Bruckstein. K-SVD: An algorithm for designing of overcomplete dictionaries for sparse representation, *IEEE Trans. on Signal Processing*, 54(11):4311–4322, November 2006.

[3] A.M. Bruckstein, D. L. Donoho, and M. Elad, From sparse solutions of systems of equations to sparse modeling of signals and images, *SIAM Review*, 51(1):34–81, 2009.

[4] O. Bryt and M. Elad, Compression of facial images using the K-SVD algorithm, *Journal of Visual Communication and Image Representation*, 19(4):270–283, May 2008.

[5] E.J. Candès, J. Romberg, and T. Tao, Robust uncertainty principles: Exact signal reconstruction from highly incomplete frequency information, *IEEE Trans. Inform. Theory*, 52:489–509, 2006.

[6] E.J. Candès and T. Tao, The Dantzig selector: Statistical estimation when p is much larger than n, *Annals of Statistics*, 35(6):2313–2351, June 2007.

[7] S.S. Chen, D.L. Donoho, and M.A. Saunders, Atomic decomposition by basis pursuit, *SIAM Journal on Scientific Computing*, 20(1):33–61 (1998).

[8] I. Daubechies, M. Defrise, and C. De-Mol, An iterative thresholding algorithm for linear inverse problems with a sparsity constraint, *Communications on Pure and Applied Mathematics*, LVII:1413–1457, 2004.

[9] G. Davis, S. Mallat, and M. Avellaneda, Adaptive greedy approximations, *Journal of Constructive Approximation*, 13:57–98, 1997.

[10] M.N. Do and M. Vetterli, The contourlet transform: an efficient directional multiresolution image representation, *IEEE Trans. Image on Image Processing*, 14(12):2091–2106, 2005.

[11] D.L. Donoho, De-noising by soft thresholding, *IEEE Trans. on Information Theory*, 41(3):613–627, 1995.

[12] D.L. Donoho, For most large underdetermined systems of linear equations, the minimal ℓ_1-norm solution is also the sparsest solution, *Communications On Pure And Applied Mathematics*, 59(6):797–829, June 2006.

[13] D.L. Donoho, For most large underdetermined systems of linear equations, the minimal ℓ_1-norm near-solution approximates the sparsest near-solution, *Communications On Pure And Applied Mathematics*, 59(7):907–934, July 2006.

[14] D.L. Donoho and M. Elad, Optimally sparse representation in general (non-orthogonal) dictionaries via l1 minimization, *Proc. of the National Academy of Sciences*, 100(5):2197–2202, 2003.

[15] D.L. Donoho and M. Elad, On the stability of the basis pursuit in the presence of noise, *Signal Processing*, 86(3):511–532, March 2006.

[16] D.L. Donoho, M. Elad, and V. Temlyakov, Stable recovery of sparse overcomplete representations in the presence of noise, *IEEE Trans. on Information Theory*, 52(1):6–18, 2006.

[17] D.L. Donoho and I.M. Johnstone, Ideal denoising in an orthonormal basis chosen from a library of bases, *Comptes Rendus del'Academie des Sciences, Series A*, 319:1317–1322, 1994.

[18] D.L. Donoho and J. Tanner, Neighborliness of randomly-projected Simplices in high dimensions, *Proceedings Of The National Academy Of Sciences*, 102(27):9452–9457, July 2005.

[19] D.L. Donoho and J. Tanner, Sparse nonnegative solutions of underdetermined linear equations by linear programming, *Proc. Natl. Acad. Sci.*, 102:9446–9451, 2005.

[20] B. Efron, T. Hastie, I.M. Johnstone, and R. Tibshirani, Least angle regression, *The Annals of Statistics*, 32(2):407–499, 2004.

[21] M. Elad, Why simple shrinkage is still relevant for redundant representations?, to appear in the *IEEE Trans. on Information Theory*.

[22] M. Elad, *Sparse and Redundant Representations: From Theory to Applications in Signal and Image Processing*, Springer, New York, 2010.

[23] M. Elad and M. Aharon, Image denoising via sparse and redundant representations over learned dictionaries, *IEEE Trans. on Image Processing* 15(12):3736–3745, December 2006.

[24] M. Elad, B. Matalon, and M. Zibulevsky, Coordinate and subspace optimization methods for linear least squares with non-quadratic regularization, *Applied and Computational Harmonic Analysis*, 23:346–367, November 2007.

[25] M. Elad and M. Zibulevsky, Iterative shrinkage algorithms and their acceleration for l1-l2 signal and image processing applications, *IEEE Signal Processing Magazine*, 27(3):78–88, May 2010.

[26] M.A. Figueiredo and R.D. Nowak, An EM algorithm for wavelet-based image restoration, *IEEE Trans. Image Processing*, 12(8):906–916, 2003.

[27] M.A. Figueiredo, and R.D. Nowak, A bound optimization approach to wavelet-based image deconvolution, IEEE International Conference on Image Processing - ICIP 2005, Genoa, Italy, 2:782–785, September 2005.

[28] I.F. Gorodnitsky and B.D. Rao, Sparse signal reconstruction from limited data using FOCUSS: A re-weighted norm minimization algorithm, *IEEE Trans. on Signal Processing*, 45(3):600–616, 1997.

[29] R. Gribonval and M. Nielsen, Sparse decompositions in unions of bases, *IEEE Trans. on Information Theory*, 49(12):3320–3325, 2003.

[30] O.G. Guleryuz, Nonlinear approximation based image recovery using adaptive sparse reconstructions and iterated denoising - Part I: Theory, *IEEE Trans. on Image Processing*, 15(3):539–554, 2006.

[31] O.G. Guleryuz, Nonlinear approximation based image recovery using adaptive sparse reconstructions and iterated denoising - Part II: Adaptive algorithms, *IEEE Trans. on Image Processing*, 15(3):555–571, 2006.

[32] K. Kreutz-Delgado, J.F. Murray, B.D. Rao, K. Engan, T-W, Lee, and T.J. Sejnowski, Dictionary learning algorithms for sparse representation, *Neural Computation*, 15(2)349–396, 2003.

[33] S.-J. Kim, K. Koh, M. Lustig, S. Boyd, and D. Gorinevsky, A method for large-scale, 1-regularized least squares problems with applications in signal processing and statistics, *IEEE J. Selected Topics Signal Processing*, 1(4):606–617, Dec. 2007.

[34] J.B. Kruskal, Three-way arrays: rank and uniqueness of trilinear decompositions, with application to arithmetic complexity and statistics, *Linear Algebra and its Applications*, 18(2):95–138, 1977.

[35] X. Liu and N.D. Sidiropoulos, Cramer-Rao lower bounds for low-rank decomposition of multidimensional arrays, *IEEE Trans. on Signal Processing*, 49(9):2074–2086, 2001.

[36] J. Mairal, M. Elad, and G. Sapiro, Sparse representation for color image restoration, *IEEE Trans. on Image Processing*, 17(1):53–69, January 2008.

[37] J. Mairal, F. Bach, J. Ponce, G. Sapiro and A. Zisserman, Non-local sparse models for image restoration, International Conference on Computer Vision (ICCV), Tokyo, Japan, 2009.

[38] S. Mallat, *A Wavelet Tour of Signal Processing*, Academic-Press, Second-Edition, 2009.

[39] S. Mallat and E. LePennec, Sparse geometric image representation with bandelets, *IEEE Trans. on Image Processing*, 14(4):423–438, 2005.

[40] S. Mallat and Z. Zhang, Matching pursuits with time-frequency dictionaries, *IEEE Trans. Signal Processing*, 41(12):3397–3415, 1993.

[41] B.K. Natarajan, Sparse approximate solutions to linear systems, *SIAM Journal on Computing*, 24:227–234, 1995.

[42] Y.C. Pati, R. Rezaiifar, and P.S. Krishnaprasad, Orthogonal matching pursuit: recursive function approximation with applications to wavelet decomposition, the twenty seventh Asilomar Conference on Signals, Systems and Computers, 1:40–44, 1993.

[43] M. Protter and Michael Elad, Image sequence denoising via sparse and redundant representations, *IEEE Trans. on Image Processing*, 18(1):27–36, January 2009.

[44] L. Rudin, S. Osher, and E. Fatemi, Nonlinear total variation based noise removal algorithms, *Physica D*, 60:259–268, 1992

[45] E.P. Simoncelli, W.T. Freeman, E.H. Adelsom, and D.J. Heeger, Shiftable multi-scale transforms, *IEEE Trans. on Information Theory*, 38(2):587–607, 1992.

[46] J.-L. Starck, E.J. Candès, and D.L. Donoho, The curvelet transform for image denoising, *IEEE Trans. on Image Processing*, 11:670–684, 2002.

[47] J.-L. Starck, M. Elad, and D.L. Donoho, Redundant multiscale transforms and their application for morphological component separation, *Advances in Imaging and Electron Physics*, 132:287–348, 2004.

[48] J.-L. Starck, M. Elad, and D.L. Donoho, Image decomposition via the combination of sparse representations and a variational approach. *IEEE Trans. on Image Processing*, 14(10):1570–1582, 2005.

[49] T. Strohmer and R. W. Heath, Grassmannian frames with applications to coding and communication, *Applied and Computational Harmonic Analysis*, 14:257–275, 2004.

[50] J.A. Tropp, Greed is good: Algorithmic results for sparse approximation, *IEEE Trans. on Information Theory*, 50(10):2231–2242, October 2004.

[51] J.A. Tropp, Just relax: Convex programming methods for subset selection and sparse approximation, *IEEE Trans. On Information Theory*, 52(3):1030–1051, March 2006.

[52] J.A. Tropp, I.S. Dhillon, R.W. Heath Jr., and T. Strohmer, Designing structured tight frames via alternating projection, *IEEE Trans. Information Theory*, 51(1):188–209, January 2005.

[53] J.A. Tropp and A.A. Gilbert, Signal recovery from random measurements via orthogonal matching pursuit, Submitted for publication, April 2005.

[54] J. Yang, J. Wright, T. Huang, and Y. Ma, Image super-resolution via sparse representation, to appear in *IEEE Trans. on Image Processing*.

Simulation of Elasticity, Biomechanics, and Virtual Surgery

J. M. Teran, J. L. Hellrung, Jr. and J. Hegemann

IAS/Park City Mathematics Series
Volume 19, 2010

Simulation of Elasticity, Biomechanics, and Virtual Surgery

J. M. Teran, J. L. Hellrung, Jr. and J. Hegemann

Introduction

This short lecture series will cover the simulation of elastic materials characteristic of biological soft tissues. The target application is virtual surgery. This relatively new field places some unique constraints on the types of algorithms needed for simulation. First and foremost, the simulation must go very fast, nearly unprecedentedly so in fact. We must run traditional scientific computing applications in real-time to provide the functionality needed for a surgical simulator. Specifically, we must update the state of the simulation every thirtieth of a second in order to provide an interactive environment. This is highly non-trivial as such steps usually require the solution of large linear systems of equations, a task that can be notoriously time consuming. Furthermore, these simulations must be abnormally robust to user input. Many of us have played video games. The presence of a bug or some source of unexpected behavior is unacceptable as it degrades the user experience and can render the environment non-interactive. Satisfying these two constraints as well as a tutorial in basic simulation of elasticity will be the primary focus of this lecture series.

UCLA Mathematics Department, Box 951555, Los Angeles, CA 90095-1555
E-mail address: jteran@math.ucla.edu

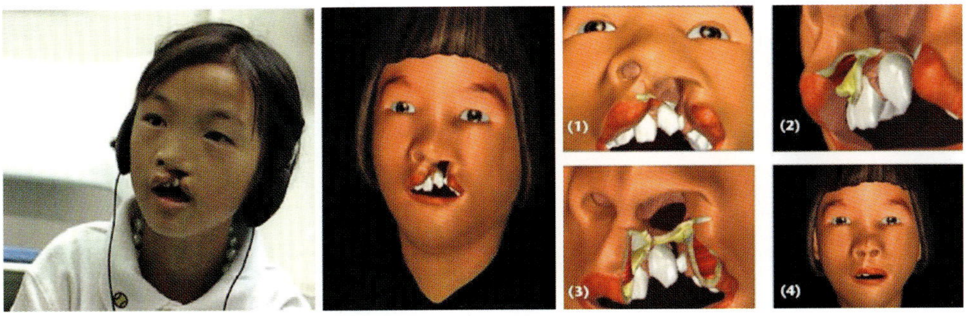

FIGURE 1. Surgical simulation will ideally be used to provide a virtual environment of prototyping procedures as well as for research and development of novel procedures. The images show a subject specific simulated cleft lip and palate repair. Elastic simulation of soft tissues is the primary algorithmic challenge in providing such technologies.

Real-time computing

Real-time simulation refers to the ability to evolve the approximation to an initial boundary value problem in less than a thirtieth of a second. In other words, less than the time it would take to observe the solution on the screen. Traditionally, computation times required for a time step in such problems have been on the order of a few minutes or even a few hours, far short of the thirtieth of a second constraint. However, such performance, should it be allowed, would provide a controllable virtual environment where a user would have the ability to change the forcing and boundary conditions of a simulation in response to real-time observation of the solution. The availability of such functionality for problems in solid and fluid mechanics engenders many new opportunities and applications. For example, imagine an interactive virtual environment where the user interacts with a finite element simulation of biomechanical soft tissues. The governing equations dictate how the tissues respond to external influence from the users. This predictive ability would allow the user to push, pull or even cut/excise portions of the tissue in real time with full confidence that the real life counterpart would behave similarly. This ability could potentially revolutionize the process of training surgical residents and medics by providing a cost effective and scientific alternative to expensive cadaver based training. Imagine the ability to know the outcome of a surgical procedure before it is performed. Reconstructive surgery for severe trauma is unpredictable by the nature of the injuries. The surgeon must design the treatment procedure on a case-by-case basis. With the aid of a predictive simulator the surgeon could perform hypothetical surgical repairs in advance to determine the most likely successful approach. This would significantly reduce complication rates and lead to drastically improved quality of life. See [**4**] for further discussion of the potential benefits of surgical simulation.

LECTURE 1

Introduction to continuum mechanics and elasticity

1. Kinematics

We will go through the basic continuum mechanics for describing the *initial boundary value problems* (IBVPs) for deformable objects as quickly as possible with an emphasis on only those details necessary for implementing a basic *finite element method* (FEM) type simulation (see the text [**1**] for more details). We will quantify the deformation of the objects of interest in terms of the mapping ϕ between an initial (or material) $\mathbf{B}_0 \in \mathbb{R}^2$ and current (or deformed) $\mathbf{B}_t \in \mathbb{R}^2$ configuration (we will just be covering 2D for simplicity). Let us introduce the convention that points in $\mathbf{B}_0$ we label as $\mathbf{X}$ and points in $\mathbf{B}_t$ we label as $\mathbf{x}$ (see Figure 1), i.e., $\mathbf{B}_0 = \{\mathbf{X}\}$ and $\mathbf{B}_t = \{\mathbf{x}\}$ so

$$\phi(\cdot, t)\colon \mathbf{B}_0 \to \mathbf{B}_t \quad \text{and} \quad \phi(\mathbf{X}, t) = \mathbf{x}.$$

It is also useful to talk about

$$\mathbf{u}(\cdot, t)\colon \mathbf{B}_0 \to \mathbb{R}^2 \quad \text{where} \quad \mathbf{u}(\mathbf{X}, t) = \phi(\mathbf{X}, t) - \mathbf{X}.$$

We will refer to this function $\mathbf{u}$ as the displacement mapping. We will also introduce a little more notation here for the derivatives of these mappings. The deformation gradient refers to $\frac{\partial \phi}{\partial \mathbf{X}}$ and is often denoted with $\mathbf{F}$:

$$\mathbf{F}(\cdot, t)\colon \mathbf{B}_0 \to \mathbb{R}^{2\times 2} \quad \text{and} \quad \mathbf{F} = \frac{\partial \phi}{\partial \mathbf{X}}.$$

Index notation for derivatives will be helpful for compactness of expressions. E.g. $F_{ij} = \frac{\partial \phi_i}{\partial X_j} = \phi_{i,j}$ and $\frac{\partial u_i}{\partial X_j} = u_{i,j}$.

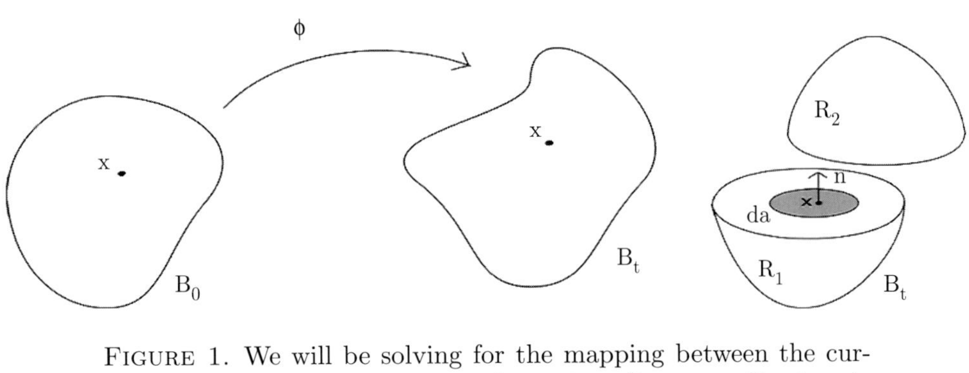

FIGURE 1. We will be solving for the mapping between the current configuration and a rest configuration. Stresses will arise via elasticity to resist changes in shape induced by this motion.

2. Basic balance laws

Again, we will only present the essential details for getting started with elastodynamics. Our governing equations of motion arise from the basic principles of continuum mechanics, ultimately $F = ma$ applied over our continuum of material $\mathbf{B}_0$. This reads as:

$$\rho_0(\mathbf{X})\frac{\partial^2 \mathbf{u}}{\partial t^2} = \nabla^{\mathbf{X}} \cdot \mathbf{P} + \mathbf{f}^{\text{ext}}.$$

Here, ρ_0 is the mass density of the material, $\mathbf{f}^{\text{ext}}$ is an externally applied force, and $\mathbf{P}$ is the first Piola-Kirchoff Stress which describes the forces of interaction in the material (specifically, $\mathbf{P}$ relates the normal of an infinitesimal surface in the material to the forces applied by one side of the material to the other; see Figure 1). We will discuss the form of this stress in the coming sections. For now, note that it is a matrix valued function over the material:

$$\mathbf{P}\colon \mathbf{B}_0 \times [0, T] \to \mathbb{R}^{2\times 2}.$$

For elasticity, we will be relating this quantity to the deformation gradient $\mathbf{F}$. It will usually make more sense for us to consider $\mathbf{P}$ as a function of $\mathbf{F}$, but we will discuss this in the constitutive modeling section. This equation is hyperbolic and roughly like the second order linear wave equation (although the specific nature of $\mathbf{P}$ for elasticity will give rise to non-linear equations). It will be useful numerically to consider both the velocities of points of material and the displacements as unknowns. We will use $\mathbf{v}(\mathbf{X}, t) = \frac{\partial \mathbf{u}}{\partial t}(\mathbf{X}, t)$ to denote the velocity of material point $\mathbf{X}$. With this convention, we can rewrite our problem with the equivalent system:

$$\begin{pmatrix} \frac{\partial \mathbf{u}}{\partial t} \\ \rho_0 \frac{\partial \mathbf{v}}{\partial t} \end{pmatrix} = \begin{pmatrix} \mathbf{v} \\ \nabla^{\mathbf{X}} \cdot \mathbf{P} + \mathbf{f}^{\text{ext}} \end{pmatrix}.$$

3. Elasticity and constitutive modeling

The equations are not yet closed because we still need to determine a relationship between the stress and the state of the material. For elastic materials, this is often referred to as the constitutive or material law. Here, we will focus on hyperelastic materials. The constitutive law for such materials arises from a so-called strain energy density function Ψ. Specifically, we define the stress as the derivative of this energy with respect to the deformation gradient $\mathbf{F}$:

$$\Psi\colon \mathbb{R}^{2\times 2} \to \mathbb{R} \quad \text{and} \quad \mathbf{P} = \frac{\partial \Psi}{\partial \mathbf{F}}.$$

This can be written in index notation as $P_{ij} = \Psi_{,ij}$. In general, the indices following the comma refer to partial differentiation.

The most simplistic model for elasticity is isotropic linear elasticity. In this case, we have

$$\epsilon_{ij} = \frac{1}{2}(F_{ij} + F_{ji}) - \delta_{ij} = \frac{1}{2}\left(\frac{\partial u_i}{\partial X_j} + \frac{\partial u_j}{\partial X_i}\right) \quad \text{and} \quad \Psi(\mathbf{F}) = \mu\epsilon_{ij}\epsilon_{ij} + \frac{\lambda}{2}\epsilon_{kk}^2.$$

We are using the convention that repeated indices are summed here and δ_{ij} is the identity matrix. Differentiating this expression with respect to $\mathbf{F}$ gives us the relationship between the stress and the state of the displaced material:

$$P_{ij} = \Psi_{,ij} = 2\mu\epsilon_{ij} + \lambda\epsilon_{kk}\delta_{ij} \quad \text{or} \quad \mathbf{P} = 2\mu\boldsymbol{\epsilon} + \lambda\,\text{tr}(\boldsymbol{\epsilon})\mathbf{I}.$$

This relationship between the stress and the displacement mapping is enough to close our governing system of equations. We will not give too much motivation for these equations, but note that $\boldsymbol{\epsilon}$ is zero when $\mathbf{F}$ is skew-symmetric. Also, note that if the displacement is spatially constant, then $\frac{\partial \mathbf{u}}{\partial \mathbf{X}} = -\mathbf{I}$ and hence $\boldsymbol{\epsilon} = \mathbf{P} = \mathbf{0}$. $\boldsymbol{\epsilon}$ is a measure of the strain (or change of shape) in the material. Ideally, we would like a measure that was invariant under rigid motions $\boldsymbol{\phi}$. Unfortunately, the motions just described are only approximately rigid. The approximation is valid if we assume that the deformation is very small. In fact, we can only say that linear elasticity is valid if the deformation is very small. Otherwise, it is not a very good model. That is, it does not accurately describe any elastic materials observed in nature unless the strain is very small. A more general expression of a rigid body motion is that the deformation gradient is orthogonal, i.e., $\mathbf{F}^T\mathbf{F} = \mathbf{I}$. In fact, more appropriate strain measures for large deformation elasticity are the right and left Cauchy-Green strain tensors: $\mathbf{G}_R = \mathbf{F}^T\mathbf{F} - \mathbf{I}$ and $\mathbf{G}_L = \mathbf{F}\mathbf{F}^T - \mathbf{I}$, respectively. Note that this measure is quadratic in the deformation. This will lead to nonlinear governing equations and considerably complicate the process of running our simulation. In other words, if we want to simulate large tissues that undergo significant deformation (as we would need in virtual surgery), we will be stuck with nonlinear governing equations.

Arguably, the most simplistic model for large deformation elasticity is the *Neo-Hookean* model. The hyperelastic strain energy density for such materials is given by the following equation:

$$\begin{aligned}\Psi(\mathbf{F}) &= \frac{\mu}{2}\left(F_{ij}F_{ji} - 2\right) - \mu\log(J) + \frac{\lambda}{2}\log(J)^2,\\ \mathbf{P}(\mathbf{F}) &= \mu\mathbf{F} + (\lambda\log(J) - \mu)\,\mathbf{F}^{-T},\end{aligned}$$

where J is the determinant of the deformation gradient:

$$J = \det(\mathbf{F}).$$

The log term in this equation is there to resist changes in volume. The λ and μ terms can be set as in linear elasticity. The most intuitive way to set these parameters is from the Young's modulus K and Poisson's ratio ν. This relationship is

$$\lambda = \frac{K\nu}{(1+\nu)(1-2\nu)} \quad \text{and} \quad \mu = \frac{K}{2(1+\nu)}.$$

The Young's modulus is a measure of the stiffness of the material and should be larger than zero (typically around 1000 for our exercises). The Poisson's ratio is a measure of the incompressibility of the material and should be between 0 and $\frac{1}{2}$ (with $\nu = \frac{1}{2}$ being the limit of an incompressible material; note that λ goes to ∞ in this case).

The models described so far are both isotropic. That is, the elastic response to deformation is the same regardless of which direction the material is stretched in. This is not very accurate for many fibrous tissues in the anatomy. In these cases, we need an auxiliary function $\mathbf{g}\colon \mathbf{B}_0 \to \mathbb{R}^2$ that describes the local fiber direction in the material configuration of the tissue (see Figure 2). The specialized response in the fiber direction can be incorporated by changing the original isotropic strain energy density to include a term based on the stretching in the fiber direction: $\mathbf{g}^T\mathbf{F}^T\mathbf{F}\mathbf{g}$. That is, the modification looks like

$$\Psi(\mathbf{F}) = \Psi_{\text{iso}}(\mathbf{F}) + \Psi_{\text{fiber}}\left(\mathbf{g}^T\mathbf{F}^T\mathbf{F}\mathbf{g}\right).$$

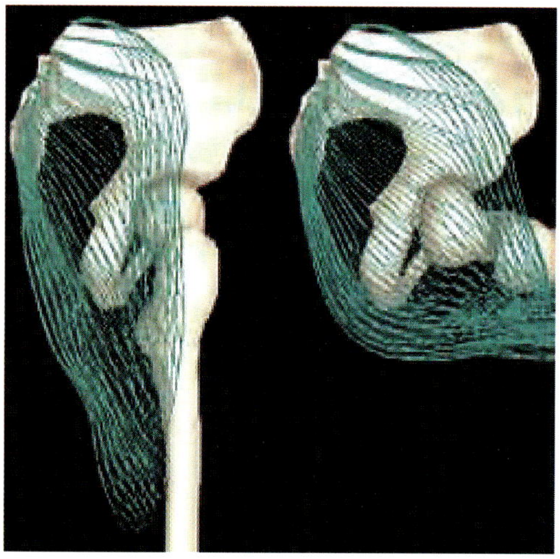

FIGURE 2. Constitutive models for muscle, tendon, ligament and many other tissues must be anisotropic to accurately reflect the natural fibrous structure of these materials. In such cases, we must represent a fiber field defined over the material configuration of the material as shown above in a rectis femoris muscle.

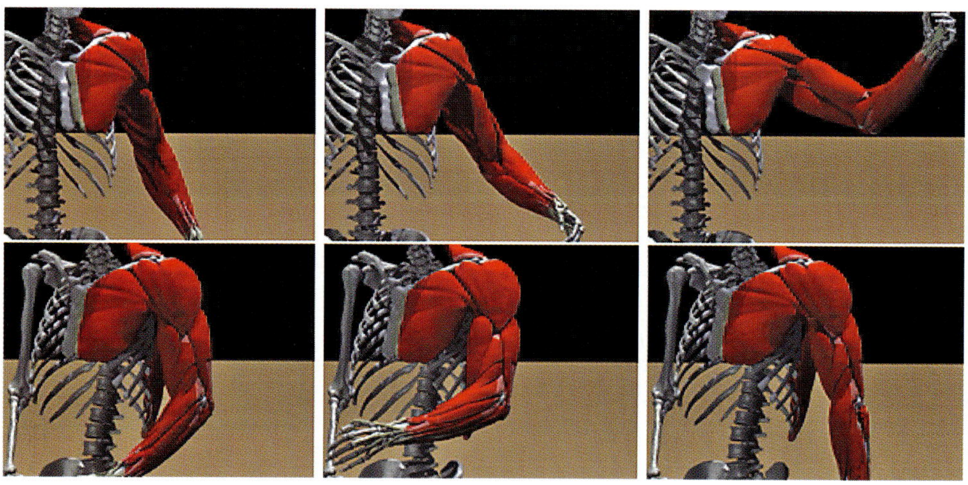

FIGURE 3. Simulation of elasticity for the musculoskeletal system.

4. Equilibrium and weak form

We can get an idea for how to treat the spatial discretization of the partial differential equation by first considering elastic equilibrium (or quasistatic) problems where the inertial terms in the governing equations are negligible (see [**6**] for more details). In this case we can solve directly for the displacement (i.e., we do not need to store the velocities as well). Ignoring inertia, our governing equations are then given by

$$-\nabla^{\mathbf{X}} \cdot \mathbf{P} = \mathbf{f}^{\text{ext}} \quad \text{or} \quad -P_{ij,j} = f_i^{\text{ext}}.$$

This can be read as the row-wise divergences of the matrix $\mathbf{P}$ balance the different components of the applied forces. We will mainly be considering *finite element method* (FEM) type discretizations in these notes, so we will first look at the weak form of the governing equations. Taking a class of test functions $\mathbf{w}\colon \mathbf{B}_0 \to \mathbb{R}^2$, we take the dot product with both sides of the governing equations and integrate over $\mathbf{B}_0$ to get

$$\begin{aligned}
&-\int_{\mathbf{B}_0} w_i P_{ij,j} d\mathbf{X} = \int_{\mathbf{B}_0} w_i f_i^{\text{ext}} d\mathbf{X} \\
&\Rightarrow -\int_{\mathbf{B}_0} \left(w_i P_{ij}\right)_{,j} - w_{i,j} P_{ij} d\mathbf{X} = \int_{\mathbf{B}_0} w_i f_i^{\text{ext}} d\mathbf{X} \\
&\Rightarrow \int_{\mathbf{B}_0} w_{i,j} P_{ij} d\mathbf{X} = \int_{\partial \mathbf{B}_0} w_i P_{ij} n_j d\mathbf{S}\left(\mathbf{X}\right) + \int_{\mathbf{B}_0} w_i f_i^{\text{ext}} d\mathbf{X}.
\end{aligned}$$

We then assume that the bottom equations hold for all suitable $\mathbf{w}$ for the weak form. Furthermore, the quantities $P_{ij} n_j$ are generally supplied as a boundary condition (this would be the external loading on the material). A FEM discretization is then done by setting up a space of interpolating functions over a mesh and considering the weak form over this finite dimensional space.

5. 1D Elasticity

To provide an introductory example, we will cover the case of 1D elasticity (we will assume some basic knowledge of FEM). Here, we will first assume the material is linearly elastic. In that case,

$$P(X) = (2\mu - \lambda) \frac{\partial u}{\partial X}$$

and the equilibrium equation is just Poisson's equation:

$$-(2\mu - \lambda) \frac{\partial^2 u}{\partial X^2} = f^{\text{ext}}.$$

Thus, we can just use a continuous piecewise linear interpolation space of a uniform discretization of our material (which we will assume for simplicity is just some interval (a, b)). In this case, we will ultimately just be solving a Poisson equation for the unknown deformation field. (This is, however, only the case in 1D.) If we assume that each node in the discrete domain has an associated interpolating function N_i, then the FEM discretized linear system is

$$\begin{aligned}
A_{ij} u_j &= F_i^{\text{ext}} + g_i \quad \text{where} \quad A_{ij} = (2\mu - \lambda) \int_a^b \frac{\partial N_i}{\partial X} \frac{\partial N_j}{\partial X} dX, \\
F_i^{\text{ext}} &= \int_a^b f^{\text{ext}} N_i dX, \quad g_i = N_i(b) P(b) - N_i(a) P(a).
\end{aligned}$$

(We assume that $P(a)$ and $P(b)$ were already given.) Now, let us consider the case of Neo-Hookean elasticity. In this case, our discretization of the weak form will give rise to a nonlinear system of equations for our displacements u_i:

$$q_i\left(\vec{u}\right) = \int_a^b \frac{\partial N_i}{\partial X} P(F(u(X))) dX - F_i^{\text{ext}} + g_i = 0$$

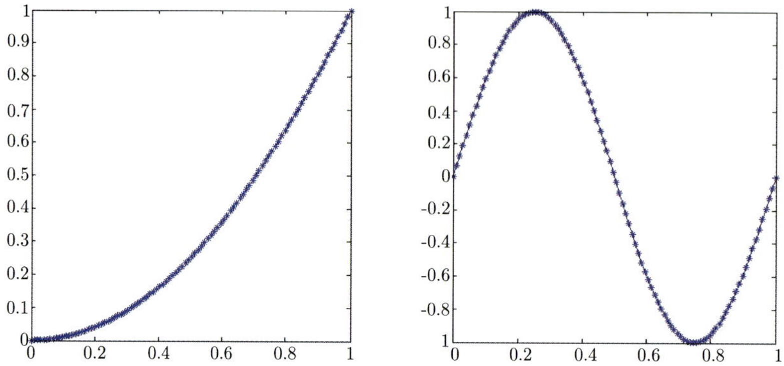

FIGURE 4. The image at the left shows an acceptable solution for the deformation ϕ (namely, $\phi(X) = X^2$). The solution at the right shows one that would produce inverted material (namely, $\phi(X) = \sin(2\pi X)$).

where $u(X) = u_i N_i(X)$ (again, where repeated indices imply summation). For Neo-Hookean materials in 1D, we have

$$P(F(u(X))) = \mu\left(\frac{\partial u}{\partial X} + 1\right) + \left(\lambda \log\left(\frac{\partial u}{\partial X} + 1\right) - \mu\right)\frac{1}{\frac{\partial u}{\partial X} + 1}.$$

We can think of $\vec{q} : \mathbb{R}^n \to \mathbb{R}^n$ as a nonlinear function of which we need to find a root. We will use Newton's method to iteratively improve our approximate solution (here, we assume $\vec{u}^k \in \mathbb{R}^n$ is our current approximation to the vector of discrete displacements) as:

$$\frac{\partial q_i}{\partial u_j}\left(\vec{u}^k\right)\Delta u_j + q_i\left(\vec{u}^k\right) = 0,$$

$$u_i^{k+1} \leftarrow u_i^k + \Delta u_i.$$

The linearization of the discrete equilibrium equations can be a little tricky, so it will be good to first consider it in 1D. The general idea is to linearize the stress with respect to the deformation gradient, then linearize the deformation gradient with respect to discrete degrees of freedom. We have

$$\frac{\partial q_i}{\partial u_j}(\vec{u}) = \int_a^b \frac{\partial N_i}{\partial X}\frac{\partial P}{\partial F}(F(\vec{u}))\frac{\partial F}{\partial u_j}dX = \int_a^b \frac{\partial N_i}{\partial X}\frac{\partial P}{\partial F}(F(\vec{u}))\frac{\partial N_j}{\partial X}dX$$

since $u(X) = u_i N_i(X)$ and $F(X) = \frac{\partial u}{\partial X}(X) + 1 = u_i \frac{\partial N_i}{\partial X}(X) + 1$. Note that the linearized system is just a spatially varying Poisson equation. However, it should be noted that for such equations we must typically require the coefficients to remain positive throughout the domain. This will not always happen and can lead to difficulties with robustness. We will discuss this issue in the next section.

6. Inversion

The deformation mapping ϕ must be a bijection (one-to-one and onto) if we are to strictly apply the principles of continuum mechanics. This precludes many solutions you might otherwise expect to see. For example, for the 1D case, the solution must never have a derivative less than or equal to zero. I.e. many functions we are

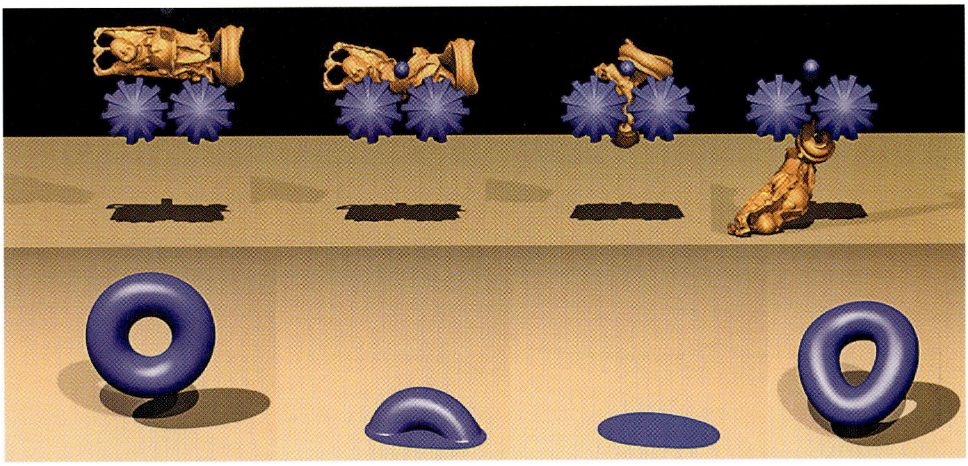

FIGURE 5. These images show the the algorithm of [**3**] in action. This algorithm is based on removing the bijection constraint from continuum mechanics in the discrete setting to provide robustness to large deformation.

used to solving for (see Figure 4) are not, strictly speaking, allowable. Unfortunately, this constraint will not always be naturally achieved by our discretization. For some models (e.g. linear elasticity) it is okay that the discrete solution does not satisfy the inversion constraint. For others like Neo-Hookean, we would end up taking the natural log of a negative number. This log term is there to prevent excessive compression. However, it has the consequence that our equations become incredibly stiff under compression. Such behavior can be very problematic numerically and must be avoided in a surgical simulator. A general approach to improving robustness to discretely inverted (i.e., non-bijective) deformation mappings was developed in [**3**]. We will not go through all of the details for the sake of time. However, we will show how we can apply the principles developed in this work to our simple Neo-Hookean model.

The problematic term in the Neo-Hookean model is the natural log. If we replace the log with another function r, we can improve the robustness of our numerical simulation to inversion. This function should look very nearly like the natural log around $J = 1$ to prevent loss of accuracy, but should not have a singularity at $J = 0$. We can do this by taking r to be the cubic Taylor expansion of the log around $J = 1$ (see Figure 6). With this choice, we have the following expressions for the Neo-Hookean hyperelastic strain energy density and first Piola-Kirchoff stresses in 2D:

$$\Psi(\mathbf{F}) = \frac{\mu}{2}(F_{ij}F_{ji} - 2) - \mu r(J) + \frac{\lambda}{2} r(J)^2,$$

$$\mathbf{P}(\mathbf{F}) = \mu \mathbf{F} + (\lambda r(J) - \mu) r'(J) \begin{pmatrix} F_{22} & -F_{12} \\ -F_{21} & F_{11} \end{pmatrix}.$$

In Figure 6, we see the effect of this modification in 1D. This new model is well-defined through inversion, but accurately reflects the original model when very near the undeformed configuration. We can make further adjustments to prevent problematic values of $\frac{\partial P}{\partial F}$ (e.g. we wanted this to be positive for our 1D example). The treatment of these terms is addressed in [**6**].

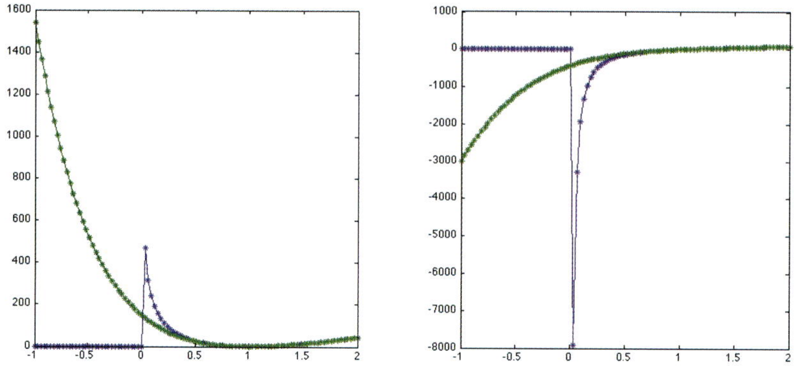

FIGURE 6. The plots at the left are of the hyperelastic energy density of the 1D Neo-Hookean model. The plots at the right are of the 1D Neo-Hookean stresses. The blue curves are the unmodified Neo-Hookean model, the green curves result from replacing the log with its cubic Taylor expansion around $J = 1$. These are preferable for the sake of robustness.

7. Time stepping

In general, our simulator will run the following loop:

for $n = 0 : N_{\text{final}}$
- get user boundary condition input
- $t \leftarrow t + \Delta t$
- solve nonlinear system for displacements and velocities
- render solution to screen

end

The key factor is that the nonlinear system must be solved in less than one thirtieth of a second to maintain an interactive simulation rate. An example of a discretization of this system that allows for inertia is the following. Let $\vec{u}^m \in \mathbb{R}^n$ represent the spatially discrete displacements at time m and let $\vec{v}^m \in \mathbb{R}^n$ represent the spatially discrete velocities. We can implement backward Euler by using

$$\vec{v}^{m+1} = \frac{1}{\Delta t}\left(\vec{u}^{m+1} - \vec{u}^m\right)$$

and

$$\rho_0 \frac{1}{\Delta t}\left(\frac{1}{\Delta t}\left(\vec{u}^{m+1} - \vec{u}^m\right) - \vec{v}^m\right) = \nabla^{\mathbf{X}} \cdot \mathbf{P}\left(\mathbf{F}\left(\vec{u}^{m+1}\right)\right) + \mathbf{f}^{\text{ext}}$$

where we solve this final system by following a similar FEM procedure to the quasistatic case for $\vec{u}^{m+1}$. Specifically, we get

$$q_i^a\left(\vec{u}^{m+1}\right) = M_{ai;bj} u_{bj}^{m+1} + \Delta t^2 \int_{\mathbf{B}_0} N_{,j}^a P_{ij} d\mathbf{X} - M_{ai;bj}\left(u_{bj}^{m+1} + \Delta t v_{bj}^m\right)$$
$$- \int_{\partial \mathbf{B}_0} N^a P_{ij} n_j d\mathbf{S}(\mathbf{X}) - \int_{\mathbf{B}_0} N^a f_i^{\text{ext}} d\mathbf{X}$$

and

$$M_{ai;bj} = \int_{\mathbf{B}_0} N^a \rho_0 N^b \delta_{ij} d\mathbf{X}.$$

LECTURE 2

Numerical solutions of the equations of elasticity

This lecture provides details and exercises pertaining to the implementation of numerical solvers for the equations of elasticity. We begin with a review of the Finite Element Method (FEM) as it applies to Poisson's equation. We will then apply this framework to numerically solve the equations of elasticity, first in dimension $d = 1$ with quasistatic evolution, then in dimension $d = 2$ with inertial effects included.

1. Numerical solution of Poisson's equation via the finite element method

Let us consider numerical solutions of (variable coefficient) Poisson's equation:

$$-\nabla \cdot (\beta \nabla u) = f \quad \in \Omega, \tag{2.1a}$$

$$u = p \quad \in \partial\Omega_d, \tag{2.1b}$$

$$\beta \nabla u \cdot \hat{\mathbf{n}} = q \quad \in \partial\Omega_n. \tag{2.1c}$$

Here, $\Omega \subset \mathbb{R}^d$ is open; $\partial\Omega_d$ and $\partial\Omega_n$ partition the boundary of Ω and define where *Dirichlet* (d) and *Neumann* (n) boundary conditions are applied, respectively; $\hat{\mathbf{n}}$ is the outward-pointing unit normal on $\partial\Omega_n$; $\beta, f \colon \Omega \to \mathbb{R}$, $p \colon \partial\Omega_d \to \mathbb{R}$, and $q \colon \partial\Omega_q \to \mathbb{R}$ are given, and β is bounded below by some positive number; and the goal is to solve for the unknown function $u \colon \Omega \to \mathbb{R}$. For now, we will ignore the details of the function spaces that u, β, f, p, and q are assumed to belong in.

For the numerical solution of (2.1), we will use a discretization based on the Finite Element Method (FEM). Briefly, an FEM discretization focuses on the weak formulation of the partial differential equation. The derivation involves multiplying the differential equation by a test function (chosen from some suitable function space), integrating by parts, and applying boundary conditions:

$$-\nabla \cdot (\beta \nabla u) = f \tag{2.2a}$$

$$\Rightarrow \int_\Omega -\nabla \cdot (\beta \nabla u)\, v = \int_\Omega f v \tag{2.2b}$$

$$\Rightarrow \int_\Omega \beta \nabla u \cdot \nabla v = \int_\Omega f v + \int_{\partial\Omega} (\beta \nabla u \cdot \hat{\mathbf{n}})\, v, \tag{2.2c}$$

$$\Rightarrow \int_\Omega \beta \nabla u \cdot \nabla v = \int_\Omega f v + \int_{\partial\Omega_n} q v \tag{2.2d}$$

where the derivation from (2.2c) to (2.2d) is possible by stipulating that $v \equiv 0$ on $\partial\Omega_d$. Clearly, then, if u satisfies (2.1), then u also satisfies (2.2d) for all appropriate v. Under suitable conditions, it turns out to be the case that the converse holds as well. Since (2.2d) involves only first order derivatives in u, its analysis is often preferable to (2.1).

We then suppose that u can be approximated by a function in a finite-dimensional function space (the *(discretized) solution space*) (e.g., continuous piecewise affine functions over some tessellation of Ω), i.e., $u \approx \tilde{u} = \sum_j u_j e_j$ for basis elements $\{e_j\}$. Likewise, v is selected from some other (often related) finite-dimensional function space (the *(discretized) test space*), thus giving a system of equations for the coefficients u_j. For example, if v is taken from the same space as $\tilde{u}$, and we let v be, in turn, each of the basis functions, then we obtain

$$\sum_j \left(\int_\Omega \beta \nabla e_j \cdot \nabla e_i \right) u_j = \int_\Omega f e_i + \int_{\partial\Omega_n} q e_i \tag{2.3}$$

for all i, which may be expressed as a linear system of equations:

$$A\vec{u} = \vec{b}, \tag{2.4}$$

where

$$A_{ij} = \int_\Omega \beta \nabla e_i \cdot \nabla e_j, \tag{2.5a}$$

$$\vec{u}_i = u_i, \tag{2.5b}$$

$$\vec{b}_i = \int_\Omega f e_i + \int_{\partial\Omega_n} q e_i. \tag{2.5c}$$

Strictly speaking, the above is only correct if v varies *precisely* throughout the same function space which $\tilde{u}$ belongs to. This is generally not the case. For example, in the presence of Dirichlet boundary conditions, $\tilde{u}$ will generally be nonzero along Ω_d, however v will be chosen from a function space which vanishes along Ω_d (hence these functions spaces will only align under homogeneous Dirichlet boundary conditions). How this manifests itself in the linear system will be investigated in the exercises.

Note that the sparsity of A is directly related to the overlap of the supports of the e_j's. Typically we choose the e_j's such that the support of one basis element overlaps the supports of only a small constant number of other basis elements, in which case A will have a number of nonzero entries proportional to the number of basis elements.

Let us now focus on dimension $d = 1$ and use *linear finite elements*. Let Ω be an interval (a, b), and let $\{x_j\}$ be grid points in (a, b), with $a = x_0 < x_1 < \cdots < x_n = b$. Often times, the x_j's will be equally spaced, which makes the analysis simpler, but it is not necessary. We will take $\tilde{u}$ (the discretized solution) and v (the discretized test function) to be continuous over $\Omega = (a, b)$ and *piecewise affine* on each segment (x_{j-1}, x_j), with $v(a)$ and/or $v(b)$ vanishing if a Dirichlet condition is specified at a and/or b, respectively. This space of continuous piecewise affine functions is spanned by the *nodal basis functions* (also called "hat" functions, because of the shape of their graphs) $\{N_j\}$, (uniquely) defined such that

$$N_j(x_k) = \delta_{jk} = \begin{cases} 1, & j = k, \\ 0, & j \neq k. \end{cases} \tag{2.6}$$

The basis $\{N_j\}$ (and its corresponding analogs in higher dimensions) is the typical basis to use for linear finite elements.

1.1. Exercises

Unless otherwise noted, assume dimension $d = 1$ linear finite elements with the basis $\{N_j\}$ given by (2.6), although these results typically extend to higher order and higher dimensional finite elements.

(1) Here is one way to incorporate the Dirichlet boundary conditions into (2.5). For concreteness, suppose $\Omega_d = \{a\}$ (with $\Omega = (a,b)$). Then we still have $\tilde{u} = \sum_{j=0}^{n} u_j N_j$, but our space of test functions $\{v\}$ is slightly smaller, span $\{N_1, \ldots, N_n\}$, to ensure that $v(a) = 0$. Thus, we obtain n equations of the form (2.3), plus one more equation from the Dirichlet boundary condition ($u_0 = u(a)$). This (ostensibly) gives enough equations to solve for the $n+1$ coefficients $\{u_j\}$.

Another conceptual way to incorporate the Dirichlet boundary conditions is to consider the function $w = u - u(a)e_0$, which satisfies a similar Poisson problem as u but now with homogeneous Dirichlet boundary conditions. Thus, the space of test functions $\{v\}$ and the solution space of w coincide, and one may use (2.5) directly.

In practice, one can actually just ignore the Dirichlet boundary conditions when initially computing the entries of A and $\vec{b}$ (i.e., just use (2.5a) and (2.5c)), then "correct" A and $\vec{b}$ to account for the Dirichlet boundary conditions afterwards. Describe a *simple* algorithm to effect this "correction" procedure. Try to keep A symmetric.

(2) Show that the linear system (2.4) has a unique solution if one specifies Dirichlet boundary conditions at either $x = a$ or $x = b$ or both (i.e., $u(a) = u_a$ and/or $u(b) = u_b$). For simplicity, you may assumed $\beta \equiv 1$. What happens if both boundary conditions are Neumann? Specifically address the conditions which $\vec{b}$ must satisfy, how this translates into a condition on f, and "how uniquely" $\tilde{u}$ is determined.

(3) Compute the integral in (2.5a) when $\beta \equiv 1$. If you wish, for simplicity, you may assume the x_j's are equally spaced, i.e., $x_j = a + jh$, where $h = (b-a)/n$. What is the structure of the matrix A (e.g., describe the sparsity)?

(4) Give an expression to reasonably approximate the integrals in (2.5a) and (2.5c) for arbitrary β, f, and q (assume β and f are sufficiently "well-behaved"). You may have to consider boundary and interior grid vertices separately.

(5) Write some code (e.g., a Matlab script) to solve the following boundary value problem using linear finite elements:

$$\begin{aligned} -\left((1+x)\,u'(x)\right)' &= (1+x)^{-2} \quad x \in (0,1), \\ u(0) &= 1, \\ (1+(1))\,u'(1) &= -1/2. \end{aligned}$$

(6) The FEM terminology for the matrix A in (2.4) is the *(global) stiffness matrix* (the meaning of the term "global" will become obvious shortly). With dimension $d = 1$ linear finite elements, it is relatively straightforward to compute the entries of the stiffness matrix, even given that one has to account for differences between boundary and interior grid vertices. However, the situation becomes more complicated with higher order or

higher dimensional finite elements, and it becomes tedious to consider all the special boundary cases. As such, it is common practice to assemble the global stiffness matrix element by element (i.e., cell by cell), with each element contributing partially to several entries of the global stiffness matrix. The collection of these contributions from a single element may be put into a small matrix, the *local stiffness matrix* for that element. The computation of the local stiffness matrix is identical to that of the global stiffness matrix upon replacing the integrations over Ω with integrations localized over the element. For example, back to dimension $d = 1$ linear finite elements, the local stiffness matrix over $I_j = (x_{j-1}, x_j)$ would consist of 2^2 (nonzero) entries (since only 2 basis functions are supported on I_j). Each of these entries would be added to accumulated global stiffness matrix entries:

$$\begin{aligned} A_{j-1,j-1} \mathrel{+}= & \int_{I_j} \beta \nabla N_{j-1} \cdot \nabla N_{j-1}, \\ A_{j-1,j}, A_{j,j-1} \mathrel{+}= & \int_{I_j} \beta \nabla N_{j-1} \cdot \nabla N_j, \\ A_{jj} \mathrel{+}= & \int_{I_j} \beta \nabla N_j \cdot \nabla N_j. \end{aligned}$$

This suggests the following procedure to build the (global) stiffness matrix A. First, initialize the sparsity pattern of A (with zeros). Then, iterate through each element and add the contributions of the element's local stiffness matrix to the corresponding entries in A.

Verify this procedure computes the same stiffness matrix as before, and modify your program from Exercise 5 to build the stiffness matrix in this fashion.

2. Neo-Hookean elasticity with quasistatic evolution in dimension 1

Let us now consider the numerical solution of the equations of elasticity with the Neo-Hookean constitutive model. To begin, for simplicity, we will consider using quasistatic evolution in dimension $d = 1$. The subsequent section will extend this into dimension $d = 2$ and add inertial effects.

2.1. Elasticity

Recall that the equations of elasticity are given by the boundary value problem

$$\rho_0 (\mathbf{X}) \frac{\partial^2 \mathbf{u}}{\partial t^2} = \nabla^{\mathbf{X}} \cdot \mathbf{P} + \mathbf{f}^{\text{ext}} \quad \in \Omega(t), \tag{2.7a}$$

$$\mathbf{u}(t, \cdot) = \mathbf{g}(t, \cdot) \quad \in \partial\Omega_d(t), \tag{2.7b}$$

$$(\mathbf{P} \cdot \hat{\mathbf{n}}) (t, \cdot) = \mathbf{h}(t, \cdot) \quad \in \partial\Omega_n(t), \tag{2.7c}$$

where $\rho_0 \colon \Omega_0 \to \mathbb{R}$ is the mass density (as a function of $\mathbf{X}$, the undeformed coordinates); $\mathbf{u}(t, \cdot) \colon \Omega_0 \to \mathbb{R}^2, \mathbf{X} \mapsto \boldsymbol{\phi}(\mathbf{X}) - \mathbf{X}$ is the unknown displacement; $\mathbf{P}$ is the first Piola-Kirchoff stress (which takes a specific form for Neo-Hookean, to be given later); $\mathbf{f}^{\text{ext}}$ is the given (external) force; and $\mathbf{g}$ and $\mathbf{h}$ specify the Dirichlet and Neumann boundary conditions, respectively. For simplicity, we will assume a uniform mass density, i.e., $\rho_0 \equiv 1$.

2.2. Neo-Hookean constitutive model

The Neo-Hookean constitutive model relates the stress $\mathbf{P}$ to the deformation gradient $\mathbf{F} := \partial\phi/\partial\mathbf{X}$ via

$$\Psi(\mathbf{F}) := \frac{\mu}{2}(F_{ij}F_{ji} - 2) - \mu \log J + \frac{\lambda}{2}\log^2 J, \tag{2.8a}$$

$$\mathbf{P}(\mathbf{F}) := \frac{\partial\Psi}{\partial\mathbf{F}} = \mu\mathbf{F} + (\lambda \log J - \mu)\mathbf{F}^{-T}. \tag{2.8b}$$

(Recall that $J := \det \mathbf{F}$.) This manifests itself in dimension $d = 1$ in terms of the (scalar) displacement u as

$$P(u) = \mu\left(\frac{du}{dX} + 1\right) + \left(\lambda \log\left(\frac{du}{dX} + 1\right) - \mu\right)\frac{1}{\frac{du}{dX} + 1}.$$

2.3. Inversion-robust neo-Hookean

Neo-Hookean as formulated above will not be robust to element inversions, due to the $\log J$ terms. To remedy this, we replace the logarithm in (2.8a) with a cubic Taylor approximation around 1:

$$\begin{aligned} r(x) &= (x-1) - \frac{1}{2}(x-1)^2 + \frac{1}{3}(x-1)^3 \\ &= -\frac{11}{6} + 3x - \frac{3}{2}x^2 + \frac{1}{3}x^3, \end{aligned} \tag{2.9a}$$

$$\begin{aligned} r'(x) &= 1 - (x-1) + (x-1)^2 \\ &= 3 - 3x + x^2, \end{aligned} \tag{2.9b}$$

$$\begin{aligned} r''(x) &= -1 + 2(x-1) \\ &= -3 + 2x. \end{aligned} \tag{2.9c}$$

This gives

$$\Psi(\mathbf{F}) = \frac{\mu}{2}(F_{ij}F_{ji} - 2) - \mu r(J) + \frac{\lambda}{2}r(J)^2,$$

$$\mathbf{P}(\mathbf{F}) = \frac{\partial\Psi}{\partial\mathbf{F}} = \mu\mathbf{F} + (\lambda r(J) - \mu)\, r'(J)\frac{\partial J}{\partial\mathbf{F}}.$$

Again, specializing this for $d = 1$ dimension, we obtain

$$P(u) = \mu\left(\frac{du}{dX} + 1\right) + \left(\lambda r\left(\frac{du}{dX} + 1\right) - \mu\right) r'\left(\frac{du}{dX} + 1\right). \tag{2.10}$$

2.4. Quasistatic evolution

We begin by studying (2.7) at equilibrium, which is the basis for quasistatic evolution. This reduces the equations to

$$-\nabla^{\mathbf{X}} \cdot \mathbf{P} = \mathbf{f}^{\text{ext}}. \tag{2.11}$$

The equivalent weak formulation is

$$\int_{\Omega_0} w_{i,j}P_{ij}d\mathbf{X} = \int_{\partial\Omega_n} w_i h_i dS(\mathbf{X}) + \int_{\Omega_0} w_i f_i^{\text{ext}} d\mathbf{X}$$

for all test functions $\mathbf{w}$. Recall that summation over repeated indices is implied, and comma'ed indices indicate differentiation.

Similarly to the derivation for Poisson's equation, we let the coordinates of $\mathbf{w}$ vary over the nodal basis functions N_i. In dimension $d = 1$, this reduces to the system of equations

$$0 = q_i(\vec{u}) := \int_a^b \frac{\partial N_i}{\partial X} P(F(u(X)))dX - b_i, \tag{2.12a}$$

$$b_i := \int_a^b N_i f^{\text{ext}} dX + [\text{Neumann boundary terms}] \tag{2.12b}$$

for each grid vertex i, where $u := u_i N_i$. The Neumann boundary terms consist of neither, one, or both of $N_i(b)h(b)$ and/or $-N_i(a)h(a)$, depending on whether b and/or a, respectively, belong to $\partial\Omega_n$.

For Neo-Hookean, P depends *non-linearly* on the displacement u, hence one must use a non-linear solver, such as Newton iteration, to solve (2.12) for $\vec{u} := (u_i)$. The Newton step looks like

$$\frac{\partial q_i}{\partial \vec{u}}(\vec{u})\,\Delta\vec{u} + q_i(\vec{u}) = 0,$$
$$\vec{u} \leftarrow \vec{u} + \Delta\vec{u},$$

where

$$\frac{\partial q_i}{\partial u_j}(\vec{u}) = \int_a^b \frac{\partial N_i}{\partial X}\frac{\partial P}{\partial F}(F(u))\frac{\partial N_j}{\partial X} dX.$$

Thus, the computation of $\Delta\vec{u}$ in each Newton iteration amounts to solving a variable coefficient Poisson problem. The coefficient is $\partial P/\partial F$, which we can express via (2.10) as

$$\frac{\partial P}{\partial F} = \mu + \lambda r'\left(\frac{du}{dX}+1\right)^2 + \left(\lambda r\left(\frac{du}{dX}+1\right) - \mu\right) r''\left(\frac{du}{dX}+1\right).$$

This gives all the necessary pieces to implement a quasistatic evolution of the equations of elasticity (2.7).

2.5. Implementation

Let us assume a regular grid on (a, b), such that we have n grid vertices $x_i := a + (i-1)\Delta x$, $\Delta x := (b-a)/(n-1)$, and i ranges from 1 to n, inclusive. We break the implementation of such a solve into several steps, outlined below and expounded upon in the subsequence subsubsections.

- Compute $\vec{b} := (b_1, \ldots, b_n)$ (from (2.12b)).
- Implement a procedure to compute $\vec{q} := (q_1, \ldots, q_n)$ (from (2.12a)).
- Implement a procedure to compute $\partial\vec{q}/\partial\vec{u}$.
- Solve for the Newton increment $\Delta\vec{u}$.

We provide pseudocode for all steps in a syntax similar to Matlab or Octave.

2.5.1. *Compute $\vec{b}$.* We begin by considering the computation of $\vec{b}$. This is a natural place to start because, first, it remains constant throughout the Newton iterations within a single time step; and second, its computation is identical to that for the right-hand side in Poisson's equation. One natural algorithm to compute $\vec{b}$ might be a vertex-based approach, where each b_i is computed explicitly and in isolation, and we consider the boundary grid vertices specially. However, as outlined in the last exercise from the previous section, an element-based approach scales better with dimension and order, hence we will focus on this approach for this and subsequent computations.

The implementation thus boils down to computing the integrals

$$\int_{x_i}^{x_{i+1}} N_j f^{\text{ext}} dX$$

where $j \in \{i, i+1\}$ (all other values of j integrate to zero). The result of this integral is then added to a running accumulation of b_j. We will assume that we have been given the values of f^{ext} at the midpoints of each interval (x_i, x_{i+1}), and approximate the integral via the midpoint rule, leading to the following implementation:

```
function b = eval_b(x1, xn, n, fext, h1, hn)
b = zeros([n 1]);
dX = (xn - x1) / (n - 1);
for i = 1:n-1
    b(i  ) += fext(i) * dX / 2; % fext(i) is fext evaluated at
    b(i+1) += fext(i) * dX / 2; %     x_{i+1/2}
end
b(1) += h1; % add Neumann boundary terms
b(n) += hn; % add Neumann boundary terms
```

2.5.2. *Computing $\vec{q}$.* We turn now toward computing $\vec{q}$ via (2.12a) given the current Newton approximation $\vec{u}$. Again, we will use an element-based approach, meaning we need to compute the integrals

$$\int_{x_i}^{x_{i+1}} \frac{\partial N_j}{\partial X} P\left(\frac{\partial u}{\partial X}\right) dX$$

where, again, $j \in \{i, i+1\}$. Note the components of the integrand are constant; indeed, on (x_i, x_{i+1}),

$$\begin{aligned} \frac{\partial N_i}{\partial X} &= -\frac{1}{\Delta X}, \\ \frac{\partial N_{i+1}}{\partial X} &= \frac{1}{\Delta X}, \\ P\left(\frac{\partial u}{\partial X}\right) &= P\left(\frac{u_{i+1} - u_i}{x_{i+1} - x_i}\right). \end{aligned}$$

It follows that

$$\int_{x_i}^{x_{i+1}} \frac{\partial N_j}{\partial X} P\left(\frac{\partial u}{\partial X}\right) dX = \pm P\left(\frac{u_{i+1} - u_i}{x_{i+1} - x_i}\right).$$

This leads to the following procedure to compute $\vec{q}$:

```
function q = eval_q(x1, xn, n, mu, lambda, b, u)
q = -b;
dX = (xn - x1) / (n - 1);
for i = 1:n-1
    du_dx = (u(i+1) - u(i)) / dX;
    r     = du_dx - du_dx^2 / 2 + du_dx^3 / 3;
    dr    = 1 - du_dx + du_dx^2;
    P     = mu * (du_dx + 1) + (lambda * r - mu) * dr;
    q(i  ) -= P; % (-) since dN_{i  }/dX = -1/dX
    q(i+1) += P; % (+) since dN_{i+1}/dX = +1/dX
```

```
end
```

2.5.3. *Computing $\partial\vec{q}/\partial\vec{u}$.* The next step is to consider the computation of $\partial\vec{q}/\partial\vec{u}$, which will be a sparse $n \times n$ matrix. The integrals we need to compute this time are

$$\int_{x_i}^{x_{i+1}} \frac{\partial N_j}{\partial X}\frac{\partial N_k}{\partial X}\frac{\partial P}{\partial F}\left(\frac{\partial u}{\partial X}\right) dX$$

where $j, k \in \{i, i+1\}$. Again, the integrand is constant, so the integration is trivial. The expression for $\partial P/\partial F$ comes from differentiating (2.10):

$$\frac{\partial P}{\partial F} = \mu + \lambda r'(F)^2 + (\lambda r(F) - \mu)\, r''(F).$$

The procedure looks as follows:

```
function dq_du = eval_dq_du(x1, xn, n, mu, lambda, u)
% create an empty tri-diagonal sparse matrix
dq_du = sparse([1:n 2:n 1:n-1], [1:n 1:n-1 2:n], 0);
dX = (xn - x1) / (n - 1);
for i = 1:n-1
    du_dx = (u(i+1) - u(i)) / dX;
    r     = du_dx - du_dx^2 / 2 + du_dx^3 / 3;
    dr    = 1 - du_dx + du_dx^2;
    d2r   = -1 + 2 * du_dx;
    dP_dF = mu + lambda * dr^2 + (lambda * r - mu) * d2r;
    dq_du(i  ,i  ) += dP_dF / dX;
    dq_du(i  ,i+1) -= dP_dF / dX;
    dq_du(i+1,i  ) -= dP_dF / dX;
    dq_du(i+1,i+1) += dP_dF / dX;
end
```

2.5.4. *Computing the Newton increment $\Delta\vec{u}$.* With the above procedures in hand, computing the Newton increment $\Delta\vec{u}$ becomes very easy. There is one thing to note regarding Dirichlet boundary conditions. When computing $\Delta\vec{u}$, we will assume that the current Newton approximation $\vec{u}$ agrees with its Dirichlet boundary conditions at the current time-step. Hence, we will want zeros in $\Delta\vec{u}$ in those entries corresponding to Dirichlet boundary conditions, which can be effected by slightly altering the matrix and right-hand side:

```
dq_du = eval_dq_du(x1, xn, n, mu, lambda, u);
q     = eval_q    (x1, xn, n, mu, lambda, b, u);
if dirichlet_bc_at_1
    dq_du(1,1) = 1;
    dq_du(1,2) = 0;
    dq_du(2,1) = 0;
    q(1) = 0;
end
if dirichlet_bc_at_n
    dq_du(n,n) = 1;
    dq_du(n,n-1) = 0;
    dq_du(n-1,n) = 0;
    q(n) = 0;
```

```
end
delta_u = dq_du \ -q;
```

2.5.5. *The full Newton method.* We now have all the pieces to solve the nonlinear equation (2.12):

```
function u = solve( ...
    x1, xn, n, ...
    mu, lambda, ...
    dirichlet_bc_at_1, g1, dirichlet_bc_at_n, gn, ...
    h1, hn, ... % Neumann boundary conditions
    fext, ...
    tol, u)
b = eval_b(x1, xn, n, fext, h1, hn);
if dirichlet_bc_at_1
    u(1) = g1;
end
if dirichlet_bc_at_n
    u(n) = gn;
end
do
    dq_du = eval_dq_du(x1, xn, n, mu, lambda, u);
    q     = eval_q    (x1, xn, n, mu, lambda, b, u);
    if dirichlet_bc_at_1
        dq_du(1,1) = 1;
        dq_du(1,2) = 0;
        dq_du(2,1) = 0;
        q(1) = 0;
    end
    if dirichlet_bc_at_n
        dq_du(n,n) = 1;
        dq_du(n,n-1) = 0;
        dq_du(n-1,n) = 0;
        q(n) = 0;
    end
    delta_u = dq_du \ -q;
    u += delta_u;
while max(abs(delta_u)) < tol
```

2.5.6. *Example problem.* We can test the code with the following example problem.

- $\Omega_0 = (0, 1)$ (so $a = 0$ and $b = 1$)
- Dirichlet boundary condition at $x = 0$: $u(0, t) = g(0, t) = \sin t$
- Neumann boundary condition at $x = 1$: $P(1, t) = 0$
- $f^{\text{ext}} \equiv 0$
- $E = 1000$ and $\nu = 0.3$; so $\mu = E/(2(1 + \nu)) = 384.$ and $\lambda = E\nu/((1 + \nu)(1 - 2\nu)) = 577.$

Since a quasistatic evolution has no inertial terms, we expect the solution displacement to simply be $u(X, t) = \sin t = g(0, t)$.

This is a pretty simple test, so feel free to experiment with other combinations of parameters.

3. Neo-Hookean elasticity with backward Euler evolution in dimension 2

Let us again recall the elasticity equations:

$$\rho_0 \frac{\partial^2 \mathbf{u}}{\partial t^2} = \nabla^{\mathbf{X}} \cdot \mathbf{P} + \mathbf{f}^{\text{ext}} \quad \in \Omega(t), \tag{2.13a}$$

$$\mathbf{u}(t,\cdot) = \mathbf{g}(t,\cdot) \quad \in \partial\Omega_d(t), \tag{2.13b}$$

$$(\mathbf{P}\cdot\hat{\mathbf{n}})(t,\cdot) = \mathbf{h}(t,\cdot) \quad \in \partial\Omega_n(t). \tag{2.13c}$$

With non-negligible inertial effects, a more sophisticated temporal evolution than quasistatic evolution must be used. One such scheme is Backward Euler, which is desirable due to its unconditional stability and dissipation of oscillatory errors. We will also see that it is not significantly more complex than quasistatic evolution. However, the transition from dimension $d = 1$ to dimension $d = 2$ does offer a fair amount of complexity, which we shall investigate.

3.1. Backward Euler

To formulate the Backward Euler time-stepping, we introduce an auxiliary variable, $\mathbf{v}$ ("velocity"), to transform (2.13) into a first-order system (in time):

$$\frac{\partial \mathbf{u}}{\partial t} = \mathbf{v},$$

$$\rho_0 \frac{\partial \mathbf{v}}{\partial t} = \nabla^{\mathbf{X}} \cdot \mathbf{P} + \mathbf{f}^{\text{ext}}.$$

The Backward Euler time discretization thus gives

$$\frac{1}{\Delta t}\left(\mathbf{u}(t+\Delta t,\cdot) - \mathbf{u}(t,\cdot)\right) = \mathbf{v}(t+\Delta t,\cdot), \tag{2.14a}$$

$$\rho_0 \frac{1}{\Delta t}\left(\mathbf{v}(t+\Delta t,\cdot) - \mathbf{v}(t,\cdot)\right) = \left(\nabla^{\mathbf{X}}\cdot\mathbf{P}\right)_{t+\Delta t} + \mathbf{f}^{\text{ext}}(t+\Delta t,\cdot). \tag{2.14b}$$

Eliminating $\mathbf{v}(t+\Delta t,\cdot)$ from (2.14) gives a (non-linear) equation which must be solved for $\mathbf{u}(t+\Delta t,\cdot)$ at each time-step. For present notational purposes, let us refer to the unknown $\mathbf{u}(t+\Delta t,\cdot)$ as simply $\mathbf{u}$, and to the known $\mathbf{u}(t,\cdot)$ and $\mathbf{v}(t,\cdot)$ as $\mathbf{u}_0$ and $\mathbf{v}_0$, respectively. Thus, (2.14) is equivalent to

$$\rho_0\mathbf{u} - \Delta t^2 \nabla^{\mathbf{X}}\cdot\mathbf{P} = \rho_0\left(\mathbf{u}_0 + \Delta t\mathbf{v}_0\right) + \Delta t^2\mathbf{f}^{\text{ext}}, \tag{2.15}$$

where $\nabla^{\mathbf{X}}\cdot\mathbf{P}$ is evaluated at $t+\Delta t$ (hence depends on the unknown $\mathbf{u}$) and $\mathbf{f}^{\text{ext}}$ is also evaluated at $t+\Delta t$.

3.2. Weak formulation and finite element discretization

Let us now derive the weak formulation of equation (2.15) to obtain a finite element discretization in space. We dot product by a test function $\mathbf{w}$, integrate over Ω_0, apply integration by parts, and simplify the integrals over $\partial\Omega_0$ by stipulating that $\mathbf{w} \equiv 0$ on $\partial\Omega_d := \partial\Omega_d(t+\Delta t)$ and substituting the Neumann condition over $\partial\Omega_n := \partial\Omega_n(t+\Delta t)$. It will be notationally convenient, at this point, to once again use index notation and implied summation. We will use subscripts to denote coordinates $(1,\ldots,d)$ and (later) superscripts to denote grid vertices.

$$\rho_0 u_i w_i - \Delta t^2 P_{ij,j} w_i = \left(\rho_0 \left((u_0)_i + \Delta t (v_0)_i\right) + \Delta t^2 f_i^{\text{ext}}\right) w_i, \tag{2.16a}$$

$$\int_{\Omega_0} \rho_0 u_i w_i - \Delta t^2 P_{ij,j} w_i = \int_{\Omega_0} \left(\rho_0 \left((u_0)_i + \Delta t (v_0)_i\right) + \Delta t^2 f_i^{\text{ext}}\right) w_i, \tag{2.16b}$$

$$\int_{\Omega_0} \rho_0 u_i w_i + \Delta t^2 P_{ij} w_{i,j} \tag{2.16c}$$

$$= \int_{\Omega_0} \left(\rho_0 \left((u_0)_i + \Delta t (v_0)_i\right) + \Delta t^2 f_i^{\text{ext}}\right) w_i + \Delta t^2 \int_{\partial\Omega_n} h_i w_i.$$

For the finite element discretization, we will suppose that $\Omega_0 \subset \mathbb{R}^d$ is tessellated by simplices $S^1, \ldots, S^N$ with grid vertices $\mathbf{X}^1, \ldots, \mathbf{X}^n \in \mathbb{R}^d$ and with S_i^α denoting the ith grid vertex of simplex S^α (we shall let "$a \in S^\alpha$" denote the relation that grid vertex a is one of the grid vertices of simplex S^α). Our finite element space will consist of continuous $\mathbb{R}^d$-valued functions which are affine over each simplex S^α. This space is spanned by the nodal basis ("hat") functions $\{N^a\}$; N^a takes the value 1 at $\mathbf{X}^a$ and the value 0 at all other grid vertices. (Strictly speaking, of course, it is really the *projection* of this function space on each coordinate that is spanned by the nodal basis functions.)

We thus take $\mathbf{w} = (w_i = \delta_{ij} N^a)$ in (2.16c), where j ranges over $1, \ldots, d$ and a ranges over $1, \ldots, n$, to obtain dn equations for $\mathbf{u}$. Likewise, we discretize each coordinate of $\mathbf{u}$ as $u_i = u_i^b N^b$, giving a (non-linear) system of equations for the coefficients $\vec{u} := \{u_i^b\}$:

$$0 = q_i^a(\vec{u}) := \left(\int_{\Omega_0} \rho_0 N^a N^b\right) u_i^b + \Delta t^2 \left(\int_{\Omega_0} P_{ij} N^a_{,j}\right) - b_i^a, \tag{2.17a}$$

$$b_i^a := \int_{\Omega_0} \left(\rho_0 \left((u_0)_i + \Delta t (v_0)_i\right) + \Delta t^2 f_i^{\text{ext}}\right) N^a + \Delta t^2 \int_{\partial\Omega_n} h_i N^a. \tag{2.17b}$$

It is worth noting that this is very similar to the non-linear system of equations which arise in a discretization of quasistatic evolution. The differences are only in the right-hand side $\vec{b} := \{b_i^a\}$ and the additional identity-like block to (2.17a).

Like in dimension $d = 1$, we shall solve (2.17) via Newton iteration:

$$\frac{\partial \vec{q}}{\partial \vec{u}}(\vec{u})\, \Delta\vec{u} + \vec{q}(\vec{u}) = 0,$$
$$\vec{u} \leftarrow \vec{u} + \Delta\vec{u}.$$

3.3. Implementation details in dimension 2

We will now consider specifically dimension $d = 2$ and go through the implementation details of the various computational steps necessary to advance one time step, from time t to time $t + \Delta t$.

3.3.1. *Computing $\vec{b}$.* Like in dimension $d = 1$, we will evaluate $\vec{b}$ via an element-based loop, requiring the evaluation of the integrals

$$\int_{S^\alpha} \left(\rho_0 \left((u_0)_i + \Delta t (v_0)_i\right) + \Delta t^2 f_i^{\text{ext}}\right) N^a.$$

(We will address the Neumann terms later.) Let us suppose we are given the value of $\mathbf{f}^{\text{ext}}$ at each grid vertex $\mathbf{X}^b$ and the value of ρ_0 for each element S^α. Then we can expand each of $(u_0)_i$, $(v_0)_i$, and f_i^{ext} as $(u_0)_i^b N^b$, $(v_0)_i^b N^b$, and $f_i^{\text{ext},b} N^b$, respectively, giving

$$\begin{aligned}
&\int_{S^\alpha} \left(\rho_0\left((u_0)_i + \Delta t(v_0)_i\right) + \Delta t^2 f_i^{\text{ext}}\right) N^a \\
&= \int_{S^\alpha} \left(\rho_0^\alpha\left((u_0)_i^b N^b + \Delta t(v_0)_i^b N^b\right) + \Delta t^2 f_i^{\text{ext},b} N^b\right) N^a \\
&= \int_{S^\alpha} \left(\rho_0^\alpha\left((u_0)_i^b + \Delta t(v_0)_i^b\right) + \Delta t^2 f_i^{\text{ext},b}\right) N^a N^b \\
&= \left(\rho_0^\alpha\left((u_0)_i^b + \Delta t(v_0)_i^b\right) + \Delta t^2 f_i^{\text{ext},b}\right) \int_{S^\alpha} N^a N^b.
\end{aligned}$$

Note that this will be nonzero only for $a, b \in S^\alpha$. It thus suffices to evaluate $\int_{S^\alpha} N^a N^b$ for $a, b \in S^\alpha$. These may be computed by a change of coordinates to the standard $\{(0,0),(1,0),(0,1)\}$ triangle, ultimately giving

$$\int_{S^\alpha} N^a N^a = \frac{1}{6}\,\text{area}\,(S^\alpha)\,, \tag{2.18a}$$

$$\int_{S^\alpha} N^a N^b = \frac{1}{12}\,\text{area}\,(S^\alpha) \quad (a \neq b)\,, \tag{2.18b}$$

and we can compute area (S^α) via

$$\text{area}\,(S^\alpha) = \frac{1}{2}\begin{vmatrix} \mathbf{X}^{S_1^\alpha} & \mathbf{X}^{S_2^\alpha} & \mathbf{X}^{S_3^\alpha} \\ 1 & 1 & 1 \end{vmatrix}.$$

In the implementation below, we assume area (S^α) and $\int_{\partial\Omega_n} \mathbf{h} N^a$ have been precomputed for each α and a, respectively.

```
function b = eval_b(tris, X, tri_areas, rho, fext, h, u0, v0, dt)
% tris(k,:)      = the 3 grid vertices within triangle k
% X(a,:)         = the coordinates of grid vertex a
% tri_areas(k)   = the area of triangle k
% rho(k)         = the mass density within triangle k
% fext(a,:)      = the external force on grid vertex a
% h(a,:)         = the Neumann boundary condition on grid vertex a
% u0(a,:)        = the previous time-step's displacement at
%                  grid vertex a
% v0(a,:)        = the previous time step's velocity at grid vertex a
% dt             = the time step increment
b = dt^2 * h;
for k = 1:size(tris,1)
    tri = tris(k,:);
    b(tri,:) += tri_areas(k) ...
             * [2 1 1;1 2 1;1 1 2]/12 ...
             * (rho(k) * (u0(tri,:) + dt * v0(tri,:)) ...
              + dt^2 * fext(tri,:));
end
```

3.3.2. *Computing* $\vec{q}$. Computation of $\vec{q}$ requires computing the integrals

$$\left(\int_{S^\alpha} \rho_0 N^a N^b\right) u_i^b + \Delta t^2 \left(\int_{S^\alpha} P_{ij} N^a_{,j}\right) - b_i^a$$
$$= \rho_0^\alpha \left(\int_{S^\alpha} N^a N^b\right) u_i^b + \Delta t^2 \left(\int_{S^\alpha} P_{ij} N^a_{,j}\right) - b_i^a,$$

where we have used the fact that ρ_0 is given element-wise. We already know the value of $\int N^a N^b$ from the previous subsubsection, so we need only address $\int P_{ij} N^a_{,j}$.

To this end, recall that inversion-robust Neo-Hookean defines $\mathbf{P}$ in terms of $\mathbf{F}$ by

$$\begin{aligned}\mathbf{P} &= \mu\mathbf{F} + (\lambda r(J) - \mu)\, r'(J) \frac{\partial J}{\partial \mathbf{F}} \\ &= \mu\mathbf{F} + (\lambda r(J) - \mu)\, r'(J) J \mathbf{F}^{-T},\end{aligned} \tag{2.19a}$$

$$r(x+1) = x - \frac{1}{2}x^2 + \frac{1}{3}x^3, \tag{2.19b}$$

$$r'(x+1) = 1 - x + x^2. \tag{2.19c}$$

We thus must compute $\mathbf{F}$ from $\mathbf{u}$. First, notice that $N^a(\mathbf{X})$ may be interpreted as the barycentric coordinate of $\mathbf{X}$ with respect to $\mathbf{X}^a$ within S^α (assuming that $a \in S^\alpha$). In other words, $N^a(\mathbf{X})$ is equal to ξ^a, where

$$\begin{pmatrix} \mathbf{X}^{S_1^\alpha} & \mathbf{X}^{S_2^\alpha} & \mathbf{X}^{S_3^\alpha} \\ 1 & 1 & 1 \end{pmatrix} \begin{pmatrix} \xi^{S_1^\alpha} \\ \xi^{S_2^\alpha} \\ \xi^{S_3^\alpha} \end{pmatrix} = \begin{pmatrix} \mathbf{X} \\ 1 \end{pmatrix}.$$

It follows that

$$\mathbf{N}^{S^\alpha}(\mathbf{X}) = \begin{pmatrix} N^{S_1^\alpha}(\mathbf{X}) \\ N^{S_2^\alpha}(\mathbf{X}) \\ N^{S_3^\alpha}(\mathbf{X}) \end{pmatrix} = \begin{pmatrix} \mathbf{X}^{S_1^\alpha} & \mathbf{X}^{S_2^\alpha} & \mathbf{X}^{S_3^\alpha} \\ 1 & 1 & 1 \end{pmatrix}^{-1} \begin{pmatrix} \mathbf{X} \\ 1 \end{pmatrix}$$

and hence

$$\frac{\partial \mathbf{N}^{S^\alpha}}{\partial \mathbf{X}} = \begin{pmatrix} \frac{\partial N^{S_1^\alpha}}{\partial \mathbf{X}} \\ \frac{\partial N^{S_2^\alpha}}{\partial \mathbf{X}} \\ \frac{\partial N^{S_3^\alpha}}{\partial \mathbf{X}} \end{pmatrix} = \begin{pmatrix} \mathbf{X}^{S_1^\alpha} & \mathbf{X}^{S_2^\alpha} & \mathbf{X}^{S_3^\alpha} \\ 1 & 1 & 1 \end{pmatrix}^{-1} \begin{pmatrix} 1 & 0 \\ 0 & 1 \\ 0 & 0 \end{pmatrix}.$$

With the ability to compute $\partial \mathbf{N}^{S^\alpha}/\partial \mathbf{X}$ in hand, and using the fact that $\mathbf{u} = \mathbf{u}^a N^a$, we can now compute $\mathbf{F}$:

$$\mathbf{F} = \frac{\partial \mathbf{u}}{\partial \mathbf{X}} + \mathbf{I} = \mathbf{u}^{S^\alpha} \frac{\partial \mathbf{N}^{S^\alpha}}{\partial \mathbf{X}} + \mathbf{I}. \tag{2.20}$$

Using (2.20) in (2.19a) will thus give $\mathbf{P}$. Finally, we can collect all $P_{ij} N^a_{,j}$ terms into a single (constant) 2×3 matrix as $\mathbf{P}\left(\partial \mathbf{N}^{S^\alpha}/\partial \mathbf{X}\right)^T$ (the rows are indexed by i, the columns by a). The implementation for computing $\vec{q}$ is then as follows.

```
function q = eval_q(tris, X, tri_areas, rho, mu, lambda, u, b, dt)
% tris(k,:)     = the 3 grid vertices within triangle k
% X(a,:)        = the coordinates of grid vertex a
% tri_areas(k)  = the area of triangle k
% rho(k)        = the mass density within triangle k
% mu            = mu
% lambda        = lambda
```

```
% u(a,:)          = the vector of u_i's at grid vertex a
% b(a,:)          = the vector of b_i's at grid vertex a
% dt              = the time step increment
q = -b;
for k = 1:size(tris,1)
    tri = tris(k,:);
    dN_dX = [X(tri,:)';1 1 1] \ [1 0;0 1;0 0];
    F     = u(tri,:)' * dN_dX + [1 0;0 1];
    J     = det(F);
    dJ_dF = [F(2,2) -F(2,1);-F(1,2) F(1,1)];
    rJ    = (J-1) - (J-1)^2 / 2 + (J-1)^3 / 3;
    drJ   = 1 - (J-1) + (J-1)^2;
    P     = mu * F + (lambda * rJ - mu) * drJ * dJ_dF;
    q(tri,:) += tri_areas(k) ...
              * (rho(k) * [2 1 1;1 2 1;1 1 2]/12 * u(tri,:) ...
               + dt^2 * (dN_dX * P'));
end
```

3.3.3. *Computing $\partial\vec{q}/\partial\vec{u}$.* It is probably not too surprising that the most complicated computation is evaluating $\partial\vec{q}/\partial\vec{u}$. To begin, we use (2.17a) and differentiate with respect to u_j^b:

$$\frac{\partial q_i^a}{\partial u_j^b} = \delta_{ij} \int_{\Omega_0} \rho_0 N^a N^b + \Delta t^2 \int_{\Omega_0} \frac{\partial P_{ik}}{\partial F_{\ell m}} \frac{\partial F_{\ell m}}{\partial u_j^b} N_{,k}^a.$$

Given that $\mathbf{F} = \mathbf{u}^{S^\alpha}\left(\partial \mathbf{N}^{S^\alpha}/\partial \mathbf{X}\right) + \mathbf{I}$ (from (2.20)), one can show that $\partial F_{\ell m}/\partial u_j^b = \delta_{j\ell} N_{,m}^b$. Substituting this into the equation above yields

$$\frac{\partial q_i^a}{\partial u_j^b} = \delta_{ij} \int_{\Omega_0} \rho_0 N^a N^b + \Delta t^2 \int_{\Omega_0} \frac{\partial P_{ik}}{\partial F_{jm}} N_{,k}^a N_{,m}^b. \tag{2.21}$$

We can already compute the first integral via (2.18), so the challenge remains to evaluate $\partial P_{ik}/\partial F_{jm}$. Using (2.19a),

$$\begin{aligned}\frac{\partial P_{ik}}{\partial F_{jm}} &= \delta_{ij}\delta_{km}\mu \\ &\quad + \left(\lambda r'(J)^2 + (\lambda r(J) - \mu)\, r''(J)\right) \frac{\partial J}{\partial F_{ik}} \frac{\partial J}{\partial F_{jm}} \\ &\quad + (\lambda r(J) - \mu)\, r'(J) \frac{\partial^2 J}{\partial F_{ik} \partial F_{jm}}.\end{aligned}$$

This gives all the information necessary to evaluate $\partial\vec{q}/\partial\vec{u}$, though the implementation is still relatively complex from all the implicitly summed indices (as a sanity check, one sees the symmetry when interchanging $a, i \leftrightarrow b, j$). We note that $\partial\vec{q}/\partial\vec{u}$ is stored as a $2n \times 2n$ matrix. Row r within the matrix refers to grid vertex $\lceil r/2 \rceil$ and coordinate $2 - r \bmod 2$. Thus, in practice, when inverting $\partial\vec{q}/\partial\vec{u}$, one needs to "flatten" the right-hand side and then "unflatten" the solution (which will turn out to be $\Delta\vec{u}$ in the Newton iteration).

```
function dq_du = eval_dq_du( ...
    tris, X, tri_areas, rho, mu, lambda, u, dt)
% tris(k,:)      = the 3 grid vertices within triangle k
```

```
% X(a,:)        = the coordinates of grid vertex a
% tri_areas(k) = the area of triangle k
% rho(k)        = the mass density within triangle k
% mu            = mu
% lambda        = lambda
% u(a,:)        = the vector of u_i's at grid vertex a
% dt            = the time step increment
n = size(X,1);
dq_du_vals = zeros([4 * 9 * size(tris,1) 1]);
dq_du_rows = dq_du_vals;
dq_du_cols = dq_du_vals;
index = 1;
for k = 1:size(tris,1)
    tri = tris(k,:);
    dN_dX = [X(tri,:)';1 1 1] \ [1 0;0 1;0 0];
    F     = u(tri,:)' * dN_dX + [1 0;0 1];
    J     = det(F);
    dJ_dF = [F(2,2) -F(2,1);-F(1,2) F(1,1)];
    rJ    = (J-1) - (J-1)^2 / 2 + (J-1)^3 / 3;
    drJ   = 1 - (J-1) + (J-1)^2;
    d2rJ  = -1 + 2 * (J-1);
    c1    = lambda * drJ^2 + (lambda * rJ - mu) * d2rJ;
    c2    = (lambda * rJ - mu) * drJ;
    for i = 1:2
        for j = 1:2
            dPik_dFjm = c1 * dJ_dF(i,:)' * dJ_dF(:,j)';
            if(i == j)
                dPik_dFjm(i,i) += mu;
            else
                dPik_dFjm(i,j) += c2;
                dPik_dFjm(j,i) -= c2;
            end
            local_dq_du = dt^2 * dN_dX * dPik_dFjm * dN_dX';
            if(i == j)
                local_dq_du += rho(k) * [2 1 1;1 2 1;1 1 2]/12;
            end
            local_dq_du *= tri_areas(k);
            for a = 1:3
                for b = 1:3
                    r = 2 * tri(a) + i - 2;
                    c = 2 * tri(b) + j - 2;
                    dq_du_vals(index) = local_dq_du(a,b);
                    dq_du_rows(index) = r;
                    dq_du_cols(index) = c;
                    ++index;
                end
            end
        end
```

```
    end
end
dq_du = sparse(dq_du_rows, dq_du_cols, dq_du_vals, 2*n, 2*n);
```

3.3.4. *Computing the Newton Increment $\Delta\vec{u}$.* As alluded to in the previous subsubsection, the stored matrix representing $\partial\vec{q}/\partial\vec{u}$ is $2n\times 2n$, requiring some "flattening" and "unflattening" (alternatively, one can use flattened vectors throughout, though the index manipulation would be somewhat more complicated). This is effected via the *reshape* function used below.

```
n = size(X,1);
dq_du = eval_dq_du(tris, X, tri_areas, rho, mu, lambda, u, dt);
q     = eval_q    (tris, X, tri_areas, rho, mu, lambda, u, b, dt);
for i = 1:length(dirichlet_vertices)
    a = dirichlet_vertices(i);
    dq_du(2*a-1,:) = 0;
    dq_du(2*a  ,:) = 0;
    dq_du(:,2*a-1) = 0;
    dq_du(:,2*a  ) = 0;
    dq_du(2*a-1,2*a-1) = 1;
    dq_du(2*a  ,2*a  ) = 1;
    q(a,:) = [0 0];
end
delta_u = reshape(dq_du \ reshape(-q', [2*n 1]), [2 n])';
```

3.3.5. *The full Newton method.* The entire solve procedure for a single time step looks as follows.

```
function u = solve( ...
    tris, X, tri_areas, ...
    rho, fext, mu, lambda, ...
    h, ...
    dirichlet_vertices, dirichlet_values, ...
    u0, v0, ...
    dt, ...
    tol, u)
b = eval_b(tris, X, tri_areas, rho, fext, h, u0, v0, dt);
for i = 1:length(dirichlet_vertices)
    u(dirichlet_vertices(i),:) = dirichlet_values(i);
end
n = size(X,1);
delta_u = inf([n 2]);
while max(max(abs(delta_u))) > tol
    dq_du = eval_dq_du( ...
                tris, X, tri_areas, rho, mu, lambda, u, dt);
    q     = eval_q( ...
                tris, X, tri_areas, rho, mu, lambda, u, b, dt);
    for i = 1:length(dirichlet_vertices)
        a = dirichlet_vertices(i);
        dq_du(2*a-1,:) = 0;
        dq_du(2*a  ,:) = 0;
```

```
        dq_du(:,2*a-1) = 0;
        dq_du(:,2*   ) = 0;
        dq_du(2*a-1,2*a-1) = 1;
        dq_du(2*a  ,2*a  ) = 1;
        q(a,:) = [0 0];
    end
    delta_u = reshape(dq_du \ reshape(-q', [2*n 1]), [2 n])';
    u += delta_u;
end
```

3.3.6. *Example problem.* We can test the code with the following example problem.

- $\Omega_0 = (-1, 1)^2$.
- Dirichlet boundary conditions at $\mathbf{X}_1 = \pm 1$: $u\,(t, \mathbf{X}) = \sin|\mathbf{X}_1|t$.
- Zero Neumann boundary conditions at $\mathbf{X}_2 = \pm 1$.
- $\rho_0 \equiv 1$.
- $\mathbf{f}^{\text{ext}} = \mathbf{0}$.
- $E = 1000$ and $\nu = 0.3$; so $\mu = E/(2(1+\nu)) = 384.$ and $\lambda = E\nu/((1+\nu)(1-2\nu)) = 577.$

LECTURE 3

Supplemental material

1. Handling inversion via diagonalization

[excerpted from [**2**]*, shortened and with minor modifications]*

1.1. Diagonalization

Since rigid body rotations do not change the physics of a deformable object, the stress $\mathbf{P}$ satisfies $\mathbf{P}(\mathbf{UF}) = \mathbf{UP}(\mathbf{F})$ for any rotation $\mathbf{U}$. (Here $\mathbf{P}(\mathbf{F})$ denotes function application.) Furthermore, if we temporarily assume an isotropic constitutive model, $\mathbf{P}$ is invariant under rotations of material space, i.e., $\mathbf{P}\left(\mathbf{F}\mathbf{V}^T\right) = \mathbf{P}(\mathbf{F})\mathbf{V}^T$. Therefore, if we diagonalize $\mathbf{F}$ via rotations $\mathbf{U}$ and $\mathbf{V}$ to obtain $\mathbf{F} = \mathbf{U}\hat{\mathbf{F}}\mathbf{V}^T$, $\mathbf{P}$ becomes

$$\mathbf{P} = \mathbf{P}(\mathbf{V}) = \mathbf{UP}\left(\hat{\mathbf{F}}\right)\mathbf{V}^T = \mathbf{U}\hat{\mathbf{P}}\mathbf{V}^T \tag{3.1}$$

where a hat superscript denotes the corresponding rotated quantity. Since the elastic energy of an isotropic material is invariant under world and material rotations, it can depend only on the invariants of $\mathbf{F}$, or equivalently on the entries of the diagonalization $\hat{\mathbf{F}}$. Therefore, the gradient of the energy, $\hat{\boldsymbol{\sigma}}$, will also be diagonal. Moreover, since the three stresses are related via $\hat{\boldsymbol{\sigma}} = (1/J)\hat{\mathbf{P}}\hat{\mathbf{F}}^T$ and $\hat{\mathbf{P}} = \hat{\mathbf{F}}\hat{\mathbf{S}}$, the diagonalization of $\mathbf{F}$ actually results in the simultaneous diagonalization of all three stresses. In particular, $\hat{\mathbf{P}}$ in equation (3.1) is diagonal for an isotropic constitutive model. For an anisotropic constitutive model, a diagonal $\hat{\mathbf{F}}$ does not result in a diagonal $\hat{\mathbf{P}}$. However, this is not restrictive, and we show examples of anisotropic constitutive models later.

The diagonalization of $\mathbf{F}$ is not unique, however. While the ordering of the entries of the diagonal matrix $\hat{\mathbf{F}}$ is unimportant, the signs of the entries determine whether the tetrahedron is inverted in a particular direction. The standard SVD convention of choosing all nonnegative entries works only when $\det \mathbf{F} \geq 0$. When $\det \mathbf{F} < 0$, the signs of the entries must be chosen carefully in order to guarantee that the forces act to uninvert the tetrahedron. In this case, $\hat{\mathbf{F}}$ has either one or three negative entries. We heuristically assume that each tetrahedron is as uninverted as possible, and thus we assume that only one entry (not three) is negative. Moreover, the entry with the smallest magnitude is chosen to be negative. This is motivated by the geometric fact that an inverted tetrahedron can be uninverted by moving any one node across the plane of the opposite face, and it is most efficient to choose the node that is closest to the opposite face.

1.2. Constitutive models

Once we have carefully diagonalized $\mathbf{F}$, we can extend our constitutive models to behave reasonably under inversion. The diagonal setting makes this quite simple.

To avoid the unnecessary stiffness associated with the neo-Hookean constitutive model, we modify the constitutive model near the origin to remove the singularity by either linearizing at a given compression limit or simply clamping the stress at some maximum value. Moreover, we extend the model past the origin into the inverted regime in order to obtain valid forces for inverted elements. These forces act to uninvert the element. Note that since we have removed both spatial and material rotations by diagonalizing, the modified model is automatically rotation invariant and isotropic.

The major strength of the diagonal setting is that these modifications can be applied to arbitrary constitutive models. This is quite natural, since the diagonal setting is also commonly used in the experimental determination of material parameters. The resulting model is identical to the physical model most of the time, and allows the simulation to continue if a few tetrahedrons invert. Furthermore, our extensions provide C^0 or C^1 continuity around the flat case, which avoids sudden jumps or oscillations which might effect neighboring elements.

Once we have computed the diagonalized stress $\hat{\mathbf{P}}$, the force computation becomes

$$\mathbf{G} = \mathbf{U}\hat{\mathbf{P}}\mathbf{V}^T\mathbf{B}_m = \mathbf{U}\hat{\mathbf{P}}\hat{\mathbf{B}}_m \tag{3.2}$$

where $\hat{\mathbf{B}}_m = \mathbf{V}^T\mathbf{B}_m$ can be computed and stored if the rotation is fixed for multiple force computations, as in some versions of Newmark time integration.

1.2.1. *Anisotropy.* If the constitutive model includes anisotropic components, it is no longer invariant under rotations of material space. However, we can continue to fully diagonalize $\mathbf{F}$, and rotate the anisotropic terms using $\mathbf{V}$. Since we still work with a diagonal $\hat{\mathbf{F}}$, the large deformation behavior of the constitutive model is still apparent and easy to modify to handle inversion. For example, if the material is stronger in a certain material direction $\mathbf{a}$, we diagonalize $\mathbf{F}$ and use $\mathbf{V}^T\mathbf{a}$ in the computation of $\hat{\mathbf{P}}$. $\hat{\mathbf{P}}$ is no longer a diagonal matrix, but we can still compute forces using equation (3.2). When constructing anisotropic constitutive models that allow inversion, we write $\hat{\mathbf{P}}$ as a diagonal matrix plus $\hat{\mathbf{F}}$ times a symmetric matrix for the anisotropic terms. Then $\hat{\mathbf{S}} = \hat{\mathbf{F}}^{-1}\hat{\mathbf{P}}$ is symmetric (preserving angular momentum) as required.

2. Constitutive model for muscle

[excerpted from **[5]***, shortened and with minor modifications]*

Muscle tissue has a highly complex material behavior—it is a nonlinear, incompressible, anisotropic, hyperelastic material and we use a state-of-the-art constitutive model to describe it with a strain energy of the following form:

$$W(I_1, I_2, \lambda, \mathbf{a}_0, \alpha) = F_1(I_1, I_2) + U(J) + F_2(\lambda, \mathbf{a}_0, \alpha),$$

where I_1 and I_2 are deviatoric isotropic invariants of the strain, λ is a strain invariant associated with transverse isotropy (it equals the deviatoric stretch along the fiber direction), $\mathbf{a}_0$ is the fiber direction, and α represents the level of activation in the tissue. F_1 is a Mooney-Rivlin rubber-like model that represents the isotropic tissues in muscle that embed the fascicles and fibers, $U(J)$ is the term associated with incompressibility, and F_2 represents the active and passive muscle fiber response. F_2 must take into account the muscle fiber direction $\mathbf{a}_0$, the deviatoric stretch in the along-fiber direction λ, the nonlinear stress-stretch relationship in muscle, and the activation level α. The tension produced in a fiber is directed along the vector tangent to the fiber direction. The relationship between the stress in the muscle and the fiber stretch has been established using single-fiber experiments and then normalized to represent any muscle fiber.

The diagonalized FEM framework is most naturally formulated in terms of a first Piola-Kirchoff stress. A stress of this type corresponding to the above constitutive model has the form

$$\mathbf{P} = w_{12}\mathbf{F} - w_2\mathbf{F}^3 + (p - p_f)\,\mathbf{F}^{-1} + 4J_{cc}T\,(\mathbf{F}\mathbf{f}_m)\,\mathbf{f}_m^T,$$

$$J_c = \det(\mathbf{F})^{-1/3}, \quad J_{cc} = J_c^2, \quad I_1 = J_{cc}\mathbf{C}, \quad \lambda = \sqrt{\mathbf{f}_m^T\mathbf{C}\mathbf{f}_m},$$

$$w_1 = 4J_{cc}\text{mat}_{c1}, \quad w_2 = 4J_{cc}^2\text{mat}_{c2}, \quad w_{12} = w_1 + I_1 w_2,$$

$$p = K\log(J), \quad p_f = \frac{1}{3}\left(w_{12}\operatorname{tr}(\mathbf{C}) - w_2\operatorname{tr}\left(\mathbf{C}^2\right) + T\lambda^2\right).$$

Here, $\mathbf{F}$ is the deformation gradient, $\mathbf{C} = \mathbf{F}^T\mathbf{F}$ is the Cauchy strain, and $\mathbf{f}_m$ is the local fiber direction (in material coordinates). mat_{c1} and mat_{c2} are Mooney-Rivlin material parameters and K is the bulk modulus. T is the tension in the fiber direction from the force length curve. Typical values for these parameters are:

$$\begin{aligned}
\text{mat}_{c1} &= 30000\text{Pa} && \text{(muscle)},\\
\text{mat}_{c1} &= 60000\text{Pa} && \text{(tendon)},\\
\text{mat}_{c2} &= 10000\text{Pa} && \text{(muscle and tendon)},\\
K &= 60000\text{Pa} && \text{(muscle)},\\
K &= 80000\text{Pa} && \text{(tendon)},\\
T &= 80000\text{Pa}.
\end{aligned}$$

This formula holds throughout both the muscle and tendon tetrahedra, however, the tendons are passive (no active stress). Note that tendon is considerably stiffer than muscle. Modeling this inhomogeneity is essential for generating muscle bulging during contraction (as well as for accurately computing the musculotendon force generating capacity). Also, large muscles, like the deltoid, trapezius, triceps, and latissimus dorsi, have multiple regions of activation. That is, muscle contraction and activation is nonuniform in the muscle. In general, the effects of varying activation within a muscle can be localized to a few contractile units in each muscle. For example, each head of the biceps and triceps receive individual activations.

3. Guaranteeing positive definiteness of the linear systems in Newton iterations

[excerpted from [**6**]*, shortened and with minor modifications]*

3.1. Finite element forces

Consider a time dependent map $\mathbf{f}$ from the undeformed material coordinates $\mathbf{X}$ to world coordinates $\mathbf{x}$. The stress at a point $\mathbf{X}$ in the material depends on the deformation gradient $\mathbf{F}(\mathbf{X}) = \partial\mathbf{x}/\partial\mathbf{X}$ of this mapping. We use constant strain tetrahedral elements where $\mathbf{F}$ is a constant 3×3 matrix in each tetrahedron. We define edge vectors for each tetrahedron in both material coordinates, $\mathbf{d}_{m1} = \mathbf{X}_1 - \mathbf{X}_0$, $\mathbf{d}_{m2} = \mathbf{X}_2 - \mathbf{X}_0$, $\mathbf{d}_{m3} = \mathbf{X}_3 - \mathbf{X}_0$, and world coordinates, $\mathbf{d}_{s1} = \mathbf{x1} - \mathbf{x0}$, $\mathbf{d}_{s2} = \mathbf{x}_2 - \mathbf{x}_0$, $\mathbf{d}_{s3} = \mathbf{x}_3 - \mathbf{x}_0$, and construct 3×3 matrices $\mathbf{D}_m$ and $\mathbf{D}_s$ using the edge vectors as columns. Then $\mathbf{F} = \mathbf{D}_s\mathbf{D}_m^{-1}$, and $\mathbf{D}_m^{-1}$ is constant and can be precomputed and stored for efficiency.

For hyperelastic materials, stress is defined as the derivative of a strain energy typically constructed from various strain invariants, and we use the first Piola-Kirchhoff stress which is the gradient of the strain energy with respect to the deformation gradient, $\mathbf{P} = \partial\Psi/\partial\mathbf{F}$. $\mathbf{P}$ maps area weighted normals in material space to forces in world space. The force on a node i due to a single tetrahedron incident to it is $\mathbf{g}_i = -\mathbf{P}\left(A_1\mathbf{N}_1 + A_2\mathbf{N}_2 + A_3\mathbf{N}_3\right)/3$, where the $A_j\mathbf{N}_j$ are the area weighted normals of the faces of the tetrahedron incident to node i. Since these do not change during the simulation, we can precompute a vector $\mathbf{b}_i$ such that $\mathbf{g}_i = \mathbf{P}\mathbf{b}_i$. For efficiency, we compute $\mathbf{g}_0 = -\left(\mathbf{g}_1 + \mathbf{g}_2 + \mathbf{g}_3\right)$ and compactly express the other three $\mathbf{g}_i$ as $\mathbf{G} = \mathbf{P}\mathbf{B}_m$ where $\mathbf{G} = (\mathbf{g}_1, \mathbf{g}_2, \mathbf{g}_3)$ and $\mathbf{B}_m = (\mathbf{b}_1, \mathbf{b}_2, \mathbf{b}_3) = -V\mathbf{D}_m^{-T}$ with V the volume of the tetrahedron in material space.

3.2. Element stiffness matrix

The global stiffness matrix in equation is constructed from the additive contributions of the element stiffness matrices, $-\partial\mathbf{f}/\partial\mathbf{x}$, which are based on contributions from individual tetrahedra. As a result of this additive decomposition, definiteness of the element stiffness matrices is a sufficient condition for definiteness of the global stiffness matrix. Motivated by this fact, we manipulate the element stiffness matrix to ensure global definiteness. Later we show that this elemental manipulation amounts to the solution of a single 3×3 symmetric eigenproblem and a few simple algebraic operations. In contrast, dealing with the global stiffness matrix directly can be prohibitively expensive, especially if eigenanalysis or Cholesky factorization of that matrix is required, as in most standard approaches to treating locally indefinite optimization problems.

In order to establish the positive definiteness of the element stiffness matrix, we must ensure that $\delta\mathbf{x}^T\left(-\partial\mathbf{f}/\partial\mathbf{x}\right)\delta\mathbf{x} = -\delta\mathbf{x}^T\delta\mathbf{f} > 0$ for any increment $\delta\mathbf{x}$. Using the formulas from the last section and some tensor manipulations yields

$$\begin{aligned}
\delta\mathbf{x}^T\delta\mathbf{f} &= \sum_{i=1}^{3}\delta\mathbf{x}_i^T\delta\mathbf{g}_i - \delta\mathbf{x}_0^T\sum_{i=1}^{3}\delta\mathbf{g}_i \\
&= \sum_{i=1}^{3}\left(\delta\mathbf{x}_i - \delta\mathbf{x}_0\right)^T\delta\mathbf{g}_i \\
&= \delta\mathbf{D}_s : \delta\mathbf{G} \\
&= \operatorname{tr}\left(\delta\mathbf{D}_s^T\delta\mathbf{G}\right) \\
&= -V\operatorname{tr}\left(\delta\mathbf{D}_s^T\delta\mathbf{P}\mathbf{D}_m^{-T}\right) \\
&= -V\operatorname{tr}\left(\mathbf{D}_m^{-T}\delta\mathbf{D}_s^T\delta\mathbf{P}\right) \\
&= -V\operatorname{tr}\left(\delta\mathbf{F}^T\delta\mathbf{P}\right) \\
&= -V\left(\delta\mathbf{F} : \delta\mathbf{P}\right).
\end{aligned}$$

Since the material element volume V is always a positive constant, the positive definiteness condition reduces to $\delta\mathbf{F} : \delta\mathbf{P} > 0$ or $\delta\mathbf{F} : \left(\partial\mathbf{P}/\partial\mathbf{F}\right) : \delta\mathbf{F} > 0$. Therefore, the positive definiteness of the element stiffness matrix is equivalent to the positive definiteness of the fourth order tensor $\partial\mathbf{P}/\partial\mathbf{F}$. This result is in direct analogy with the energy based formulation of the Newton-Raphson iteration system, since by definition $\mathbf{P} = \partial\Psi/\partial\mathbf{F}$ and thus $\partial\mathbf{P}/\partial\mathbf{F} = \partial^2\Psi/\partial\mathbf{F}^2$.

3.3. Diagonalization

Testing and enforcing positive definiteness of the fourth order tensor $\partial\mathbf{P}/\partial\mathbf{F}$ directly can be rather unwieldy. Instead, we start by rotating both stresses and deformations into diagonal space (transforming our configuration using the rotation matrices that diagonalize the current $\mathbf{F}$ and $\mathbf{P}$). In order to do this, first note that $\delta\mathbf{P} = (\partial\mathbf{P}/\partial\mathbf{F})|_{\mathbf{F}} : \delta\mathbf{F}$ where we explicitly stress the dependency of $\mathbf{P}$ on $\mathbf{F}$ with $\mathbf{P}(\mathbf{F})$. We can manipulate this equality into

$$\delta\mathbf{P} = \left.\frac{\partial\mathbf{U}\mathbf{P}\left(\mathbf{U}^T\mathbf{F}\mathbf{V}\right)\mathbf{V}^T}{\partial\left(\mathbf{U}^T\mathbf{F}\mathbf{V}\right)}\right|_{\mathbf{F}} : \delta\left(\mathbf{U}^T\mathbf{F}\mathbf{V}\right), \tag{3.3}$$

$$= \mathbf{U}\left(\left.\frac{\partial\mathbf{P}\left(\mathbf{F}\right)}{\partial\mathbf{F}}\right|_{\mathbf{U}^T\mathbf{F}\mathbf{V}} : \mathbf{U}^T\delta\mathbf{F}\mathbf{V}\right)\mathbf{V}^T, \tag{3.4}$$

$$= \mathbf{U}\left(\left.\frac{\partial\mathbf{P}}{\partial\mathbf{F}}\right|_{\hat{\mathbf{F}}} : \mathbf{U}^T\delta\mathbf{F}\mathbf{V}\right)\mathbf{V}^T, \tag{3.5}$$

where the first equality comes from equation

$$\mathbf{P}\left(\mathbf{F}\right) = \mathbf{U}\mathbf{P}\left(\mathbf{U}^T\mathbf{F}\mathbf{V}\right)\mathbf{V}^T = \mathbf{U}\mathbf{P}\left(\hat{\mathbf{F}}\right)\mathbf{V}^T$$

and replacing $\delta\mathbf{F}$ with a rotated version, the second comes from a change of variables and the fact that $\mathbf{U}$ and $\mathbf{V}$ are chosen independent of $\mathbf{F}$, and the third comes from choosing $\mathbf{U}$ and $\mathbf{V}$ to be the rotation matrices that diagonalize the initial value of $\mathbf{F}$, i.e., where we evaluate $\partial\mathbf{P}/\partial\mathbf{F}$ to linearize for iteration. Also in the last equality, we drop the explicit dependence of $\mathbf{P}$ on $\mathbf{F}$.

Equation (3.3) provides all the information we need for solving the Newton-Raphson iteration system using a conjugate gradient solver, since the nodal force differentials can readily be computed from the stress differentials as $\delta\mathbf{G} = \delta\mathbf{P}\mathbf{B}_m$. Furthermore we have

$$\begin{aligned}\delta\mathbf{P} : \delta\mathbf{F} &= \delta\mathbf{U}\left(\left.\frac{\partial\mathbf{P}}{\partial\mathbf{F}}\right|_{\hat{\mathbf{F}}} : \mathbf{U}^T\delta\mathbf{F}\mathbf{V}\right)\mathbf{V}^T : \delta\mathbf{F} \\ &= \mathbf{U}^T\delta\mathbf{F}\mathbf{V} : \left.\frac{\partial\mathbf{P}}{\partial\mathbf{F}}\right|_{\hat{\mathbf{F}}} : \mathbf{U}^T\delta\mathbf{F}\mathbf{V}\end{aligned}$$

illustrating that the condition for definiteness, $\delta\mathbf{P} : \delta\mathbf{F} > 0$, derived in the previous subsection is equivalent to positive definiteness of $(\partial\mathbf{P}/\partial\mathbf{F})|_{\hat{\mathbf{F}}}$. We might expect that applying the rotations that diagonalize the current deformation $\mathbf{F}$ to $\delta\mathbf{P}$ and $\delta\mathbf{F}$ would induce a simple structure for the tensor $(\partial\mathbf{P}/\partial\mathbf{F})|_{\hat{\mathbf{F}}}$. In fact this tensor turns out to have a block diagonal structure in the case of isotropic materials.

3.4. Enforcing positive definiteness

In order to reveal the block diagonal structure of $(\partial\mathbf{P}/\partial\mathbf{F})|_{\hat{\mathbf{F}}}$, we rewrite the $3\times3\times3\times3$ fourth order tensor as a 9×9 matrix. To do this, we consider the rearrangement of a 3×3 matrix S into the 9×1 vector $(s_{11}, s_{22}, s_{33}, s_{12}, s_{21}, s_{13}, s_{31}, s_{23}, s_{32})$. We can then represent $(\partial\mathbf{P}/\partial\mathbf{F})|_{\hat{\mathbf{F}}}$ as the 9×9 matrix that maps the vector equivalent of $\delta\mathbf{F}$ to the vector equivalent of $\delta\mathbf{P}$. For isotropic materials this matrix is block diagonal with diagonal components $\mathbf{A}$, $\mathbf{B}_{12}$, $\mathbf{B}_{13}$ and $\mathbf{B}_{23}$ where

$$\mathbf{A} = \begin{pmatrix} \alpha_{11}+\beta_{11}+\gamma_{11} & \gamma_{12} & \gamma_{13} \\ \gamma_{12} & \alpha_{22}+\beta_{22}+\gamma_{22} & \gamma_{23} \\ \gamma_{13} & \gamma_{23} & \alpha_{33}+\beta_{33}+\gamma_{33} \end{pmatrix},$$

$$\mathbf{B}_{ij} = \begin{pmatrix} \alpha_{ij} & \beta_{ij} \\ \beta_{ij} & \alpha_{ij} \end{pmatrix}.$$

Here,

$$\alpha_{ij} = 2\Psi_I + 4\left(\sigma_i^2+\sigma_j^2\right)\Psi_{II},$$

$$\beta_{ij} = 4\sigma_i\sigma_j\Psi_{II} - \frac{2III\Psi_{III}}{\sigma_i\sigma_j},$$

$$\gamma_{ij} = \begin{pmatrix} 2\sigma_i & 4\sigma_i^3 & 2III\sigma_i^{-1} \end{pmatrix} \frac{\partial^2\Psi}{\partial\left(I, II, III\right)^2} \begin{pmatrix} 2\sigma_j \\ 4\sigma_j^3 \\ 2III\sigma_j^{-1} \end{pmatrix} + \frac{4III\Psi_{III}}{\sigma_i\sigma_j},$$

where $\Psi = \Psi\left(I, II, III\right)$ is the strain energy written in terms of the invariants $I = \operatorname{tr}(\mathbf{C})$, $II = \mathbf{C}:\mathbf{C}$ and $III = \det\mathbf{C}$ with $\mathbf{C} = \mathbf{F}^T\mathbf{F}$ and subscripts representing partial derivatives. Also, σ_1, σ_2 and σ_3 are the diagonal components that constitute $\hat{\mathbf{F}}$.

Positive definiteness of $(\partial\mathbf{P}/\partial\mathbf{F})|_{\hat{\mathbf{F}}}$ is equivalent to positive definiteness of each of the blocks $\mathbf{A}$, $\mathbf{B}_{12}$, $\mathbf{B}_{13}$ and $\mathbf{B}_{23}$. For $\mathbf{A}$ a simple 3×3 diagonalization is required, followed by the clamping of all negative eigenvalues to zero. For the 2×2 matrices $\mathbf{B}_{12}$, $\mathbf{B}_{13}$ and $\mathbf{B}_{23}$ no eigenanalysis is necessary since the negative eigenvalue, if present, can be clamped to zero analytically.

Our algorithm computes the stress differential $\delta\mathbf{P}$ as outlined in equation (3.3). First we compute the rotated deformation differential $\mathbf{U}^T\delta\mathbf{F}\mathbf{V}$, and then convert this 3×3 second order tensor into a 9×1 vector and multiply it by the 9×9 matrix for $(\partial\mathbf{P}/\partial\mathbf{F})|_{\hat{\mathbf{F}}}$ to carry out the contraction. Of course, we use the clamped positive definite version of $(\partial\mathbf{P}/\partial\mathbf{F})|_{\hat{\mathbf{F}}}$. The result is then converted from a 9×1 vector back to a 3×3 second order tensor, before being premultiplied by $\mathbf{U}$ and postmultiplied by $\mathbf{V}^T$.

Since we clamp eigenvalues to zero, the element stiffness matrices are only positive semi-definite, not positive definite, which raises the issue of whether the resulting global stiffness matrix could be semi-definite or ill-conditioned itself. In practice, the additive contributions of neighboring elements and boundary conditions always lead to a positive definite global stiffness matrix, even for extreme configurations. (Note that one could clamp to a small positive value as well.) The effect of boundary conditions on the definiteness of the stiffness matrix is analogous to that observed in the matrix resulting from the discretization of the Poisson equation. When all Neumann boundary conditions are specified, the resulting matrix is positive semi-definite. In this case a special version of Conjugate Gradients is still applicable, since an analytic description of the null space is available and, similarly, the global stiffness matrix of an elastic object has a null space corresponding to global translation and linearized rotation. Specification of one or more Dirichlet boundary conditions makes the Poisson matrix strictly positive definite, with positional constraints having the same effect on the definiteness of the global stiffness matrix for elasticity.

Bibliography

1. Javier Bonet and Richard D Wood, *Nonlinear continuum mechanics for finite element analysis*, Communications in Numerical Methods in Engineering **24** (2008), no. 11, 1567–1568.
2. G. Irving, J. Teran, and R. Fedkiw, *Invertible finite elements for robust simulation of large deformation*, Proceedings of the 2004 ACM SIGGRAPH/Eurographics symposium on Computer animation (Aire-la-Ville, Switzerland, Switzerland), SCA '04, Eurographics Association, 2004, pp. 131–140.
3. ———, *Tetrahedral and hexahedral invertible finite elements*, Graph. Models **68** (2006), 66–89.
4. Eftychios Sifakis, Jeffrey Hellrung, Joseph Teran, Aaron Oliker, and Court Cutting, *Local flaps: a real-time finite element based solution to the plastic surgery defect puzzle.*, Stud Health Technol Inform **142** (2009), 313–8.
5. Joseph Teran, Eftychios Sifakis, Silvia S. Blemker, Victor Ng-Thow-Hing, Cynthia Lau, and Ronald Fedkiw, *Creating and simulating skeletal muscle from the visible human data set*, IEEE Transactions on Visualization and Computer Graphics **11** (2005), no. 3, 317–328.
6. Joseph Teran, Eftychios Sifakis, Geoffrey Irving, and Ronald Fedkiw, *Robust quasistatic finite elements and flesh simulation*, Proceedings of the 2005 ACM SIGGRAPH/Eurographics symposium on Computer animation (New York, NY, USA), SCA '05, ACM, 2005, pp. 181–190.